4th ed 2023
✓ ing on ebooked.
GRL
12/2023

Occupational Safety and Health Law Handbook

Third Edition

authors

Melissa A. Bailey
Donelle R. Buratto
Matthew C. Cooper
Frank D. Davis
William K. Doran
John B. Flood
Margaret S. Lopez
John F. Martin

Marshall Lee Miller
Gwendolyn K. Nightengale
Shontell Powell
Phillip B. Russell
Arthur G. Sapper
Kenneth B. Siepman
Zachary S. Stinson
R. Lance Witcher

of:

Ogletree, Deakins, Nash, Smoak & Stewart, P.C.
McDermott Will & Emery LLP
Baise & Miller, P.C.

Lanham • Boulder • New York • London

Published by Bernan Press
An imprint of The Rowman & Littlefield Publishing Group, Inc.
4501 Forbes Boulevard, Suite 200, Lanham, Maryland 20706
www.rowman.com
800-865-3457; info@bernan.com

Unit A, Whitacre Mews, 26-34 Stannary Street, London SE11 4AB

The first edition of this book was previously cataloged by the Library of Congress as follows:

Occupational safety and health law handbook / authors, Lesa L. Byrum ... [et al.].
   p. cm.
   Includes bibliographical references and index.,
   1. Industrial safety—Law and legislation—United States. 2. Industrial hygiene—Law and legislation—United States.
   KF3570 .O844 2001
   344.73/0465   2001040320

ISBN 978-1-59888-678-8 (pbk. : alk. cloth)—ISBN 978-1-59888-679-5 (electronic)

⊖™ The paper used in this publication meets the minimum requirements of American National Standard for Information Sciences—Permanence of Paper for Printed Library Materials, ANSI/NISO Z39.48-1992.

Printed in the United States of America

Occupational Safety and Health Law Handbook

Third Edition

Summary Contents

Contents

3. The Duty to Comply with Standards 70

4. The General Duty Clause 90

8. Hazard Communication: Implementation of the Globally Harmonized System in the 21st Century...178

9. Voluntary Safety and Health Self-Audits 199

12. Criminal Enforcement of Violations 250

13. Judicial Review of Enforcement Actions 261

Preface

Nothing in the workplace is more important than the safety and health of the employees. Good safety and health conditions and practices are maintained through the dedication and teamwork of employees and their supervisors, management, and safety professionals.

A major component of safety and health in a workplace is compliance with the Occupational Safety and Health Act of 1970 (OSH Act) and the laws and policies that the Occupational Safety and Health Administration (OSHA) issues pursuant to the OSH Act. This handbook provides guidance regarding these legal requirements and the enforcement programs administered by OSHA.

The authors of the handbook have a wealth of knowledge and experience in occupational safety and health law. Collectively, the authors have used their knowledge and expertise to provide comprehensive analysis on the key issues that may arise when managing a safety and health program in the workplace.

The Third edition of this handbook provides updates on significant legal decisions impacting workplace safety in health and new OSHA standards, regulations, and policies including the new hazard communication requirements and the new injury and illness reporting requirements issued since the last edition.

As in the previous edition, this edition begins with an overview of the OSH Act in Chapter 1 and then addresses who is covered under the OSH Act, the development of safety and health laws, and the employer's duty to comply with occupational safety and health laws in Chapters 2 through 4. Chapter 5 provides a detailed discussion of the recording and reporting rules for work-related injuries and illness as well other recordkeeping requirements under specific safety and health standards, such as OSHA's Bloodborne Pathogen standard. Chapters 6 and 7 focus on the rights of employers and employees, including employees' right against unlawful discrimination related to protected safety activity. In Chapter 8, the authors provide a detailed discussion of the new hazard communication provisions that conform with the Globally Harmonized System. Chapter 9 covers the use of voluntary safety and health self-audits that employers may use to identify and correct safety and health hazards. Chapters 10 and 11 discuss OSHA's primary enforcement tools — inspections and investigations and citations issued following an inspection or investigation. These chapters also contain insight on steps employers may need to take during an inspection or investigation. Chapter 12 explains the procedures for challenging a citation before the Occupational Safety and Health Review Commission and federal court of appeals. Chapter 13 covers criminal enforcement of OSHA standards. The last chapter covers imminent domain inspections.

These chapters are intended to serve as a useful guide for practitioners in the occupational safety and health field who are responsible for addressing safety and health concerns in the workplace.

Shontell Powell
Ogletree, Deakins, Nash, Smoak & Stewart, P.C.

About the Authors

Melissa A. Bailey

Melissa Bailey has an extensive background in government relations and labor and employment law, including substantial litigation experience in occupational safety and health matters. Ms. Bailey provides counseling and litigates occupational safety and health cases before federal agencies and courts, represents clients in government inspections and investigations, and performs audits and compliance counseling. Ms. Bailey represents employers in a wide range of industries, including chemical, manufacturing, mining, construction, automotive, health and hospitality services. Ms. Bailey advises employers and trade associations about strategies to minimize liability during government inspections, and she assists employers in developing effective compliance plans.

Donelle R. Buratto

Donelle Buratto develops practical and adaptable solutions for overall workforce management and specific employee issues in the areas of workplace safety and employment counseling and litigation. She works directly with corporate counsel, safety compliance personnel, and human resources professionals to preemptively address potential hazards, formally contest agency enforcement actions, provide strategic advice and management of investigations and incident responses, and respond to employment discrimination claims based on safety activity. She supports employers in a wide variety of industries, including automotive suppliers, pipeline, tree service, packaged food products, and household products.

Matthew C. Cooper

At the time of authorship, Mr. Cooper was an attorney with Ogletree Deakins, where he practiced both Health & Safety and Labor & Employment law out of its Denver office. Mr. Cooper's practice consists of helping employers navigate regulatory inspections, investigations, accidents, and other crises management, as well as litigating a wide array of government enforcement actions, including both MSHA and OSHA actions. As an employment attorney, he also litigates and counsels employers on a comprehensive range of workplace matters, including wage and hour compliance, agency audits and investigations, employment agreements, policy drafting and implementation, and discrimination, retaliation, and whistleblower claims. Mr. Cooper currently practices law in the Denver office of Squire Patton Boggs..

Frank D. Davis

Frank Davis represents management exclusively in the area of labor and employment law throughout the United States. His extensive experience includes the handling of representation proceedings before the National Labor Relations Board, labor arbitrations, unfair labor practice charges, union campaigns, employment contracts, dis- crimination, harassment, retaliation, wage and hour compliance, leave issues, and workplace safety. Mr. Davis also has obtained summary judgments or dismissals on behalf of clients in dozens of cases filed in state and federal courts.

William K. Doran

William Doran's practice is concentrated in safety and health law, labor law, employment law, and litigation. Mr. Doran represents and advises companies with respect to issues arising under federal and state safety and health statutes. Mr. Doran represents companies in accident investigations, special investigations, audits, and discrimination investigations conducted by OSHA, MSHA, and state agencies. He represents companies in litigation be- fore the Occupational Safety and Health Review Com- mission, the Federal Mine Safety and Health Review Commission and the federal courts. He also represents companies and trade associations on local and national issues within the purview of federal agency administrators.

John B. Flood

John B. Flood is an experienced litigator who practices in the areas of safety and health law and employment litiga- tion. Mr. Flood represents clients in litigation before the Mine Safety and Health Administration, the Occupational Safety and Health Administration, as well as in fed- eral and state courts at the trial and appellate levels. He represents clients in a wide variety of industries, including defense contracting, mining, real estate, and retail sales.

Margaret S. Lopez

Margo Lopez is the Managing Shareholder of the Washington, D.C. office and a long time workplace safety lawyer representing mining companies in the coal, metal, stone, sand and gravel industries. She is a highly successful litigator and regularly presents company and industry cases before Administrative Law Judges, Federal Review Commissions and the United States Court of Appeals. In a recent high profile Court of Appeals case, Margo effectively saved the mining industry from severe limitations on its rights to manage its workforce, involving potentially millions of dollars, by successfully arguing that temporary reinstatement of miners in discrimination cases must end when MSHA decides not to take the case. Margo represents mine operators in federal investigations and cases arising out of mine incidents and agency enforcement actions under the Federal Mine Safety and Health Act. She has been lead counsel in the aftermath of serious mine accidents and she represents companies and witnesses in MSHA special.

John F. Martin

John Martin focuses his practice on occupational safety and health compliance and litigation. He serves as national OSHA counsel for three publicly-traded companies, and has over 15 years of experience in defending employers in federal court and before the Occupational Safety and Health Review Commission (OSHRC). John has defended clients in 18 states and counsels clients on developing safety programs to eliminate and reduce workplace injuries. John also consults employers and industry groups on federal and state safety regulations and ongoing rulemaking and legislation. He has represented clients in federal and state courts all across the country, up to the Supreme Court of the United States. John represents employers in a wide range of industries, including health care, oil and gas exploration, drilling, well servicing, manufacturing, software, alcohol distribution, and professional sports.

Marshall Lee Miller

Marshall Miller is a partner in the Washington, D.C., office of the law firm Baise & Miller, where he specializes in the areas of environmental law, occupational health and safety, and international transactions. Mr. Miller was previously special assistant to the first administrator of the U.S. Environmental Protection Agency, chief EPA judicial officer, associate deputy attorney general in the U.S. Department of Justice, and deputy administrator and acting head of the Occupational Safety and Health Administration. He was educated at Harvard, Oxford, Heidelberg, and Yale.

Gwendolyn K. Nightengale

Gwendolyn Nightengale's practice is concentrated in workplace safety and health law. Ms. Nightengale represents companies in litigation before the Federal Mine Safety and Health Review Commission. She also assists companies with government investigations and inspections, conferences with regulatory officials, formal contests of agency enforcement actions, and employment discrimination claims based on safety activity.

Shontell Powell

Shontell focuses her practice on occupational safety and health law, assisting employers in enforcement matters before the Occupational Safety and Health Review Commission and state plan review boards across the United States. She also counsels employers on complex OSHA compliance issues. Prior to joining Ogletree Deakins, Shontell worked in the Office of General Counsel at the Occupational Safety and Health Review Commission for over six years as an attorney-advisor. Shontell represents employers in a wide range of industries including tree care, freight services, electric utility, health care, manufacturing, telecommunications, oil and gas exploration, drilling, and well servicing.

Phillip B. Russell

Phillip Russell is a practical labor and employment lawyer who represents businesses in a wide range of labor and employment law matters, including workplace safety and health (OSHA), traditional labor relations (staying union-free), unfair competition and trade secrets litigation, employment litigation defense, and advising and counseling employers on workplace legal compliance issues. Phillip represents clients in the construction, staffing, technology, manufacturing, banking, and other industries. He is also a nationally recognized speaker and author on labor and employment law issues. In 2015, Phillip is celebrating his 20th year of practicing law. He is a peer-review AV-Preeminent® Rated Attorney (by Martindale-Hubbell), rated among the top 5% of lawyers by Florida Super Lawyers (Thompson Reuters), and is listed in Best Lawyers® by U.S. News and World Report. He earned his law degree from Stetson University and has B.S. Management and M.S. Economics degrees from Georgia Tech.

Arthur G. Sapper

Arthur G. Sapper is a partner in the OSHA Practice Group of McDermott, Will & Emery in Washington, D.C. Mr. Sapper regularly represents employers and major trade associations in OSHA cases before the federal appellate courts and the Occupational Safety and Health Review Commission and has participated in ground- breaking cases in the field. For nine years, he was an adjunct professor at Georgetown University Law Center, where he taught a graduate course in OSHA law. Mr. Sapper was a member of the Committee on Model Agency Procedural Rules of the Administrative Conference of the United States. Previously, he was the deputy general counsel of the Occupational Safety and Health Review Commission, and the special counsel to the Federal Mine Safety and Health Review Commission. Mr. Sapper has several times testified before Congress on occupational safety and health issues, has published numerous articles in the field of OSHA law, and is a contributing editor of Occupational Hazards magazine. Mr. Sapper was awarded a J.D. from Georgetown University Law Center.

Kenneth B. Siepman

Mr. Siepman has represented and advised public and private employers in virtually all areas of labor and employment law for fifteen years. Mr. Siepman regularly serves as lead counsel for employers in arbitrations, federal and state trial and appellate courts, and administrative agencies such as the Equal Employment Opportunity Com- mission, the National Labor Relations Board, the Occupational Safety and Health Administration, and the Indiana and United States Departments of Labor. He has considerable experience representing public employers in defense of the first and fourteenth amendments and other constitutional and civil rights cases.

Zachary S. Stinson

Zachary Stinson is an associate in the firm's Washington, D.C., office. His practice includes counseling and representing employers in a wide range of employment litigation matters. Mr. Stinson focuses his practice on discrimination, whistleblower, retaliation, trade secrets, restrictive covenants, and employment torts in state and federal courts. In addition to his litigation practice, Mr. Stinson counsels clients on compliance issues, such as drug testing and other pre-employment screening, as well as drafting and reviewing handbooks, policies, and contracts to ensure compliance with state and federal law. Before joining Ogletree Deakins, Mr. Stinson served as the law clerk to the Honorable Joan Zeldon and the Honorable Rufus G. King, III, of the Superior Court of the District of Columbia. In law school, Mr. Stinson was the managing editor of Volume 58 of the Catholic University Law Review. He also served as an extern at the U.S. Court of Federal Claims to the Honorable Edward J. Damich, and was a summer associate at an industry-leading intellectual-property law firm, where he gained experience in trademark and other intellectual-property matters.

R. Lance Witcher

Lance Witcher has represented employers in a wide range of employment-related litigation in both state and federal court. In addition to representing companies against private litigants in class and single plaintiff cases, Lance regularly litigates against the EEOC and the U.S. Department of Labor. While Lance concentrates his practice primarily in litigation, he regularly provides proactive HR counseling to avoid or reduce the risks associated with employment-related litigation. Additionally, Lance regularly handles matters before OSHA and the Mine Safety Health Administration, and has tried numerous cases before the Occupational and Mine Safety and Health Review Commissions. Lance has appeared before the U.S. Supreme Court, Eighth Circuit Court of Appeals and the Missouri Court of Appeals, and has obtained favorable results for numerous employers before the EEOC, the Missouri Commission on Human Rights, other state and local Fair Employment Practice agencies, the U.S. Department of Labor, and the National Labor Relations Board.

Chapter 1

Occupational Safety and Health Act

Marshall Lee Miller, Esq.
Baise & Miller, P.C.
Washington, D.C.

1.0 Overview

The U.S. Occupational Safety and Health Administration (OSHA) was once called the most unpopular agency in the federal government. It was criticized for its confusing regulations, chronic mismanagement, and picayune enforcement. With somewhat less accuracy, business groups likened it to an American gestapo, while labor unions denounced it as ineffective, unresponsive, and bureaucratic.

Most damning of all, OSHA was often simply ignored. It no longer is. Although OSHA still has its weaknesses and many of its standards are sadly outmoded, its penalties have sharply increased in severity. This has caught the attention of labor and management alike. Moreover, the agency has gradually improved its general reputation. This led the prestigious Maxwell School of Government at Syracuse University a few years ago to grade OSHA a B-minus, the same mark it gave the U.S. EPA.[1] Sadly, this grade is far too generous. While OSHA does deserve considerable credit for shaking off an initially dreadful image, its failure to expand or even update its forty-year-old health standards is indefensible. Enforcement cannot be effective if there is nothing to enforce.

It is not often recognized, however, that OSHA is also perhaps the most important environmental health agency in the government. Even EPA, with far greater resources and public attention, deals with a smaller range of much less hazardous exposures than does OSHA. After all, individuals are more likely to be exposed to high concentrations of dangerous chemicals in their workplaces than in their backyards.

1.1 Comparison of OSHA and EPA

There are several distinct differences between OSHA and EPA, besides the obvious occupational jurisdiction.

First, OSHA has major responsibility over safety in the workplace as well as health. Second, OSHA is essentially an enforcement organization, with a majority of its employees as inspectors, performing tens of thousands of inspections a year. This "highway patrol" function, inspecting and penalizing thousands of businesses large and small, has been the major reason for OSHA's traditional unpopularity. At EPA, on the other hand, inspections and enforcement are a relatively smaller part of the operation.

Third, whereas EPA is an independent regulatory agency, albeit headed by presidential appointees, OSHA is a division of the Department of Labor. This organizational arrangement not only provides less prestige and less independence for OSHA, but also has posed an internal conflict of whether OSHA should be primarily a health (and safety) or a labor-oriented agency. Nevertheless, OSHA and EPA regulate different aspects of so many health issues—asbestos, vinyl chloride, carcinogens, hazard labeling, and others—that it is reasonable to regard them as overlapping environmental organizations.[2]

1.2 OSHA, the Organization

OSHA has a staff of 2,200 throughout the country in ten regional offices and scores of area offices. Almost exactly half of the personnel are safety and health inspectors. Around 600 workers are located at OSHA headquarters in Washington, D.C., near Capitol Hill. The annual budget is over half a billion dollars.

The organization is headed by the assistant secretary of labor for occupational safety and health, a presidential appointee. Under the Barack Obama administration, the current OSHA administrator is David Michaels, a former professor at George Washington University who previously served in the Bill Clinton administration as head of the environmental and occupational safety program at the Department of Energy. The previous appointee, under the George W. Bush administration, was (from February 2006) Edwin G. Foulke, Jr., a South Carolina labor lawyer who was also chairman of the quasi-judicial Occupational Safety and Health Review Commission in the 1990s. His predecessor until the end of 2004 was John L. Henshaw, an official in the North Carolina Department of Labor and the department's chief lobbyist on occupational safety and health matters with the state legislature.

The head of OSHA has traditionally been aided by one to three deputy assistant secretaries, as well as by a number of other senior personnel who head offices such as health standards, safety standards, enforcement, policy planning, and federal programs.[3]

This chapter emphasizes the health aspects of OSHA, because most press attention and the agency's own public emphasis since the mid-1970s has been on toxic hazards. Nevertheless, OSHA is predominantly an occupational safety organization. The two parts of the organization are quite distinct: There are separate inspectors and standards offices for each, and the two groups are different in terms of background, education, and age. There are also far more safety inspectors than health inspectors.

In the most recent fiscal year, the agency conducted 40,000 inspections and proposed penalties of around $100 million. Over half of the inspections were in the construction area, and a quarter were in manufacturing. The number of inspections is only about half of what it was in some earlier years, but this fact alone is not a particularly reliable indicator of agency effectiveness. State OSHA inspections average a little fewer than 60,000 a year, but with only $50 million in proposed penalties.

2.0 Legislative Framework

OSHA was created in December 1970—the same month as EPA—with the enactment of the Occupational Safety and Health Act (OSH Act)[4] and officially began operation in April 1971. Compared with other environmental acts, the OSH Act is very simple and well drafted. This does not mean that one necessarily agrees with the provisions of every section, but it is clearly and concisely written so that details can be worked out in implementing regulations. And unlike the other environmental laws that have been amended several times, becoming more tangled each time, the OSH Act has scarcely been amended or modified since its original passage.[5]

2.1 Purpose of the Act

The Act sets an admirable but impossible goal: to assure that "*no* employee will suffer material impairment of health or functional capacity" from a lifetime of occupational exposure.[6] It does not require—or even seem to allow—a balancing test or a risk-benefit determination.[7] The supplementary phrase in the OSH Act, "to the extent feasible," was not meant to alter this. This absolutist position, comparable only to one provision in the Clean Air Act,[8] reflects Congress's displeasure at previous overly-permissive state standards, which traditionally seemed always to be resolved against workers' health. In fact, the concession to *feasibility* was added almost as an afterthought.

Business groups did obtain two provisions in the law as their price for support. First, industry insisted that states should be encouraged to assume primary responsibility for implementation, in order to minimize the role of the federal OSHA. Second, because of their distrust for the allegedly pro-union bias of the Department of Labor, responsibility for first-level adjudication of violations would be vested in an independent Occupational Health and Safety Review Commission (OHSRC) with a three-member panel of judges named by the president and approved by the Senate.

Congress did reject an industry effort to separate the standard-setting authority from the enforcement powers of the new organization, but it gave a special role to the National Institute for Occupational Safety and Health (NIOSH), located in another government department, the Department of Health and Human Services, in the standard-setting process.

Thus, the three main roles of OSHA are

1. setting of safety and health standards,

2. their enforcement through federal and state inspectors, and

3. employer and employee education and consultation.

2.2 Coverage of the Act

In general, coverage of the Act extends to all employers and their employees in the fifty states and all territories under federal government jurisdiction.[9] An *employer* is defined as any "person engaged in a business affecting commerce who has employees but significantly does not include the United States or any State or political subdivision of a State."[10] Coverage of the Act was clarified by regulations published in the *Federal Register* in January 1972.[11] These regulations interpret coverage as follows:

1. The term *employer* excludes the United States and states and political subdivisions.

2. Any employer employing one or more employees is under its jurisdiction, including professionals, such as physicians and lawyers; agricultural employers; and nonprofit and charitable organizations.

3. Self-employed persons are not covered.

4. Family members operating a farm are not regarded as employees.

5. To the extent that religious groups employ workers for secular purposes, they are included in the coverage.

6. Domestic household employment activities for private residences are not subject to the requirements of the act.

7. Workplaces already protected by other federal agencies under other federal statutes (discussed later) are also excluded.

In total, OSHA directly or indirectly covers more than 100 million workers in six million workplaces.

2.3 Exemptions from the Act

The OSH Act and regulations exempt a number of different categories of employees. The most important exemption is for workplaces employing 10 or fewer workers. What often is not recognized is that this exemption is only partial; these smaller establishments are still subject to accident and worker complaint investigations and the hazard communication requirements (discussed below).

Federal and state employees are also exempted from direct coverage by OSHA. As discussed below, however, the former are subject to OSHA rules under OSH Act Section 19 and several presidential executive orders, and most states with their own state OSHA plans also cover their state and local government workers.

Workers are also exempted if they are covered under other federal agencies, such as railroad workers under the Federal Railroad Administration or maritime workers subject to Coast Guard regulations. This exemption has sometimes generated intergovernmental friction where the other agency has general safety and health regulations but not the full coverage of OSHA regulations. In other words, is the exemption absolute or only proportional?

Under OSH Act Section 9, OSHA is supposed to defer to the other agency if it can better protect the workers and, similarly, the other agency is expected to recede when the situation is reversed. Of course, considerations of turf and politics are often paramount.[12]

2.4 Telecommuting and Home Workplaces

Workplaces are workplaces, even if they are in a private home. That was at least the principle OSHA relied on in 1999 to attempt to exert its authority over the growing number of white collar workers who use their modems rather than their motor cars to commute to work. This is a good example of the type of political furor OSHA can create, often unintentionally.

Of course, OSHA had always claimed (if rarely exercised) jurisdiction over "sweat shops" and other industries, even if operated from someone's home. Therefore, when OSHA was asked for a simple interpretation about its coverage of home office workers, it applied the same logic. In an interpretative ruling from the Office of Compliance Programs in November 1999, the agency stated that OSHA would hold employers responsible for injuries to employees at home.[13] This triggered a political explosion.

The National Association of Manufacturers declared, "We see this as the long arm of OSHA coming into people's homes."[14] The chairman of a powerful congressional committee warned that the policy would put "home workers in the position of having to comply with thousands of pages of OSHA regulations."[15]

What OSHA had failed to realize was these new workers were not someone's employees needing protection from exploitive bosses. They were their own bosses, or they certainly saw themselves as such. And they saw OSHA intervention not as protective but intrusive.

On January 5, 2000, the Secretary of Labor, Alexis Herman, announced the cancellation of the short-lived OSHA policy.

3.0 Scope of OSHA Standards

To give the reader an idea of the areas covered by the standards, the following is a subpart listing from the Code of Federal Regulations, Part 1910, Occupational Safety and Health Standards. Note that the listings are mostly safety standards. The health standards are all contained in Subpart Z, except for Subparts A, C, G, K, and R, which cover both categories.

3.1 Areas Covered by the OSHA Standards

• Subpart A: General (purpose and scope, definitions, applicability of standards, etc.)

• Subpart B: Adoption and Extension of Established Federal Standards (construction work, ship repairing, long shoring, etc.)

• Subpart C: General Safety and Health Provisions (preservation of records)

• Subpart D: Walking-Working Surfaces (guarding floor and wall openings, portable ladders, requirements for scaffolding, etc.)

• Subpart E: Means of Egress (definitions, specific means by occupancy, sources of standards, etc.)

• Subpart F: Powered Platforms, Manlifts, and Vehicle-Mounted Work Platforms (elevating and rotating work platforms, standards, organizations, etc.)

• Subpart G: Occupational Health and Environmental Control (ventilation, noise exposure, radiation, etc.)

• Subpart H: Hazardous Materials (compressed gases, flammables, storage of petroleum gases, effective dates, etc.)

• Subpart I: Personal Protective Equipment (eye/face, respiratory, electrical devices, etc.)

• Subpart J: General Environmental Controls (sanitation, labor camps, safety color code for hazards, etc.)

• Subpart K: Medical and First Aid (medical services, sources of standards)

• Subpart L: Fire Protection (fire suppression equipment, hose and sprinkler systems, fire brigades, etc.)

• Subpart M: Compressed Gas and Compressed Air Equipment (inspection of gas cylinders, safety relief devices, etc.)

• Subpart N: Materials Handling/Storage (powered industrial trucks, cranes, helicopters, etc.)

• Subpart O: Machinery and Machine Guarding (requirements for all machines, woodworking machinery, wheels, mills, etc.)

• Subpart P: Hand and Portable Powered Tools and Other Hand-Held Equipment (guarding of portable power tools)

• Subpart Q: Welding, Cutting, and Brazing (definitions, sources of standards, etc.)

• Subpart R: Special Industries (pulp, paper and paperboard mills, textiles, laundry machinery, telecommunications, etc.)

• Subpart S: Electrical (application, National Electrical Code)

• Subpart T: Commercial Diving Operations (qualification of team, pre- and postdive procedures, equipment, etc.)

• Subpart U–Y: [Reserved]

• Subpart Z: Toxic and Hazardous Substances (air contaminants, asbestos, vinyl chloride, lead, benzene, etc.)

3.2 Overview of Standards

When OSHA was created, Congress realized that the new agency would require years to promulgate a comprehensive corps of health and safety standards. The OSH Act therefore provided that for a two-year period ending in April 1972, the agency could adopt as its own the standards of respected professional and trade groups. These are the *consensus standards* issued under Section 6(a) of the statute.[16] Nobody could have imagined that over four decades later, these imperfect and outdated standards would still form the overwhelming majority of OSHA regulations.

3.3 Overview of Health Standards

Health issues, notably environmental contaminants in the workplace, have increasingly become a national concern over the past few years. Health hazards are much more complex, more difficult to define, and—because of the delay in detection—perhaps more dangerous to a larger number of employees. Unlike safety hazards, the effects of health hazards may be slow, cumulative, irreversible, and complicated by nonoccupational factors.

If a machine is unequipped with safety devices and maims a worker, the danger is clearly and easily identified and the solution usually obvious.

However, if workers are exposed for several years to a chemical that is later found to be carcinogenic, there may be little help for those exposed.

In the nation's workplaces, there are tens of thousands of toxic chemicals, many of which are significant enough to warrant regulation. Yet OSHA only has a list of fewer than 500 substances, and these are mostly simple threshold limits adopted under Section 6(a) from the recommended lists of private industrial hygiene organizations back in the 1960s and early 1970s. This list is being updated now but with glacial slowness.

The promulgation of health standards involves many complex concepts. To be complete, each standard needs medical surveillance requirements, recordkeeping, monitoring, and multiple physical reviews, just to mention a few. At the present rate, promulgation of standards on every existing toxic substance could take centuries.

Ironically, an attempt to update the health standards for hundreds of substances in one regulatory action by borrowing newer figures from respected health professional organizations was opposed by the labor unions (and industry) and struck down by an appellate court in 1992.[17]

3.4 Overview of Safety Standards

Safety hazards are those aspects of the work environment that, in general, cause harm of an immediate and sometimes violent nature, such as burns, electrical shock, cuts, broken bones, loss of limbs or eyesight, and even death. The distinction from health hazards is usually obvious; mechanical and electrical are considered safety problems, while chemicals are considered health problems. Noise is difficult to categorize; it is classified as a health problem.

The Section 6(a) adoption of national consensus and other federal agency standards created chaos in the safety area. It was one thing for companies to follow industry or association guidelines that, in many cases, had not been modified in years; it was another thing for those guidelines actually to be codified and enforced as law. In the two years that the Act provided for OSHA to produce standards derived from these existing rules, the agency should have examined these closely, simplified them, deleted the ridiculous and unnecessary ones, and promulgated final regulations that actually identified and eliminated hazards to workers. But in the commotion of organizing an agency from scratch, it did not happen that way.

Nor did affected industry groups register their objections until later. During the entire two-year comment period, not a single company or association filed an objection with OSHA.

Almost all of the so-called Mickey Mouse standards were safety regulations, such as the requirements that fire extinguishers be attached to the wall exactly so many inches above the floor. Undertrained OSHA inspectors often failed to recognize major hazards while citing industries for minor violations "which were highly visible, but not necessarily related to serious hazards to workers' safety and health."[18]

Section 6(g) of the OSH Act directs OSHA to establish priorities based on the needs of specific "industries, trades, crafts, occupations, businesses, workplaces, or work environments." The Senate report accompanying the OSH Act stated that the agency's emphasis initially should be put on industries where the need was determined to be most compelling.[19] OSHA's early attempts to target inspections, however, were sporadic and, for the most part, unsuccessful. The situation has improved somewhat in recent years, for both health and safety, in part because of the recent requirement that some priority scheme be used that could justify search warrants. But, as we shall see, that has brought its own problems.

4.0 Standard Setting

Setting standards can be a complex and protracted process. There are thousands of chemical substances, electrical problems, fire hazards, and many other dangerous situations prevalent in the workplace for which standards needed to be developed.

To meet the objectives defined in the Act, three different standard-setting procedures were established:

1. Consensus Standards, under Section 6(a)

2. Permanent Standards, under Section 6(b)

3. Emergency Temporary Standards, under Section 6(c)

4.1 Consensus Standards: Section 6(a)

Congress realized that OSHA would need standards to enforce while it was developing its own. Section 6(a) allowed the agency, for a two-year period that ended on 25 April 1973, to adopt standards developed by other federal agencies or to adopt consensus standards of various industry or private associations.[21] This resulted in a list of around 420 common toxic chemicals with maximum permitted air concentrations specified in parts per million (ppm) or in milligrams per cubic meters (mg/M^3).

There are several problems inherent in these standards. First, these threshold values are the only elements in the standard. There are no required warning labels, monitoring, or medical recordkeeping, and they do not generally distinguish between the quite different health effects in eight-hour, 15-minute, peak, annual average, and other periods of exposure.

Second, being thresholds, they are based on the implicit assumption that there are universal *no-effect levels*, below which a worker is safe. For carcinogens, this assumption is quite controversial.

Third, most of the standards were originally established not on the basis of firm scientific evidence but, as the name implies, from existing guidelines and limits of various industry, association, and governmental groups. Before OSHA's creation, they were intended to be general, nonbinding guidelines, and had been in circulation for a number of years with no urgency to keep them current. Consequently, neither industry nor labor bothered to comment when OSHA first proposed the consensus standards. Many of these "interim" standards were out of date by the time they were adopted by OSHA, and they are now frozen in time until OSHA goes through the full Section 6(b) administrative rulemaking process.

Fourth, OSHA consensus standards often involve "incorporation by reference," especially in the safety area. In some cases, these pre-1972 publications were not standards or even formal association guidelines but mere private association pamphlets that are no longer in print and not easily obtainable. For example, the general regulation on compressed gases merely states that the cylinders should be in safe condition and maintained "in accordance with Compressed Gas Association pamphlet P-1-1965" and several similar documents.[22]

Fifth, not all of these "toxics" are on the list because they really pose a health hazard. Although that has been the unquestioned assumption of certain later rulemakings, such as the requirement for Material Safety Data Sheets, some chemicals, such as carbon black, were listed because of "good housekeeping" practices—a facility with even a small amount of this intrusive black substance will look filthy—and not because it was hazardous at the levels set.

Nevertheless, Congress was undoubtedly correct in requiring the compilation of such a list. Otherwise, there would have been no OSHA health standards at the beginning. There are virtually no others even now.

4.2 Standards Completion and Deletion Processes

The agency has attempted to deal with one of the shortcomings of the consensus standards by what is called the Standards Completion Process. Over a number of years, OSHA has taken some threshold standards and added various medical, monitoring, and other requirements.[23] At least a broader range of protection is offered to exposed workers.

The agency has also sought to reduce the number of safety standards. This is done by eliminating the so-called "Mickey Mouse" standards that accomplish little but impose voluminous requirements. More often, the simplification has come by removing redundant sections and cross-references. This eliminates pages but not a lot more.

Nevertheless, OSHA is proud of its compliance with the presidential directive that federal agencies review and remove duplicative or repetitive regulations.[24]

4.3 Permanent Standards: Section 6(b)

Permanent standards must now be developed pursuant to Section 6(b). This is the familiar standard-setting and rule-making process followed by most other federal agencies under the Administrative Procedure Act.[25]

Permanent standards may be initiated by a well-publicized tragedy, court action, new scientific studies, or the receipt of a criteria document from NIOSH, an organization described later in this chapter. The criteria document is a compilation of all the scientific reports on a particular chemical, including epidemiological and animal studies, along with a recommendation to OSHA for a standard. The recommendation, based supposedly only on scientific health considerations, includes suggested exposure limits (eight-hour average, peaks, etc.) and appropriate medical monitoring, labeling, and other proscriptions.

Congress assumed that NIOSH would be the primary standard-setting arm of OSHA, although the two are in different government departments—Health and Human Services (HHS) and Labor, respectively. According to this model, OSHA would presumably take the scientific recommendations from NIOSH, factor in engineering and technical feasibility, and then promulgate as similar a standard as possible. However, the system has never worked this way. Instead, OSHA's own standards office has generally regarded NIOSH's contribution as just one step in the process—and not one entitled to a great deal of deference.[26]

Following receipt of the criteria document or some other initiating action, OSHA will study the evidence and then possibly publish a proposed standard. Most candidate standards never get this far: The hundreds of NIOSH documents, labor union petitions, and other serious recommendations have resulted in only a few new health standards since 1970.[27]

The proposed standard is then subjected to public comment for (typically) a 90-day period, often extended, after which the reactions are analyzed and informal public hearings are scheduled. In a few controversial instances, there may be more than one series of hearings and comments. Then come the posthearing comments, which are perhaps the most important presentations by the parties. After considerable further study, a final standard is eventually promulgated. The entire process might theoretically be accomplished in under a year, but in practice it takes a minimum of several years or, as with asbestos, even decades.

The following is a list of some of the final health standards that OSHA has promulgated to date:

1. Asbestos

2. Fourteen carcinogens

—4-Nitrobiphenyl	—benzidine
—alpha-nephthylamine	—ethyleneimine
—methyl chloromethyl ether	—beta-propiolactone
—3,3'-dichlorolenzidine	—2-acetylaminofluorene
—bis-chloromethyl ether	—4-dimethylaminozaobenzene
—beta-naphthylamine	—N-nitrosodimethylamine
—4-aminodiphenyl	—(MOCA stayed by court action)

3. Vinyl chloride

4. Inorganic arsenic

5. Lead

6. Coke-oven emissions

7. Cotton dust

8. 1,2-dibromo-3-chloropropane (DBCP)

9. Acrylonitrile

10. Ethylene oxide

11. Benzene

12. Field sanitation

And, most recently, in February 2006,

13. Hexavalent chromium

The list is obviously incredibly short for three and a half decades of OSHA standard setting.

4.4 Emergency Temporary Standards

The statute also provides for a third standard-setting approach, specified for emergency circumstances where the normal, ponderous rulemaking procedure would be too slow. Section 6(c) gives the agency authority to issue an emergency temporary standard (ETS) if necessary to protect workers from exposure to *grave danger* posed by substances "determined to be toxic or physically harmful or from new hazards."[28]

Such standards are effective immediately upon publication in the *Federal Register*. An ETS is only valid, however, for six months. OSHA is thus under considerable pressure to conduct an expedited rulemaking for a permanent standard before the ETS lapses. For this reason, a quest for an emergency standard has been the preferred route for labor unions or other groups seeking a new OSHA standard. These ETSs have not fared well, however, when challenged in the courts; virtually all have been struck down as insufficiently justified.

4.5 General Duty Clause, 5(a)(1)

There is actually a fourth type of enforceable standard, one that covers situations for which no standards currently exist.

Since OSHA has standards for only a few hundred of the many thousands of potentially dangerous chemicals and workplace safety hazards, there are far more situations than the rules cover. Therefore, inspectors have authority under the General Duty Clause to cite violations for unsafe conditions even where specific standards do not exist.[29] Agency policy has shifted back and forth between encouraging the use of "Section 5(a)(1)," as the clause is often termed, since this ensures that unsafe conditions will be addressed, and discouraging its use on the theory that employers should be liable only for compliance with specific standards of which they are given knowledge.

However, the agency has acknowledged that many of the standards that do exist are woefully out of date and thus cannot be relied upon for adequate protection of worker safety and health. The traditional notion was that compliance with an existing specific standard—even if demonstrably unsafe—precluded an OSHA citation.[30] This has been called into question by the courts. In April 1988, a federal appellate court allowed OSHA to cite for violations of the General Duty Clause even where a company was in full compliance with a specific numerical standard on the precise point in question.[31] Bare compliance with the standards on the books, therefore, might not be responsible management.

4.6 Feasibility and the Balancing Debate

There has been a continuing debate over feasibility and balancing in OSHA enforcement. The important issues include the following:

• Can OSHA legally consider economic factors in setting health or safety standards levels?

• If so, is this consideration limited only to extreme circumstances?

• Does the Occupational Safety and Health Act provide for a balancing of costs and benefits in setting standards?

• Can OSHA mandate engineering controls although they alone would still not attain the standard?

• Can OSHA require engineering controls even if personal protective equipment (such as ear plugs) could effectively, if often only theoretically, reduce hazards to a safe level and at a much lower cost?

These questions have been extensively litigated before the Occupational Safety and Health Review Commission (OSHRC) and the courts. Most of the debate has been over the interpretation of *feasibility* in Section 6(b)(5) of the Act.

One must remember that OSHA legislation was originally seen by Congress in rather absolutist terms: Any standard promulgated should be one "which most adequately assumes . . . that no employee will suffer material impairment of health." Only late in the congressional debate was the Department of Labor able to insert the phrase "to the extent feasible" into the text. This was intended to prevent companies having to close because unattainable standards were imposed on them, but it was not spelled out to what extent economic as well as technical feasibility was included.[32]

Since the term *feasibility* was not clearly defined, there has been much confusion over how to interpret what Congress intended, as the earlier cases show. In *Industrial Union Department, AFL v. Hodgson*, the D.C. Circuit accepted that economic realities affected the meaning of *feasible*, but only to the extent that "a standard that is prohibitively expensive is not 'feasible.'"[33] It was Congress's intent, the court added, that this term would prevent a standard unreasonably "requiring protective devices unavailable under existing technology or by making financial viability generally impossible." The court warned, however, that this doctrine should not be used by companies to avoid needed improvements in their workplaces:

> Standards may be economically feasible even though, from the standpoint of employers, they are financially burdensome and affect profit margins adversely. Nor does the concept of economic feasibility necessarily guarantee the continued existence of individual employers.[34]

A similar view was adopted in 1975 by the Second Circuit in *The Society of the Plastics Industry v. OSHA*, written by Justice Clark, who cited approvingly the case above.[35] He held that *feasible* meant not only that which is attainable technologically and economically now, but also that which might reasonably be achievable in the future. In this case, which concerned strict emissions controls on vinyl chloride, he declared that OSHA may impose "standards which require improvements in existing technologies or which require the development of new technology, and . . . is not limited to issuing standards based solely on devices already fully developed."[36]

Neither court undertook any risk-benefit analysis, such as attempting to compare the hundreds of millions of dollars needed to control vinyl chloride with the lives lost to angiosarcoma of the liver. Those who have attempted to develop such equations have generally concluded the task is undoable, at least for most such chronic health effects.[37]

A third federal appeals court, however, took a strongly contrary position in a case involving noise. In *Turner Co. v. Secretary of Labor*, the Seventh Circuit Court of Appeals decided that the $30,000 cost of abating a noise hazard should be weighed against the health damage to the workers, taking into consideration the availability of personal protective equipment to mitigate the risk.[38]

This holding is not unreasonable, but is based on a highly tenuous interpretation of the law. The court, without providing any clear rationale for its view, held that "the word 'feasible' as contained in 29 C.F.R. § 1910.95(6)(1) must be given its ordinary and common sense meaning of 'practicable.'" (This may be so, but is of no analytical value.) From this the court concluded:

> Accordingly, the Commission erred when it failed to consider the relative cost of implementing engineering controls . . . versus the effectiveness of an existing personal protective equipment program utilizing fitted earplugs.[39]

This interpretation does not follow from the analysis. In fact, since the Turner Company had both the financial resources and the technical capability to abate the noise problem, compliance with the regulation would appear to be "practicable." The court, however, considered this term to mean that a cost-benefit computation should be made.

In 1982, the U.S. Court of Appeals for the Ninth Circuit, in the case of *Donovan v. Castle & Cooke Foods and OSHRC*,[40] also held that the Noise Act and the regulations permit consideration of relative costs and benefits to determine what noise controls are feasible.

OSHA gave the plant a citation on the grounds that, although Castle & Cooke required its employees to wear personal protective equipment, its failure to install technologically feasible engineering and administrative controls[41] constituted a violation of the noise standard, and that the violation could only be abated by the implementation of such controls. OSHA argued that engineering and administrative controls should be considered economically infeasible only if their implementation would so seriously jeopardize the employer's economic condition as to threaten continued operation.

On appeal, OSHA argued that neither the OSHRC nor the courts are free to interpret *economic feasibility*, because its definition is controlled by the Supreme Court's decision in *American Textile Manufacturers Institute, Inc. v. Donovan*.[42] The appeals court, however, decided that the Supreme Court's interpretation of the term *feasible* made in *American Textile* was not deemed controlling for the noise standards. It also affirmed that economic *feasibility* should be determined through a cost-benefit analysis, and that in the case of *Castle & Cooke* the costs of economic controls did not justify the benefit that would accrue to employees. Thus, the decision to vacate the citation was upheld.

5.0 Variances

Companies that complain that OSHA standards are unrealistic are often not aware that they might be able to create their own version of the standards. The alternative proposed has to be at least as effective as the regular standard, but it can be different.

5.1 Temporary Variances

Section 6(b)(6)(A) of the OSH Act establishes a procedure by which any employer may apply for a "temporary order granting a variance from a standard or any provision thereof." According to the Act, the variance will be approved when OSHA determines that the requirements have been met and establishes:

> • that the employer is unable to meet the standard "because of unavailability of professional or technical personnel or of materials and equipment," or because alterations of facilities cannot be completed in time;

> • that he is "taking all available steps to safeguard" his workers against the hazard covered by the standard for which he is applying for a variance; and

> • that he has an "effective program for coming into compliance with the standard as quickly as practicable."[43]

This temporary order may be granted only after employees have been notified and, if requested, there has been sufficient opportunity for a hearing. The variance may not remain in effect for more than one year, with the possibility of only two six-month renewals.[44] The overriding factor an employer must demonstrate for a temporary variance is good faith.[45]

5.2 Permanent Variances

Permanent variances can be issued under Section 6(d) of the OSH Act. A permanent variance may be granted to an employer who has demonstrated "by a preponderance" of evidence that the "conditions, practices, means, methods, operations or processes used or proposed to be used" will provide a safe and healthful workplace as effectively as would compliance with the standard.

6.0 Compliance and Inspections

OSHA is primarily an enforcement organization. In its early years both the competence of its inspections and the size of the assessed fines were pitifully inadequate; they were the primary reason OSHA was not taken seriously by either labor unions or the business community. That picture has now changed significantly.

6.1 Field Structure

The Department of Labor (DOL) has divided the territory subject to the OSH Act into ten federal regions, the same boundaries that EPA also uses. Each region contains from four to nine area offices. When an area office is not considered necessary because of a lack of industrial activity, a district office or field station may be established. Each region is headed by a regional administrator, each area by an area director. In the field, compliance officers represent area offices and inspect industrial sites in their vicinity, except in situations where a specialist or team might be required.

6.2 Role of Inspections

The only way to determine compliance by employers is inspections, but inspecting all the workplaces covered by the OSH Act would require decades. Each year there are tens of thousands of federal inspections, and as many or more state inspections, but there are several million workplaces.

Obviously, a priority system for high-hazard occupations is necessary, along with random inspections just to keep everyone "on his toes."

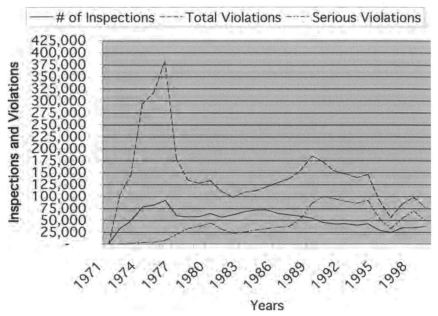

Figure 1.1 Inspections and Violations

Inspections are supposed to be surprises; there are criminal penalties for anyone alerting the sites beforehand. The inspections may occur in several ways: They may be targeted at random, triggered by worker complaints, set by a priority system based on hazardous probabilities, or brought on by events such as a fatality or explosion. Inspectors expect admittance without search warrants, but a company has the constitutional right to refuse admittance until OSHA obtains a search warrant from a federal district court.[46] Such refusal is frankly not a good idea except in very special circumstances, such as when the additional delay would allow a quick cleanup of the workplace to bring it into compliance.

6.3 Training and Competence of Inspectors

There has been a major problem with OSHA inspectors in the past—the training program did not adequately prepare them for their tasks, and the quality of the hiring was uneven. In the early days there was tremendous pressure from the unions to get an inspection force on the job as soon as possible, so recruitment was often hurried and training was minimal.

Inspectors would walk into a plant where, for example, pesticide dust was so thick workers could not see across the room, yet, because there was no standard as such, the inspectors would not think there was a problem.[47] Yet, had there been a fire extinguisher in the wrong place, and had the inspector been able to see it through the haze, he would have cited the plant for a safety violation. This early bumbling was the source of much of the animus against OSHA that persists even today.

Competence among staff has markedly improved since the early days of the program. Both in-house training efforts by OSHA and increased numbers of professional training programs conducted by colleges and universities have contributed to these improvements. There is also a greater sensitivity towards workers and their representatives.[48]

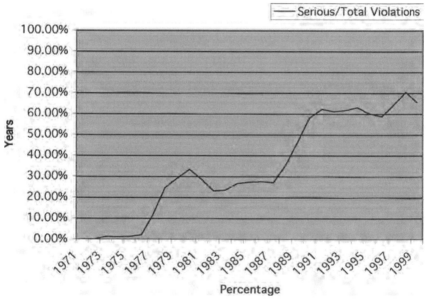

Figure 1.2 Serious Violations as a Proportion of Total Violations

6.4 Citations, Fines, and Penalties

If the inspector discovers a hazard in the workplace, a citation and a proposed fine may be issued. Citations can be serious, nonserious, willful, or repeated. By one count, there are at least nine types of penalty findings under the OSH Act. They are as follows:

• De minimis—These are technical violations, but they pose insignificant risk and for which no monetary penalty is warranted.

• Nonserious—This is the basic type of penalty. No risk of death or serious injury is posed, but the violation might still cause some harm.

• Serious—The hazard could lead to death or serious injury.

• Failure to correct—Violations when found must be remediated within a certain period of time. If a subsequent reinspection finds this has not been done, or the violation has been allowed to recur, this fairly serious citation is in order.

• Repeated—These are continuous violations, discussed below.

• Willful—These are intentional violations, discussed below.

• Criminal—These violations are applicable under the OSH Act only for cases involving death.[49]

• Egregious—These are supposedly heinous situations, discussed below.

• Section 11(c)—These are penalties for company retaliation against complainers and whistle-blowers, discussed at length below.

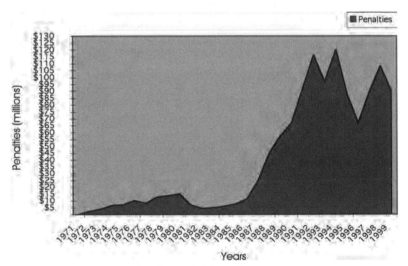

Figure 1.3 Penalties

6.5 OSHA Citation and Penalty Patterns

OSHA now averages over 40,000 inspections a year.[50] These are focused on the industries and sectors where statistics indicate greater potential hazards. Contrary to the common assumption that most inspections are in manufacturing, in fact that sector accounts for only about one-fourth of the inspections. Over half are in the construction industry, with another quarter distributed over all other types of workplaces.

Specially targeted sectors in the manufacturing area, with four-digit Standard Industrial Classification (SIC) codes, have most recently been designated[51]:

- Plastic products (3089)

- Sheet metalwork (3444)

- Fabricated structural metal (3441)

- Metal stampings (3469)

- Fabricated metal products (3499)

- Motor vehicle parts (3714)

- Construction machinery (3431)

- Shipbuilding and repair (3731)[52]

6.6 Communicating and Enforcing Company Rules

Many accidents—arguably even most—are due to human negligence, often involving an act that is contrary to company policy. Merely claiming a company policy, however, is not enough, for OSHA does not look very favorably upon this defense. For employers to plead employee misconduct as a defense to an OSHA citation, the company must first demonstrate three things:

- First, of course, is to prove the existence of such rules.[53]

- Second, an employer must prove that these rules were effectively communicated to the employees. Proof can include written instructions, evidence of required attendance at education sessions, the curriculum of training programs, and other forms that should be documented.[54]

• Third, many companies that can demonstrate the above two principles fall short on the third, namely that there should be evidence the policies are effectively enforced.[55] For this, evidence of disciplinary action taken against infractions of the rules, though not necessarily the precise rule that would have prevented the accident under investigation, is necessary. The closer to the actual circumstance, of course, the more that proof of active company enforcement is dispositive.[56]

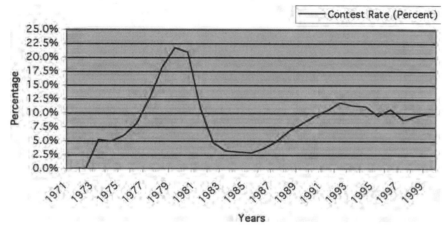

Figure 1.4 Contest Rate

If the above three principles can all be demonstrated, they constitute a reasonable defense to charges of violating the regulations, even in cases of death or serious injury.

Note that this defense is not limited to the misconduct of a low-ranking employee. Misconduct of a supervisor, although it may suggest inadequate company policy and direction, can also be shown as an isolated and personal failing. According to an appellate court, the proper focus of a court is on the effectiveness of the employer's implementation of his safety program and not on whether the unforeseeable conduct was by an employee or by supervisory personnel.[57]

6.7 Warrantless Inspections: The Barlow Case

Litigants have challenged OSHA's constitutionality on virtually every conceivable grounds, from the First Amendment to the Fourteenth.[58] The one case that has succeeded has led to the requirement of a search warrant, if demanded, for OSHA inspectors.

The Supreme Court in *Marshall v. Barlow's Inc.*,[59] decided that the Fourth Amendment to the Constitution, providing for search warrants, was applicable to OSHA, thereby declaring unconstitutional Section 8(a) of the act, in which Congress had authorized warrantless searches.[60]

While the court held that OSHA inspectors are required to obtain search warrants if denied entry to inspect, it added that OSHA must meet only a very minimal *probable cause* requirement under the Fourth Amendment in order to obtain them. As Justice White explained:

> Probable cause in the criminal sense is not required. For purposes of an administrative search such as this, probable cause justifying the issuance of a warrant may be based not only on specific evidence of an existing violation but also on a showing that "reasonable legislative or administrative standards for conducting an . . . inspection are satisfied with respect to a particular [establishment]."[61]

Moreover, if too many companies demanded warrants, so that the inspection program was seriously impaired, the Court indicated that it might reconsider its ruling. This ironically would make enjoyment of a Constitutional right partly contingent on few attempting to exercise it. It is therefore not surprising that commentators, both liberals and conservatives, were critical of the decision. Conservative columnist James J. Kilpatrick declared flatly:

> If the Supreme Court's decision in the *Barlow* case was a "great victory," as Congressman George Hansen proclaims it, let us ask heaven to protect us from another such victory anytime soon.[62]

7.0 Recordkeeping

For an agency that seems grounded in practical workplace realities, OSHA's regulations increasingly emphasize recordkeeping and paperwork requirements. Moreover, recent OSHA enforcement efforts have been directed heavily toward paperwork violations.

7.1 Accident Reports

Any workplace accident requiring treatment or resulting in lost work time must be recorded within six working days on an OSHA Form 300. This is officially entitled the Log and Summary of Recordable Occupational Injuries and Illnesses, although no one uses that longer term. This document is supposed to provide insight into accident types and causes for both the company and OSHA inspectors. It must be retained for five years. Criminal penalties apply to any "knowing false representation" on these and other required records.[63]

There is a new document, Form 300A, which provides additional information and supposedly makes it easier for employers to calculate injury incidence rates.

A third document is the OSHA Form 301, which describes in detail the nature of each of the recorded accidents. All the supporting information does not have to be on this one form, provided that the material is available in the file. This form is officially called the Supplementary Record of Occupational Injuries and Illnesses.

A fourth required document is the Annual Summary of accidents and illnesses, statistics based on the Form 300 data. This summary must be signed by a responsible corporate official and posted in some conspicuous place by the following 1st of February each year.[64]

7.2 Monitoring and Medical Records

OSHA's health standards increasingly contain provisions calling for medical records, monitoring of pollution, and other information. Safety, as well as health standards, may also require periodic inspections of workplaces or equipment, and these inspections must be recorded. These medical and exposure records must be retained for a staggering *30 years*. A company going out of business or liquidating must transfer these records to NIOSH.[65]

For example, the OSHA noise standards mandate baseline and periodic hearing tests,[66] the lead standard requires measuring of blood-lead levels and other data that can be the basis for removal from the workplace until the levels go down; and the ionizing radiation regulation requires careful recording of exposure and absorption information.

A host of safety (and some health) regulations requires (1) written safety programs, (2) specified training, or (3) documented routine inspections, or combinations of all three.

There is no clear pattern to these requirements; they must be checked separately for each regulation. For example, the safety standard on derricks does not call for the first but does call for the second and third requirements, while cranes require only the third.[67] Some safety standards, such as fire protection, lockout/tagout, process safety management, and employee alarms, require all three.[68] The health standards tend to require all three as well, including those for bloodborne pathogens and for hazard communications.[69]

OSHA has increasingly levied substantial fines for failure to comply with these recordkeeping regulations. For example, in October 2000 a Texas steelmaker was fined $1.7 million, much of it for "purposefully" not recording workplace injuries and illnesses.[71]

7.3 Hazard Communication

The OSHA hazard communication (hazcom) program, which is described more fully later in this chapter, requires companies making or using hazardous chemicals to provide information to their workers on possible exposure risks. The program provides for these measures:

1. Toxic chemical labeling,

2. Warning signs and posters,

3. Material Safety Data Sheets (MSDSs) on hazardous chemicals,

4. A written policy setting forth the company's handling of issues under the hazcom program; and

5. A list of hazardous chemicals on premises.[71]

7.4 Access to Records

Employees and their designated legal or union representatives have the right to obtain access to their records within 15 working days. They may not be charged for duplication or other costs. Former employees are also given this access.

There are certain limited exceptions to disclosure dealing with psychiatric evaluation, terminal illness, and confidential informants. Otherwise the view is that even the most secret chemical formulas and business information must be revealed to the employees or former employees if they are relevant to exposure and toxicity. This could be a godsend for industrial espionage, but so far there have been few claims that this is a practical problem.

OSHA inspectors also have access to these records. From time to time some company challenges this access as a violation of the Fourth Amendment, but an inspector has little difficulty in obtaining a search warrant.

7.5 Programmatic Standards

OSHA is giving more attention to programmatic standards. The controversial ergonomics proposed standard, for example, was based on companies providing evidence that they have set up a specific program rather than having OSHA dictate what the detailed content of that program should be. Although these are perhaps not recordkeeping in the absolute sense, their reliance on paperwork and documentation merits their mention in this recordkeeping section.

8.0 Refusal to Work and Whistle-blowing

Employees have a right to refuse to work when they believe conditions are unsafe. OSHA rules protect them from discrimination based on this refusal. And if employees see unsafe or unhealthy workplace conditions, they have a right to report them to OSHA without fear of reprisals or discrimination.

8.1 Refusal to Work

OSHA has ruled, and the Supreme Court has unanimously upheld, the OSHA principle that workers have the right to refuse to work in the face of serious injury or death.[72] The leading case was a simple one. Two workers refused to walk on the thin wire mesh screens suspended high above the workplace through which several workers had fallen partway through and, two weeks before, another worker had fallen to his death. When reprimanded, the workers complained to OSHA. The Supreme Court had no difficulty in finding that the workers had been improperly discriminated against by their employer in this case.

How a court would rule in less glaring circumstances is harder to predict. Interestingly, there has not been a swarm of such cases in the two decades since this decision, despite dire predictions of wholesale refusal and consequent litigation.

8.2 Protection of Whistle-blowing

If a worker is fired or disciplined for complaining to governmental officials about unsafe work conditions, he has a legal remedy under the OSH Act for restoration of his job or loss of pay.[73] Similar provisions, administered also by OSHA's "11(c)" staff, have been inserted into 13 other federal statutes, including EPA's Emergency Planning and Community Right-to-Know Act (EPCRA) in the Superfund legislation, as well as ones that have little to do with environmental or occupational protection.[74]

Congress assumed that the employees in a given workplace would be best acquainted with the hazards there. It therefore statutorily encouraged prompt OSHA response to worker complaints of violations.[75] Since this system could be undermined if employers penalized complaining employees, the Act in Section 11(c) provides sanctions against such retaliation or discrimination:

> No person may discharge or in any manner discriminate against any employee because such employee has filed any complaint or instituted or caused to be instituted any proceeding under or related to this Act or has testified or is about to testify in any such proceeding or because of the exercise by such employee on behalf of himself or others of any right afforded by this Act.[76]

If discrimination occurs, particularly if an employee is fired, a special OSHA team intervenes to obtain reinstatement, back wages, or—if return to the company is undesirable—a cash settlement for the worker. If agreement cannot be reached, the agency resorts to litigation.

This entire system has not worked quite as expected. First, worker complaints have surprisingly not been a very fruitful source of health and safety information. Far too many of the complaints came in bunches, coinciding with labor disputes in a particular plant.[77] OSHA therefore finally had to abandon its policy of trying to investigate every single complaint.[78]

Second, the Title 11(c) process has worked slowly and uncertainly, so even though an employee may receive vindication, the months (or even years) of delay and anguish are a strong disincentive for workers to report hazards.

Third, it is often difficult to determine whether a malcontented worker was fired for informing OSHA or for a number of other issues that might cloud the employer-employee relationship. Does the complaint have to be the sole cause of dismissal or discrimination, or can some (fairly arbitrary) allocation be made?

Fourth, there is continuing controversy over whether 11(c) should protect workers complaining of hazards to those other than OSHA, even if the direct or indirect result is an OSHA inspection.

In the famous Kepone tragedy of 1975, an employee complained of hazardous chemicals to his supervisor, was fired, and only then went to OSHA. Not only was he declared unprotected by the Act, but his complaint, no longer a worker complaint, was not even investigated at the time. Although agency officials have sworn not to repeat that mistake, the issue of what triggers 11(c) protection, either (a) a complaint of unsafe workplace conditions, or (b) reporting that matter to OSHA, is a continuing one.

A related current issue is whether an employee who reports a hazard to the press, whose ensuing publicity triggers an OSHA investigation, is protected by 11(c). In one notable instance, OSHA regional officials decided in favor of the worker and won the subsequent litigation in federal district court. The solicitor of labor, however, disagreed and attempted to withdraw the agency from a winning position.[79]

Still, in one recent case, a Brooklyn bookstore worker supposedly dismissed for whistle-blowing in March 2006 was reinstated with a small cash settlement by August. More typical was an airline worker in Puerto Rico who took four years to litigate through the district court and court of appeals before being awarded a somewhat larger amount.[80] So a worker can never really be sure how he might fare if he does complain.

9.0 Federal and State Employees

The exclusion of federal and state employees has been the topic of much discussion and debate.

9.1 Federal Agencies

Federal employees are not covered directly by OSHA, at least not to the extent that federal agencies are subject to fines and other penalties. However, the presumption was that the agencies would follow OSHA regulations in implementing their own programs. Section 19 of the OSH Act designates the responsibility for providing safe and healthful working conditions to the head of each agency. A series of presidential executive orders has emphasized that this role should be taken seriously. Nevertheless, many commentators feel the individual agencies' programs are inadequate and inconsistent.

In 1980 the leading presidential executive order[81] was issued, which broadened the responsibility of federal agencies for protecting their workers, expanded employee participation in health and safety programs, and designated circumstances under which OSHA will inspect federal facilities. In the operation of their internal OSHA programs, agency heads have to meet requirements of basic program elements issued by the Department of Labor and comply with OSHA standards for the private sector unless they can justify alternatives.

9.2 State Employees

The OSH Act excludes the employees of state governments. Virtually all states with their own OSHA programs—about half—however, cover their state and local employees. Some labor unions believe this exclusion of state workers is one of the most serious gaps in the OSH Act, and several congressional bills have sought in vain to remedy the perceived omission. In light of recent Supreme Court decisions, however, such bills even if enacted might not be constitutional.

10.0 State OSHA Programs

The federal OSHA program was intended by many legislators and businesses only to fill the gaps where state programs were lacking. The states were to be the primary regulatory control. It has not happened that way, of course, but approximately two dozen state programs are still important.[82]

10.1 Concept

The OSH Act requires OSHA to encourage the states to develop and operate their own job safety and health programs, which must be "at least as effective as" the federal program.[83] Until effective state programs were approved, federal enforcement of standards promulgated by OSHA preempted state enforcement,[84] and continue to do so where state laws have major gaps. Conversely, state laws remain in effect when no federal standard exists.

Before approving a submitted state plan, OSHA must make certain that the state can meet criteria established in the Act.[85] Once a plan is in effect, the Secretary may exercise "authority . . . until he determines, on the basis of actual operations under the State plan, that the criteria set forth are being applied."[86] But he cannot make such a determination for three years after the plan's approval. OSHA may continue to evaluate the state's performance in carrying out the program even after a state plan has been approved. If a state fails to comply, the approval can be withdrawn, but only after the agency has given due notice and opportunity for a hearing.

10.2 Critiques

The program has not developed as anticipated into an essentially state-oriented system, although almost half the states have their own system.

Organized labor has never liked the state concept, both because of its poor experience with state enforcement in the past and because it realized that its strength could more easily be exercised in one location—Washington, D.C.—than in all fifty states and the territorial capitals, many of which are traditionally hostile to labor unions. This has meant, ironically, that some of the better state programs, in areas where unions had the most influence, were among the first rejected by state legislators under strong union pressure.

Industry has cooled to the local concept, which requires multistate companies to contend with a variety of state laws and regulations instead of a uniform federal plan. Furthermore, state OSHAs are often considerably larger than the local federal force, so there can be more inspections.

It was therefore never clear what incentive a state had to maintain its own program, since a governor could always terminate his state's plan and save the budgetary expenses, knowing that the federal government would take up the slack. California's Governor George Deukmejian, for example, came to this conclusion in 1987 and terminated the state Cal-OSHA. However, the idea did not stick; California's state program was soon reestablished and, surprisingly, the notion did not spread.

Organized labor and industry are not alone in their criticism of the state programs. Health research organizations, OSHA's own national advisory committee (NACOSH), and some of the states themselves have also voiced disapproval of the state program policy. Ineffective operations at the state level, disparity in federal funding, and the lack of the necessary research capability are just a few of the criticisms lodged.[87]

There is some defense of state control, however: "To the extent that local control increases the responsiveness of programs to the specific needs of people in that area, this [a state plan] is a potentially good policy."[88] But reevaluation and revision will be necessary in the next several years if OSHA's policy for state programs is to be accepted by all the factions involved.

11.0 Consultation

Employers subject to OSHA regulation, particularly small employers, would benefit from on-site consultation to determine what must be done to bring their workplaces into compliance with the requirements of the OSH Act. This was particularly true during the agency's formative years. Although OSHA's manpower and resources are limited, this assistance, where rendered, should be free from citations or penalties.

OSHA typically conducts around 30,000 consultative visits a year to small and medium companies, in addition to a million or so telephone enquiries to its telephone service and many millions of hits on the internet service. Larger companies have been aided by programs such as the Safety and Health Achievement a recognition Program (SHARP).

As in so many other areas of OSHA regulation, there has been a great deal of controversy surrounding the consultation process. Union leaders have always feared that OSHA could become merely an educational institution rather than one with effective enforcement. But Section 21(c) of the Act does mandate consultation with employers and employees "as to effective means of preventing occupational injuries and illnesses."[90]

11.1 Education

Along with the consultation provisions, the statute provides for "programs for the education and training of employers and employees in the recognition, avoidance, and prevention of unsafe or unhealthful working conditions in employments covered" by the Act.[91] OSHA produces brochures and films to educate employees about possible hazards in their workplaces. But there are problems at every stage of the information process, from generation to utilization.

Back in 1979, OSHA began experimenting with a New Directions Training and Education Program, which made available millions in grants to support the development and strengthening of occupational safety and health competence in business, employee, and educational organizations. This program supported a broad range of activities, such as training in hazard identification and control; workplace risk assessment; medical screening and recordkeeping; and liaison work with OSHA, the National Institute for Occupational Safety and Health, and other agencies. "The goal of the program was to allow unions and other groups to become financially self-sufficient in supporting comprehensive health and safety programs."[89] This program, criticized by some as a payoff to constituent groups, especially labor unions, was a natural target of the budget cutters during the Reagan administration, but the concept of increased consultation has been given even greater emphasis.

There is also a provision that state plans may include on-site consultation with employers and employees to encourage voluntary compliance.[93] The personnel engaged in these activities must be separate from the inspection personnel, and their existence must not detract from the federal enforcement effort. These consultants not only point out violations, but also give abatement advice.

11.2 Alliances

Much of OSHA's focus over the past decade has been in arranging "alliances" with trade associations, businesses, professional groups, and even universities. The present program, initiated in March 2002, attempts to enlist other organizations in the safety fight and also serves to make the public image of OSHA less confrontational. Whether this will result in better workplace safety and health than, say, reviving the near-dead standard-setting effort remains to be seen.

12.0 Overlapping Jurisdiction

There are other agencies involved with statutory responsibilities that affect occupational safety and health. These agencies indirectly regulate safety and health matters in their attempt to protect public safety.

One example of an overlapping agency is the Department of Transportation and its constituent agencies, such as the Federal Railroad Administration and the Federal Aviation Administration. These agencies promulgate rules concerned with the safety of transportation crews and maintenance personnel, as well as the traveling public, and consequently overlap similar responsibilities of OSHA.

Section 4(b)(1) of the OSH Act states that when other federal agencies "exercise statutory authority to prescribe or enforce standards or regulations affecting occupational safety or health," the OSH Act will not apply to the working conditions addressed by those standards. MOUs between these agencies and OSHA have eliminated much of the earlier conflict.

The Environmental Protection Agency is the organization that overlaps most frequently with OSHA. When a toxic substance regulation is passed by EPA, OSHA is affected if that substance is one that appears in the workplace. For instance, both agencies are concerned with pesticides, EPA with the general environmental issues surrounding the pesticides and OSHA with some aspects of the agricultural workers who use them. During the early 1970s, there was a heated interagency conflict over field reentry standards for pesticides, a struggle that spilled over into the courts and eventually had to be settled by the White House in EPA's favor.[94] The OSHA-EPA MOU of 1990 and similar such agreements hopefully will prevent repetitions of such problems.

Thus, although the health regulatory agencies generally function in a well-defined area, overlap does occur. As another example, there are toxic regulations under Section 307 of the Federal Water Pollution Control Act, Section 112 of the Clean Air Act, and under statutes of the Food and Drug Administration (FDA) and Consumer Product Safety Commission (CPSC). These regulatory agencies realized the need for coordination, particularly when dealing with something as pervasive as toxic substances, and under the Carter administration combined their efforts into an interagency working group called the Interagency Regulatory Liaison Group (IRLG). Although the IRLG was abolished at the beginning of the Reagan administration, the concept of interagency working groups is a good one. The federal agencies involved in regulation should rid themselves of the antagonism and rivalry of the past and cooperate with one another to meet the needs of the public.

13.0 Occupational Safety and Health Review Commission

The OSH Act established the Occupational Safety and Health Review Commission (OSHRC) as "an independent quasi-judicial review board"[95] consisting of three members appointed by the president to six-year terms. Any enforcement actions of OSHA that are challenged must be reviewed and ruled upon by the Commission.[96]

13.1 OSHRC Appeal Process

Any failure to challenge a citation within fifteen days of issuance automatically results in an action of the Review Commission to uphold the citation. This decision by default is not subject to review by any court or agency. When an employer challenges a citation, the abatement period, or the penalty proposed, the Commission then designates a hearing examiner: an administrative law judge who hears the case; makes a determination to affirm, modify, or vacate the citation or penalty; and reports his finding to the Commission.[97] This report becomes final within thirty days unless a Commission member requests that the Commission itself review it.

The employer or agency may then seek a review of the decision in a federal appeals court.

13.2 Limitations of the Commission

One of the major problems with the Review Commission is the question of its jurisdiction: "The question has arisen of the extent to which the Commission should conduct itself as though it were a court rather than a more traditional administrative agency."[98] The Commission cannot look to other independent agencies in the government for a resolution of this problem "because its duties and its legislative history have little in common with the others."[99] It cannot conduct investigations, initiate suits, or prosecute; therefore, it is best understood as an administrative agency with the limited duty of "adjudicating those cases brought before it by employers and employees who seek review of the enforcement actions taken by OSHA and the Secretary of Labor."[100]

Another problem inherent in the organization of the Commission is the separation from the president's administration. There has been a question of where the authority of the administration ends and the authority of the Commission begins. Because of the autonomous nature of the Review Commission, it cannot always count on the support of the Executive agencies. In fact, OSHA has generally ignored Review Commission decisions, and few inspectors are even aware of the Commission interpretations on various regulations.

14.0 National Institute for Occupational Safety and Health

The standard-recommending arm of OSHA is actually in a totally separate government department. This procedure has never worked well, even at best during the late 1970s, and has now ceased to work at all.

14.1 In Theory

Under the OSH Act, the Bureau of Safety and Health Services in the Health Services and Mental Health Administration was restructured to become the National Institute for Occupational Safety and Health (NIOSH), so as to carry out HEW's responsibilities under the Act.[101] (HEW—the Department of Health, Education, and Welfare—has since become the Department of Health and Human Services, HHS.) NIOSH reports, illogically, to the Centers for Disease Control (CDC), and the two organizations have headquarters in Atlanta.

From the beginning NIOSH has claimed the training and research functions of the Act, along with its primary function of recommending standards. For this latter task, NIOSH provides recommended standards to OSHA in the form of criteria documents for particular hazards. These are compilations and evaluations of all available relevant information from scientific, medical, and (occasionally) engineering research.

The order of hazards selected for criteria development is determined several years in advance by a NIOSH priority system based on severity of response, population at risk, existence of a current standard, and advice from federal agencies (including OSHA) as well as involved professional groups.[102] The criteria documents may actually have some value apart from the role in standards-making. Even though they do not have the force of law, they are widely distributed to industry, organized labor, universities, and private research groups as a basis to control hazards. The criteria documents also serve as a "basis for setting international permissible limits for occupational exposures."[103]

14.2 In Practice

To the extent that certain criteria documents may be deficient, as discussed earlier, this expansive role for them among laymen poses a real problem. This problem may unfortunately become worse if NIOSH declines in both funds and morale. Nevertheless, there is arguably some small benefit in having the two organizations separate. NIOSH has on occasion criticized OSHA for regulatory decisions that the former believed were scientifically untenable.

A review of NIOSH criteria projects shows that virtually all date from the 1970s and still await OSHA action. Almost no new documents have emerged in the past two decades.

Since OSHA does not react to its recommendations, the agency has tried to reinvent itself as a guide to the public. For example, in 2006, the U.S. Chemical Safety and Hazard Investigation Board (CSB) proposed that NIOSH include research on chemical production safety in the National Occupational Research Agenda (NORA). This suggestion would focus the organization on ways to prevent, for example, the series of deadly explosions at oil refineries in Texas and elsewhere.[104]

Not under consideration is the idea of just eliminating the moribund agency.

15.0 Hazard Communication Regulations

OSHA's output of health standards has never been impressive. In recent years, it has tried three new approaches to get around this bottleneck. The first was the "federal" cancer policy designed to create a template for dealing in an expedited fashion with a number of hazardous chemicals. The second was the wholesale review initiated in 1988 of all the Z-1 list consensus standards—an effort struck down by the courts.

The third, characterized by one OSHA official as the agency's most important rulemaking ever, is the hazard communication (hazcom) regulation issued in November 1983.[105]

15.1 Reason for the Regulation

This standard, sometimes known as the *worker right-to-know rule*, provides that hazardous chemicals must be labeled, Material Safety Data Sheets (MSDSs) on hazards be prepared, and workers and customers should be informed of potential chemical risks.

How could a rule with such far-reaching consequences be issued from an administration that so stressed deregulation and deliberately avoided issuing other protective regulations? The answer lies in an almost unprecedented grassroots movement at the state and municipal level to enact their own "worker right-to- know" laws that, many businessmen felt, could be a considerable burden on interstate commerce. They therefore lent their support to OSHA in its confrontation with the Office of Management and Budget (OMB) at the White House. A federal regulation on this subject would arguably preempt the multiplicity of local laws.

The rule was originally presumed to apply to only a few hundred, perhaps a thousand, particularly hazardous chemicals. The individual employers would evaluate the risk and then decide for themselves which products merited coverage. Most employers were unable or unwilling to make such scientific determinations. Within a year or two, this limited program expanded into universal coverage.

15.2 Scope and Components

Published on 25 November 1983, OSHA's Hazard Communication or "Right-to-Know" Standard[106] went into effect two years later, in November 1985, for chemical manufacturers, distributors, and importers, and in May 1994 for manufacturers that use chemicals. It required that employees be provided with information concerning hazardous chemicals through labels, Material Safety Data Sheets, training and education, and lists of hazardous chemicals in each work area. Originally it covered only manufacturing industries classified in SIC codes 20–39, but by court order in 1987, it was extended to virtually all employers.[107]

Every employer must assess the toxicity of chemicals it makes, distributes, or uses based on guidelines set forth in the rule. Then it must provide this material downstream to those who purchase the chemicals through MSDSs.[108] The employers are then required to assemble a list of the hazardous materials in the workplace, label all chemicals, provide employees with access to the MSDSs, and provide training and education. While all chemicals must be evaluated, the "communication" provisions apply—in theory—only to those chemicals known to be present in the workplace in such a way as to potentially expose employees to physical or health hazards.

Special provisions apply to the listing of mixtures that constitute health hazards. Each component that is itself hazardous to health and that comprises one percent or more of a mixture must be listed. Carcinogens must be listed if present in quantities of 0.1% or greater.

The Hazard Communication Standard is a performance-oriented rule. While it states the objectives to be achieved, the specific methods to achieve those objectives are at the discretion of the employer. Thus, in theory, employers have considerable flexibility to design programs suitable for their own workplaces. However, this may mean the employers will have questions about how to comply with the standard.

The purpose of labeling is to give employees an immediate warning of hazardous chemicals and a reminder that more detailed information is available. Containers must be labeled with identity, appropriate hazard warnings, and the name and address of the manufacturer. The hazard warnings must be specific, even as to the endangered body organs. For example, if inhalation of a chemical causes lung cancer, the label must specify that and cannot simply say "harmful if inhaled" or even "causes cancer." Pipes and piping systems are exempt from labeling, as are those substances required to be labeled by another federal agency.

MSDSs, used in combination with labels, are the primary tools for transmitting detailed information on hazardous chemicals. An MSDS is a technical document that summarizes the known information about a chemical. Chemical manufacturers and importers must develop an MSDS for each hazardous chemical produced or imported and pass it on to the purchaser at the time of the first shipment. The employer must keep these sheets where employees will have access to them at all times.

The purpose of employee information and training programs is to inform employees of the labels and MSDSs and to make them aware of the actions required to avoid or minimize exposure to hazardous chemicals. The format of these programs is left to the discretion of the individual employer. Training programs must be provided at the time of initial assignment and whenever a new hazard is introduced into the workplace.

15.3 Hazard Evaluation

Chemical manufacturers are required to evaluate all chemicals they sell for potential health and physical hazards to exposed workers. Purchasers of these chemicals may rely on the supplier's determination or may perform their own evaluations.

There are really no specific procedures to follow in determining a hazard. Testing of chemicals is not required, and the extent of the evaluation is left to the manufacturers and importers of hazardous chemicals. However, all available scientific evidence must be identified and considered. A chemical is considered hazardous if it is found to be so by even a single valid study.

Chemicals found on the following master lists are automatically deemed hazardous under the standard:

• The International Agency for Research on Cancer (IARC) monograph

• The *Annual Report on Carcinogens* published by the National Toxicology Program (NTP)

• OSHA's Subpart Z list, found in Title 29 of the Code of Federal Regulations, Part 1910 or

• Threshold Limit Values for Chemical Substances and Physical Agents in the Work Environment, published by the American Conference of Governmental Industrial Hygienists

If a substance meets any of the health definitions in Appendix A of the standard, it is also to be considered hazardous. The definitions given are for a carcinogen, a corrosive, a chemical that is highly toxic, an irritant, a sensitizer, a chemical that is toxic, and target organ effects.

Appendix B of the standard gives the principal criteria to be applied in complying with the hazard determination requirement. First, animal as well as human data must be evaluated. Second, if a scientific study finds a chemical to be hazardous, the effects must be reported whether or not the manufacturers or importers agree with the findings.

Appendix C of the standard gives a lengthy list of sources that may assist in the evaluation process. The list includes company data from testing and reports on hazards, supplier data, MSDSs or product safety bulletins, scholarly text books, and government health publications.

In practice, as noted above, companies have begun requiring MSDSs from manufacturers for all chemicals they purchase, so the evaluation aspect of the standard has become unimportant.

15.4 Trade Secrets

Although there is agreement that there must be a delicate balance between the employee's right to be free of exposure to unknown chemicals and the employer's right to maintain reasonable trade secrets, the exact method of protection has been considerably disputed.

Under the standard, a trade secret is considered to be defined as in the *Restatement of Torts*, that is, something that is not known or used by a competitor. However, OSHA had to revise its definition to conform to a court ruling that said that a trade secret may not include information that is readily discoverable through reverse engineering.

Although the trade secret identity may be omitted from the MSDS, the manufacturer must still disclose the health effects and other properties about the chemical. A chemical's identity must immediately be disclosed to a treating physician or nurse who determines that a medical emergency exists.

In nonemergency situations, any employee can request disclosure of the chemical's identity if he demonstrates through a written statement a *need to know* the precise chemical name and signs a confidentiality agreement. The standard specifies all purposes that OSHA considers demonstrate the need to know a specific chemical identity.

The standard initially limited this access to health professionals, but in 1985, the U.S. Court of Appeals for the Third Circuit ruled that trade secrets protections must be narrowed greatly, allowing not only health professionals, but also workers and their designated representatives the same access as long as they follow the required procedures.[109] In response, OSHA issued a final rule on trade secrets in September 1986[110] that narrows the definition of *trade secret*. It denies protection to chemical identity information that is readily discoverable through reverse engineering. It also permits employees, their collective bargaining representatives, and occupational nurses access to trade secret information.

Upon request, the employer must either disclose the information or provide written denial to the requester within 30 days. If the request is denied, the matter may be referred to OSHA, whereupon evidence to support the claim of trade secret and alternative information that will satisfy the claim are needed.

15.5 Federal Preemption Controversy

Several states and labor groups have filed suits challenging state laws that are more protective. New Jersey, for example, has enacted the toughest labeling law in the nation, requiring industry to label all its chemical substances, whether they are hazardous or not, and supply the information to community groups and health officials, as well as to workers.

They were also concerned that, because the original OSHA standard only covered the manufacturing sector, more than 50 percent of the workers (such as those workers in the agricultural and construction fields) would be unprotected, and OSHA did not cover (and still does not) such groups as state employees and consumers. Moreover, they argued that OSHA would be incapable of enforcing worker protection because of the staff cuts made by the Reagan administration.

The chemical industry, on the other hand, favored a uniform federal regulation because they believed it would be less costly and easier to comply with one federal rule as opposed to several state and local rules that would often conflict or be confusing.

In October 1985, the U.S. Court of Appeals for the Third Circuit ruled that the federal Hazard Communication Standard does not preempt all sections of New Jersey's right-to-know laws designed to protect workers and the public from chemical exposure—only those that apply to groups the agency's rules covered, which were then only in the manufacturing sector.[111] Thus, while some parts of a state law may be preempted, other provisions may not be.

In September 1986, the Third Circuit also found that the federal Hazard Communication Standard did not entirely preempt requirements under Pennsylvania's right-to-know act pertaining to worker protection in the manufacturing industry where the state rules relate to public safety generally and for protection of local government officials with police and fire departments. However, five days later, also in September 1986, the U.S. Court of Appeals for the Sixth Circuit ruled that a right-to-know ordinance enacted by the city of Akron, Ohio, is preempted by the federal standard in manufacturing sector workplaces.

In 1992, the Supreme Court came down strongly on the side of preemption. The *Gade v. National Solid Waste Management Association* case, although it involved OSHA's so-called HAZWOPER regulations[112] rather than hazard communication, involved a state law requiring additional training for heavy equipment operators on hazardous waste sites. The high court found that the OSHA regulations preempted the state despite arguments that the federal rules only set a minimum that the state could exceed—the situation in most environmental laws—and the more transparent claim that the state laws had a dual purpose in protecting the public as well as workers.[113]

In 1997 a unanimous federal appeals court, relying on *Gade*, held that the OSHA hazard communication rule preempted California's famous Proposition 65 requirements (the public must be warned of carcinogens and other harmful substances, including buildings).

16.0 Ergonomics Issues

The ergonomics issue has been one of the few championed by OSHA in recent years. Ironically, when the standard finally emerged, it ran into a buzz saw of hostility that forced OHA to retreat. Opposition ranged from criticism of it as a defective standard, to the concern that ergonomic issues were highly particular to a given workplace. This controversy aptly illustrates the interplay of factors affecting OSHA's performance.

16.1 Background

For almost three decades OSHA officials have worked towards developing a standard on ergonomics. The original impetus was a series of reports from Midwest poultry and meatpacking plants that workers were developing "carpal tunnel syndrome" (CTS). This condition develops from repetitive motion of the hand and wrist, which irritates the nerve running through a bone channel near the thumb. Because similar conditions can develop from repetitive motion or strain in other parts of the body, such as with "tennis elbow," the malady was relabeled as "cumulative trauma syndrome," also conveniently abbreviated CTS, then changed to "repetitive motion syndrome," and so on until the more sweeping term "ergonomics" was adopted.

Along the way, OSHA was hitting offending companies with fines in the millions of dollars, some of the biggest in the agency's history. All of this had to be done under OSHA's "general duty clause," the famous Section 5(a)(1) of the act, because there was no specific standard that addressed this particular condition. With the congressional rejection of OSHA's Ergonomic Standard, OSHA may again fall back to the general duty clause to deal with clear cases of abuse.

16.2 Scope of the Problem

Ergonomics was not a new word or a new concept. It had long been used in Europe to denote arrangements of workers and tools that maximized productivity with a minimum of wasted effort. This was based on, ironically, the American-developed "time and motion studies" from almost a century ago. The *erg* in ergonomics, after all, is from the Greek word meaning "work." The term also came to be used in furniture and office design with the connotation of comfortable and well laid out.

The workplace collision came when the concept of mass production—with each worker repeating a number of simple steps all day—clashed with the possible physical irritation caused to certain parts of the body. The better companies sought to deal with the problem, though most were concerned with the boredom and carelessness aspects of endless repetition rather than with possible deleterious effects on the body. The remedies, however, tended to be very specific to each worksite or even each job. So how could a general standard be developed?

A very different problem was raised for OSHA. Considering the host of unregulated chemicals, life-threatening workplace hazards, and a pathetically slow agency pace for dealing with them, is this where OSHA should be putting its priorities for at least a decade?

Congress did not think so, and for a number of years put a "rider" on OSHA's appropriation bills that such an omnibus ergonomic standard should not be developed. Due to a congressional slip-up, however, and the confusion in the year right before the 2000 presidential elections, OSHA was able to slide out a proposal in November 2000, due to take effect just days before the new president and Congress took office and could do anything about it. Once in effect, it was legally much more difficult to overturn it, except for denying appropriations for enforcement, and a closely divided House and Senate had more pressing issues.[114]

16.3 Scope of the Standard

The new standard that emerged in 2000 was designed to reduce the incidence of musculoskeletal disorders (MSDs) by requiring that companies establish programs to prevent them. In other words, the standard is not prescriptive but procedural.

The standard applied to all general industry workplaces under OSHA but not, for technical legal reasons, to the construction, maritime, agricultural, or most railroad operations. They were eventually supposed to have their own standards once the legal steps were completed. Being subject to the standard, however, does not mean that it automatically applies in its entirety. Some actions have to be taken, and others need to occur only after an *action trigger*. The trigger was, in Western parlance, a hair-trigger that would go off very easily. Therefore, most workplaces expected to fall under its provision sooner rather than later.

Certain specified *initial actions* had to be taken by every employer: Every employee had to be given, first, basic information about MSDs, including symptoms and reporting obligations; second, a summary of the requirements of the Act; and third, a written notice in a conspicuous place or by electronic communication.

An action trigger occurs when an employee reports a work-related MSD that rises above a certain threshold, namely when (1) the disorder requires days away from work, restricted work, or medical treatment beyond first aid; or (2) when the symptoms last for more than seven consecutive days. The trigger is then met if the employee's job "routinely involves, on one or more days a week, exposure to one or more relevant risk factors at the levels described in the Basic Screening Tool in Table W-1." In making this determination an employer could seek assistance from a health care professional (HCP), who plays a key role in implementation of the subsequent program.

17.0 Legislation

The OSH Act has remained virtually untouched since its passage in 1970. With the inauguration of a Democratic president, William Clinton, in 1993 and Democratic control of both houses of Congress, the expectation was that the labor unions would secure the passage of the first significant revisions in the law.

Under the circumstances, the proposed legislation was surprisingly innocuous. It included verbose and often unnecessary sections on enforcement, refusal to work, and other issues. Among them was a seemingly innocuous section providing for labor-management safety committees in the workplace. Both employers and employees have found these committees quite useful, but some manufacturers' organizations criticized the language as forcing a much greater role for labor unions.

With the Republican election victories in the House and Senate in late 1994, not only did these Democratic legislative plans collapse, but the victors prepared their own onslaught on the OSH Act. To the surprise of many, the draconian Republican plans to curtail or even eliminate OSHA got no further than the previous Democratic plans: "Organized labor counted its victories in this year's Congress by the number of bills defeated rather than enacted."[115] However, the Republicans' hostile scrutiny of OSHA paralyzed the agency leadership and led to a sharp decline in both enforcement and standard setting.

Congressional oversight has been most intense regarding OSHA's proposed ergonomics standard. As discussed above, Congress used the appropriations process to order the agency not to issue the standard, while the Clinton administration eventually refused to sign the legislation that included these prohibitions on the eve of the 2000 presidential elections.

However, legislation was also introduced to direct OSHA to encourage safer medical needles, give small businesses more input into agency regulatory proceedings, bar home office inspections, and (signed into law) expand federal compensation for radiation-related exposure.[116] Since then, except for brief outbursts, usually following some regrettable accident, serious efforts to introduce new OSHA legislation have been, well, not.

Notes

1. "Report Card In on Government Agencies," Associated Press (AP), 2 February 1999. Nevertheless, in an- other study, OSHA tied with the Internal Revenue Service for the lowest ranking among federal agencies in terms of customer approval. University of Michigan Business School, "American Customer Satisfaction Re- port," 15 December 1999. Not everything has improved. A detailed critique of OSHA prepared by an out-going senior official a quarter-century ago could regrettably be reissued today with relatively few changes. *See* "Report on OSHA: Regulatory and Administrative Efforts to Protect Industrial Health," January 1977, 108 pp., by the author of this chapter.

2. To prevent this overlap from causing jurisdictional confusion, the two agencies developed a Memorandum of Understanding (MOU) in 1990 to delineate and coordinate their respective activities. OSHA-EPA MOU, 23 November 1990.

3. In the fifteen-month interval between Henshaw and Foulke, a Deputy Assistant Secretary, Texas lawyer Jonathan Snare, served as acting Assistant Secretary.

4. Occupational Safety and Health Act of 1970, PL 91-596, 84 Stat. 1590.

5. This lack of change could obviously also be considered a negative factor, but a comparison with some of EPA's ponderously detailed legislation shows the benefits of keeping the basic statute simple. OSHA annual appropriations legislation, however, has been modified several times to restrict OSHA authority over small businesses, farming, hunting, and other subjects.

6. OSH Act § 6(b)(5); emphasis added.

7. This issue will be discussed in detail in section 4.6 of this chapter.

8. Clean Air Act § 112, 42 U.S.C. § 1857, the National Emission Standards for Hazardous Air Pollutants (NESHAP).

9. OSH Act § 4(a)–4(b)(2).

10. OSH Act § 3(5). Congress's annual appropriations language has excluded several "peripheral" categories of employers in the past few years.

11. 37 FR 929, 21 January 1972, codified at 29 Code of Federal Regulations (C.F.R.) § 1975.

12. EPA learned this lesson back in 1984 when Deputy Administrator James Barnes quite properly deferred to OSHA on certain asbestos workplace matters. Congressional

critics, who believed OSHA would not treat the matter seriously or competently, raised such furor that EPA retained jurisdiction. Even earlier, in 1973, OSHA and EPA had an acrimonious dispute over which agency should have primary jurisdiction over protecting farm workers from pesticides. EPA won.

13. Richard Fairfax, director of the Office of Compliance Programs, opinion letter to CSC Credit Services of Houston, Texas, 15 November 1999. Lest one think this was merely a hasty OSHA response, note that the company's request for an opinion was submitted in August 1997, 27 months before.

14. Jenny Krese, director of NAM's employment policy, Bureau of National Affairs (BNA), *OSHA Reporter*, 6 January 2000, p. 5.

15. Rep. Pete Hoekstra (R-Mich.), chairman of the House Oversight and Investigations Subcommittee of the House Education and Workforce Committee, *id.*, 13 January 2000, p. 22.

16. These consensus standards are discussed in sections 3.4 and 4.1 of this chapter.

17. The American Federation of Labor and Congress of Industrial Organizations (AFL-CIO) v. OSHA, 965 F.2d 962 (11th Cir. 1992).

18. Statement of Basil Whiting, Deputy Assistant Secretary of Labor for OSHA, before the Committee on La- bor and Human Resources, U.S. Senate, 21 March 1980, pp. 5–6.

19. For the legislative history of the Act, *see especially* the Conference Report 91-1765 of 16 December 1970, as well as H.R. 91-1291 and S.R. 91-1282.

20. *See Marshall v. Barlow's Inc.*, 436 U.S. 307 (1978), which is discussed later in this chapter.

21. 39 FR 23502, 27 June 1974.

22. 29 C.F.R. § 1910.101.

23. Since the 6(a) process ended in April 1972, the standards promulgated thereunder cannot be modified or revised without going through the notice and comment administrative procedures under Section 6(b).

24. *See*, for example, the OSHA press release of 19 June 1998: "OSHA Eliminates Over 1,000 Pages of Regulations to Save Employers Money, Reduce Paperwork, and Maintain Protection."

25. 5 U.S.C. § 553 et seq.

26. NIOSH criteria documents vary considerably in quality, depending in part on to whom they were subcontracted, but another problem is that too often they are insufficiently discriminating in evaluating question- able studies. That is, one scientific study is regarded as good as any other study, without regard to the quality of the data or the validity of the protocols. Of course, another factor in OSHA's attitude just might be the "not invented here" syndrome. This is discussed in detail in section 14 of this chapter.

27. This meager number of chemicals does *not* reflect OSHA's scientific judgment that the other candidates are unworthy or that the agency has sharply different priorities,

although these may be partial factors. More important reasons are poor leadership, technical inexperience, and a bit of politics.

28. OSH Act § 6(c)(1).

29. OSH Act § 5(a)(1); 29 U.S.C. § 654(a)(1).

30. This is exemplified by *Phelps Dodge Corporation* (OSHRC Final Order, 1980), 9 OSHC 1222, which found no violation of the Act to expose workers to "massive amounts of sulphur dioxide for short periods of time" since there was no maximum ceiling value in the standard and the employer was complying with the eight- hour average value required in the specific sulphur dioxide regulation. The citation for violation of § 5(a)(1) was therefore vacated.

31. *International Union, UAW v. General Dynamics Land System Division*, 815 F.2d 1570, 13 OSHC 1201 (CADC 1988). The Court held that employer's knowledge was the crucial element; if he knew that the OSHA standard was not adequate to protect workers from a hazard, he could not claim he was maintain- ing a safe workplace within the meaning of § 5(a)(1), even if he were adhering to a standard he knew was outmoded. The Court thereby dismissed the argument that the employer would not know what is legally expected of him; he was expected to maintain a safe workplace, specific regulations notwithstanding. There was no specific provision in the statute that prevented a general duty citation when a specific standard existed. Note, however, that no other court has since used this rationale.

32. This account of the behind-the-scenes machinations is based on this author's personal discussions with the late Congressman William Steiger (R-Wisc.), a principal author of the Act, and Judge Lawrence Silberman, now of the Court of Appeals for the District of Columbia Circuit, who was then solicitor of labor. The legislative history is relatively unhelpful on this subject. *See*, for example, hearings before the Select Sub- committee on Labor, Committee on Education and Labor, "Occupational Safety and Health Act of 1969," two vols., 1969.

33. 499 F.2d 467, 1 OSHC 1631 (D.C. Cir., 1974).

34. 1 OSHC 1631 at 1639.

35. 509 F.2d 1301, 2 OSHC 1496 (2nd Cir., 1975), *cert. den.* 421 U.S. 922.

36. 509 F.2d at 1309, 2 OSHC at 1502 (2nd Cir., 1975).

37. *See*, for example, the conclusions of the National Academy of Sciences report, "Government Regulation of Chemicals in the Environment," 1975.

38. 561 F.2d 82, 5 OSHC 1970 (7th Cir., 1977). The Occupational Safety and Health Review Commission (OSHRC) decisions on *Turner* and the related *Continental Can* case can be found at 4 OSHC 1554 (1976) and 4 OSHC 1541 (1976), respectively.

39. 5 OSHC 1790 at 1791.

40. 692 F.2d 641, 10 OSHC 2169 (1982).

41. Engineering controls are those that reduce the sound intensity at the source of the noise. This is achieved by insulation of the machine, by substituting quieter machines

and processes, or by isolating the machine or its operator. Administrative controls attempt to reduce workers' exposure to excess noise through use of variable work schedules, variable assignments, or limiting machine use. Personal protective equipment in- cludes such devices as ear plugs and ear muffs provided by the employer and fitted to individual workers.

42. 101 S. Ct. 2478, 9 OSHC 1913 (17 June 1981). In this case, representatives of the cotton dust industry challenged proposed regulations limiting permissible exposure levels to cotton dust. Section 6(b)(5) of the Act requires OSHA to "set the standard which most adequately assures, to the extent feasible . . . that no employee will suffer material impairment of health...." The industry contended that OSHA had not shown that the proposed standards were economically feasible. However, the Supreme Court upheld the cotton dust regulations, holding that the "plain meaning of the word 'feasible' is capable of being done, executed, or effected," and that a cost-benefit analysis by OSHA is not required.

43. OSH Act § 7(b)(6)(A).

44. Id.

45. E. Klein, Variances, in Proceedings of the Occupational Health and Safety Regulations Seminar (Washington, D.C.: Government Institutes, 1978), p. 74.

46. See a later section in this chapter on the Supreme Court's Barlow's decision interpreting the Fourth Amend- ment to the U.S. Constitution.

47. This happened with kepone in the notorious Hopewell, Virginia, incident in 1975, and with asbestos for years at a plant in Tyler, Texas.

48. Statement of Lane Kirkland, president, AFL-CIO, before the Senate Committee on Labor and Human Re- sources on Oversight of the Occupational Safety and Health Act, 1 April 1980.

49. This is discussed in a separate chapter later in this book.

50. In 2006 there were 38,579 federal OSHA inspections resulting in 83,913 violations, of which 61,337 (73.1 percent) were categorized as "Serious": state occupational safety and health agencies did another 58,058 in- spections finding 127,284 violations, of which 45.3 percent was classified as Serious.

51. The SIC codes are officially being supplanted by something called the North American Industry Classifi- cation System (NAICS), but the transition is a gradual one.

52. "Top Ten Federal OSHA Targeted SIC Codes," *Manufacturing Sector*, 4th Quarter 1998, OSHA.

53. *The Carborundum Company* (OSHRC Judge, 1982), 10 OSHC 1979.

54. *Schnabel Associates, Inc.* (OSHRC Judge, 1982), 10 OSHC 2109. Moreover, employers should retain copies of training curriculum, tests, and other evidence of the educational program, recommends Susan M. Olan- der, counsel for the Federated Rural Electric Insurance Exchange. BNA, *OSHA Reporter*, 19 October 200, pp. 933–934.

55. *Galloway Enterprises, Inc.* (OSAHRC Judge, 1984), 11 OSHC 2071.

56. *Bethlehem Steel Corporation, Inc.* (OSAHRC Judge, 1985), 12 OSHC 1606. *Dover Electric Company, Inc.* (OSAHRC Judge, 1984), 11 OSHC 2175.

57. *Brock v. L. E. Myers Company*, 818 F.2d 1270, 13 OSHC 1289 (6th Cir., 1987).

58. A good, if dated, summary of these challenges is found in Volume I of *A Practical Guide to the Occupational Safety and Health Act*, by Walter B. Connolly and David R. Cromwell, II, (New York: New York Law Jour- nal Press, 1977).

59. 436 U.S. 307 (1978).

60. There are non-OSHA circumstances in which warrants are not required, such as federal inspection of liquor dealers, gun dealers, automobiles near international borders, and in other matters with a long history of fed- eral involvement.

61. *Marshall v. Barlow's Inc.*, *supra*, quoting *Camara v. Municipal Court*, 387 U.S. 523 at 538 (1967).

62. Washington Star, 2 June 1978.

63. OSH Act § 17(g).

64. Recordkeeping requirements are set forth generally in 29 C.F.R. 1904.

65. 29 C.F.R. § 1910.20.

66. OSHA's noise monitoring and recordkeeping requirements for hearing loss and *standard threshold shift* (STS, previously "significant threshold shift") are particularly complex and have been subject to considerable liti- gation.

67. 29 C.F.R. § 1910.181, and § 1910.178–179.

68. 29 C.F.R. §§ 1910.156 et seq., 1910.147, 1910.119, and 1910.165.

69. 29 C.F.R. § 1910.1030, and 29 C.F.R. § 1200.

70. BNA, *OSHA Reporter*, 26 October 2000, p. 951.

71. 29 C.F.R. 1910.1200.

72. 29 C.F.R. § 1977.12 (1979); *Whirlpool Corp. v. Marshall*, 445 U.S. 1, 8 OSHC 1001 (1980). This case also stands as a textbook example of when not to appeal a lower court's ruling.

73. OSH Act § 11(c), 29 U.S.C. § 660; 29 C.F.R. § 1977.

74. Title III of the Superfund Amendment and Reauthorization Act of 1986 (SARA). These amendments are designed to "prevent future Bhopals" (the 1984 chemical disaster that killed thousands of residents of the city of Bhopal, India) by informing community fire and emergency centers what chemicals a company has on site. Dominique Lapierre and Javier Moro, *Five Past Midnight in Bhopal: The Epic Story of the World's Deadliest Industrial Disaster* (New York: AOL Time Warner, 2002).

75. OSH Act § 8(f)(1).

76. OSH Act § 11(c)(1).

77. A contrary view by Peg Seminario, AFL-CIO's director of Health and Safety, is that "Historically OSHA inspections conducted as a result of a complaint produce just as significant results in identifying serious vi- olations and uncovering hazards as the general scheduled inspections." Quoted in BNA, *OSHA Reporter*, 16 March 2000, p. 202.

78. OSHA has nevertheless strengthened the workers' role in the on-site consultation process. 29 C.F.R. § 1908 (December 2000).

79. *Washington Post*, "About Face Considered in OSHA Suit," 20 October 1982.

80. *Occupational Hazard*, news for 2 August 2006 and 10 March 2006, respectively, at www.occupationalhazard.com.

81. Executive Order 12196, signed 26 February 1980, 45 FR 12769, superseding E.O. 11807 of 28 September 1974.

82. There are 21 states and one territory with complete state plans for the private and public sectors, and three states and one territory that cover only public employees. Around 10 others have withdrawn their programs over the past three decades, but the number has never been more than half the states.

83. OSH Act §§ 2(b)(11) and 18(c)(2).

84. OSH Act § 18(a).

85. OSH Act § 18(c)(1–8).

86. OSH Act § 18(c).

87. *Robert Hayden, "Federal and State Rules" in Proceedings of the Occupational Health and Safety Regulation Seminar* (Washington, D.C.: Government Institutes, 1978), pp. 9–10.

88. Nicholas A. Ashford, Crisis in the Workplace: Occupational Disease and Inquiry (Boston: MIT Press, 1976), p. 231.

89. "OSHA: 2005, 2006 and Beyond," in *Occupational Hazards*, news for 26 January 2006, at www.occupa- tionalhazards.com.

90. OSH Act § 21(c)(2).

91. OSH Act § 21(c)(1).

92. U.S. Department of Labor, "OSHA News," 12 April 1978.

93. 29 C.F.R. § 1902.4(c)(2)(xiii).

94. *Florida Peach Growers Assn. v. Dept. of Labor*, 489 F.2d 120 (5th Cir., 1974). To avoid this type of con- frontation, in 1976 Congress provided in Section 9 of the Toxic Substances Control Act for detailed coor- dination procedures to be followed when jurisdictional overlap occurs.

95. Ashford, *Crisis*, p. 145.

96. OSH Act § 12(a)–(b).

97. OSH Act § 12(j).

98. Ashford, *Crisis*, p. 145.

99. *Id.*, pp. 281-82.

100. *Id.*

101. OSH Act § 22(a).

102. John F. Finklea, "The Role of NIOSH in the Standards Process," in *Proceedings of the Occupational Health and Safety Regulation Seminar*, p. 38.

103. *Id.*, p. 39.

104. "CSB Proposes New Ideas for Updated NIOSH Agenda", in *Occupational Hazards*, news for 7 March 2006, at www.occupationalhazards.com.

105. 49 FR 52380, 25 November 1983.

106. 48 FR 53280; 29 C.F.R. § 1200.

107. 52 FR 31852, 24 August 1987, in response to United Steelworkers of America, AFL-CIO v. Pendergrass, 819 F.2d 1263 (3rd Cir., 1987).

108. There is some legal question whether OSHA, which has jurisdiction over employer-employee health and safety relations, has authority over the relationship between a company and its downstream customers.

109. *United Steelworkers of America, AFL-CIO-CLC v. Auchter, et al.*, 763 F.2d 728; 12 OSHC 1337 (3rd Cir., 1985).

110. 51 FR 34590.

111. *New Jersey State Chamber of Commerce v. Hughey*, 774 F.2d 587, 12 OSHC 1589 (3rd Cir., 1985).

112. Hazardous Waste Operations and Emergency Response (HAZWOPER) regulations in 29 C.F.R. § 1910.120.

113. *Gade v. National Solid Wastes Management Assoc.*, 112 S. Ct. 2374 (1992).

114. OSHA ergonomics program standard final rule, 65 FR 68261 (14 November 2000), taking effect 60 days later.

115. "GOP Labor Bills Make Little Headway," AP, 28 October 1998.

116. In the 106th Congress, H.R. 987, H.R. 4577, and S.1070 restricting OSHA on issuing an ergonomics stan- dard; H.R. 5178 on needle stick prevention, S. 1156 amending the Small Business Regulatory Enforcement Fairness Act (SBREFA) of 1996; H.R. 4098 barring home office inspections; and S. 1515 signed into law by President Clinton on 10 July 2000 as P.L. 106-245.

Chapter 2

The Rulemaking Process

Shontell Powell
Ogletree, Deakins, Nash, Smoak & Stewart, P.C.
Washington, D.C.

1.0 Overview

Congress enacted the Occupational Safety and Health Act (OSH Act or Act) to "assure so far as possible . . . safe and healthful working conditions" for every person employed in the United States.[1] In order to facilitate this goal, the OSH Act bestowed upon the Secretary of Labor (Secretary) the power to issue rules that govern safety and health in the workplace. The Secretary of Labor (Secretary) has delegated this rulemaking authority to the Occupational Safety and Health Administration (OSHA). OSHA is also responsible for enforcing the occupational safety and health standards and regulations promulgated pursuant to the OSH Act.

The OSH Act authorizes the promulgation of two types of rules: (1) occupational health and safety standards and (2) regulations. Occupational safety and health standards prescribe conditions or practices that employers must have in place to provide a safe and healthy working environment.[2] These are the rules with which employers typically are most concerned, as these rules are intended to address the safety and health hazards that may exist in the workplace. OSHA standards can be categorized by industry: general,[3] construction,[4] agricultural,[5] and maritime.[6] Section 5 (a)(1) of the OSH Act, often referred to as the general duty clause, covers serious hazards to which no these specific OSHA standard applies.[7]

OSHA regulations, issued pursuant to Section 8 of the Act, effectuate other administrative statutory purposes such as recordkeeping requirements, inspections, and the conduct during administrative proceedings.[8]

2.0 The Rulemaking Process

The OSH Act authorizes the Secretary to promulgate standards in three ways. First, the Act allowed the Secretary to adopt any then-existing "national consensus standard" or "established Federal standard" for the two years following the effective date of the OSH Act, without following the notice and comment procedures mandated by the Act and the Administrative Procedure Act (APA).[9] Under the second method of promulgating a standard — the most common — the Secretary may promulgate, modify, or revoke safety standards after full notice and comment proceedings as provided for in the APA.[10] The third method allows the Secretary, in limited circumstances, to quickly issue emergency temporary standards by circumventing the usual rulemaking procedures under the APA. After issuing the emergency temporary standard, the Secretary must engage in notice and comment rulemaking to issue a permanent standard no later than six months after issuing the emergency standard.[11]

2.1 Petitions for Rulemaking

Any person or organization interested in having the Secretary promulgate, modify, or revoke a particular standard may petition OSHA to engage in the appropriate rulemaking.[12] The petition must include a draft of the proposed rule and a statement explaining the reason for the promulgation, modification, or revocation and the effect of that action. If OSHA denies the petition, the petitioner may seek review of the denial in federal court.

Courts have generally been reluctant to require OSHA to promulgate a standard. For example, in *UAW v. Chao*,[13] the International Union of United Automobile, Aerospace & Agriculture Implement Workers of America (collectively UAW) petitioned OSHA to promulgate a standard that protects workers from "the health effects of occupational exposure to machining fluids."[14] After its petition had been pending with OSHA for more than 10 years, the UAW brought suit in the United States Court of Appeals for the Third Circuit, requesting that the court review OSHA's "unreasonable delay" in issuing the standard. The Third Circuit held that while there is "little doubt, and it is not disputed here, that exposure to [metalworking fluids] can have debilitating health effects," it had no basis to order OSHA to issue a standard. In the decision, the court noted that "OSHA never decided to regulate [machine fluids], much less formally initiated rulemaking proceedings with the publication of a proposed rule."[15]

2.2 National Institute for Occupational Safety and Health

Section 22 of the OSH Act created the National Institute for Occupational Safety and Health (NIOSH) to conduct research and make recommendations to the Secretary for the prevention of work-related injuries and illnesses.[16]

NIOSH, which is part of the Centers for Disease Control and Prevention in the U.S. Department of Health and Human Services (HHS), does not have authority to promulgate or enforce OSHA standards and regulations. For rulemaking initiatives, NIOSH gathers information on potential hazards to the health and safety of workers and makes recommendations to OSHA. The first step in this process involves gathering and analyzing all available information from the scientific literature, which can take years. The second step typically involves conducting interviews and investigations to gather additional information on the potential hazards under study. NIOSH may conduct its own interviews or may choose to enlist the help of other organizations to gather this information. Once the information from the literature has been gathered and the interviews and investigations are complete, NIOSH will generally prepare and send its recommendations to the Secretary. Collecting the information and preparing the recommendations may also take several months or several years to complete. Therefore, the initial stages of the rulemaking process conducted by NIOSH can take a significant amount of time to complete.

2.3 Advisory Committees

The OSH Act provides that the Secretary may establish advisory committees to assist with rulemaking.[17] An advisory committee may be a standing committee or an ad hoc committee. A standing committee is permanent or long-standing and formed to address a variety of occupational safety and health issues. An ad hoc committee is temporary and created to research and investigate a specific safety or health concern. Once an ad hoc committee has completed its mission, it is dissolved.[18]

2.3.1 National Advisory Committee on Occupational Safety and Health

The National Advisory Committee on Occupational Safety and Health (NACOSH) is a standing committee with twelve members that advises the Secretary of Labor and the Secretary of HHS on occupational safety and health, including rulemaking and policy under the OSH Act.[19] The

NACOSH members, appointed by the Secretary of Labor, consists of representatives from management, labor, occupational health professions, and the public. Two of the health representatives and two of the public members are designated by the Secretary of HHS.[20]

2.3.2 Federal Advisory Council on Occupational Safety and Health

The Federal Advisory Council on Occupational Safety and Health (FACOSH), established in 1971, advises the Secretary on matters relating to the health and safety of federal government employees.[21] FACOSH consists of 16 members appointed by the Secretary, with half representing management of federal departments and agencies and half representing federal government labor organizations.[22]

2.3.3 Advisory Committee on Construction Safety and Health

The Contract Work Hours and Safety Standards Act ("Construction Safety Act") established the Advisory Committee on Construction Safety and Health (ACCSH).[23] OSHA must consult with ACCSH when promulgating a construction industry standard.[24] ACCSH, a 15-member committee appointed by the Assistant Secretary of OSHA, is comprised of representative for employers, employees, state safety and health agencies, the public, and one Secretary of HHS designee.[25]

2.3.4 Maritime Advisory Committee for Occupational Safety and Health

The Secretary established the Maritime Advisory Committee for Occupational Safety and Health to provide advice on occupational safety and health issues in the maritime industry.[26] MACOSH, initially established in 1995 and re-established in 2012, is comprised of 15 members appointed by the Secretary. The appointees represent management, labor, safety and health professional organizations, government organizations, and the public.[27]

2.4 Regulatory Agenda

OSHA issues a semiannual regulatory agenda that prioritizes and provides a timeline for any standards and regulations under development or review. The agenda indicates what stage of the rulemaking process that a developing rule is at and a target issuance date for each rule. The activities included in OSHA's agendas are primarily those currently planned to have an Advance Notice of Proposed Rulemaking (ANPRM), a Notice of Proposed Rulemaking (NPRM), or a Final Rule issued within the next 12 months. However, OSHA's regulatory agenda usually includes a long-term section for rules it expects will have the next regulatory action more than 12 months after publication of its agenda.

2.5 Request for Information & Advanced Notice of Rulemaking

OSHA may solicit information from the public to determine whether it is necessary to promulgate a new rule. To solicit information from the public, OSHA will publish a Request for Information (RFI) or an Advanced Notice of Rulemaking in the *Federal Register.*

2.6 Notice of Proposed Rulemaking

Once the potential safety or health hazard has been identified and researched, OSHA will decide whether to initiate the rulemaking process under Section 6(b) of the Act and the APA. This process involves a number of steps to provide adequate notice of the proposed rule and to provide interested persons with an opportunity to submit comments and suggestions to OSHA.

OSHA must publish a Notice of Proposed Rulemaking (NPRM) in the *Federal Register* that includes the proposed language for the rule. The preamble to the proposed rule must explain the rationale for the rule, including the legal basis for the rule. The preamble may also contain a review of the scientific basis for the rule; a discussion of the expected impact and economic cost of the rule for the regulated community; enforcement guidance; and a discussion of comments received in response to the Request for Information or other sources.

The OSH Act requires the Secretary to provide a period of at least 30 days for interested persons to submit written comments on the proposed rule.[28] However, the Secretary usually provides a longer comment period. Written comments (and any oral testimony at hearings) are a critical part of the rulemaking process, as OSHA must consider the comments before issuing a final rule and the comments become part of the official rulemaking record.

In addition to submitting written comments, interested persons also have the right to request a hearing regarding the rule. The Act provides that 30 days after the comment period has ended, the Secretary must publish a notice specifying "the occupational safety or health standard to which objections have been filed and a hearing requested."[29]

2.7 Hearings

If objections are made to a proposed rule, interested parties will have an opportunity to participate in an oral hearing conducted by a Department of Labor administrative law judge.[30] The oral hearings are informal, like a legislative hearing, but permit cross-examination on crucial issues. Also, a transcript of the hearing must be made available to interested parties.[31]

OSHA's regulations state that "fairness may require an opportunity for cross-examination on crucial issues" at the hearing.[32] The presiding judge has discretion to either permit or deny cross-examination of any witness.[33] The goal of cross examination at the hearing is to "provide an opportunity for effective oral presentation by interested persons which can be carried out with expedition and in the absence of rigid procedures which might unduly impede or protract the rulemaking process."[34] OSHA usually provides an opportunity for interested parties to submit written comments and briefs after the hearing.

2.8 The Final Rule

The OSH Act provides that within 60 days after the written notice and comment period has ended (with no hearing) or within 60 days after the certification of the record (if a hearing has been held), the final rule "shall" be published in the *Federal Register*.[35] However, typically, the period before promulgation of the final rule is substantially longer than this specified timeframe. If the Secretary decides not to promulgate a rule, the determination that such a rule will not be issued must also to be published in the *Federal Register*.[36] If a determination is made that a rule should not be issued, the Secretary may solicit additional data or information from interested persons involved.[37] If the Secretary decides that a rule should be promulgated after this additional comment period, the rule is to be issued within 60 days after that comment period ends.[38]

Section 6(e) of the Act, requires OSHA to publish a "statement of reasons" with the final rule.[39] The statement operates as an internal check to ensure that the agency has determined that its reasons for issuing a rule are not arbitrary but based on documented research and information. It also "makes possible informed public criticism of a decision by making known its underlying rationale; and it facilitates judicial review of agency action by providing an important part of the record of the decision."[40] In

Dry Color Manufactuers' Assoc. v. Department of Labor,[41] the court held that to "satisfy subsection 6(e), the statement of reasons should indicate which data in the record is being principally relied on and why that data suffices to show that the substances covered by the standard are harmful and pose a grave danger of exposure to employees." The statement of reasons does not necessarily have to include any findings of fact, but it must demonstrate the significant issues that were addressed and articulate the rationale for the rule.[42]

3.0 Negotiated Rulemaking

In recent years, there has been increased interest in "negotiated rulemaking." Under negotiated rulemaking, interested parties work closely with OSHA to negotiate the content of a proposed rule that both OSHA and the affected parties believe appropriately and adequately addresses a safety or health hazard, thus minimizing the potential for litigation challenging the final rule. For example, OSHA charted the Cranes and Derricks Negotiating Rulemaking Advisory Committee to negotiate revisions to the standard for cranes and derricks in construction.

4.0 Other Applicable Statutes Concerning Rulemaking

In addition to the OSH Act and the APA, there are other statutes that apply to OSHA's rulemaking. The Paperwork Reduction Act (PRA)[43] is designed to reduce the burden of complying with federal statutes mandating recordkeeping and reporting and ensure that federal regulations limit required information collection to that which is truly necessary. Under the PRA, OSHA, like other federal rulemaking agencies, is required to review the paperwork burden for all proposed rules and to provide a rationale for the information that will be collected under the rule. The PRA also requires OSHA to prepare an Information Collection Request (ICR) for each existing rule requiring collection of information for approval to the Office of Management and Budget (OMB). Generally, an ICR must be resubmitted to OMB every three years.

The Regulatory Flexibility Act (RFA), as amended by the Small Business Regulatory Enforcement Fairness Act (SBREFA), is intended to address the particular needs of small businesses to not be overwhelmed by costly and burdensome federal regulations. The RFA requires agencies conducting rulemaking to evaluate the impact that a proposed rule will have on small businesses. If the agency determines that the rule will have a significant economic impact on a substantial number of small businesses, the agency is required to publish with the proposed rule a Regulatory Impact Analysis (RIA). Affected entities will have an opportunity to submit comments on the RIA and the impact of the proposed rule on their businesses for the agency's consideration on further rulemaking. With the final rule, the agency is required to publish a final RIA addressing the comments received and explaining the agency's determination as to the economic impact of the rule on small businesses. SBREFA provides for judicial review of the agency's determination. It also requires a proposed rule affecting small businesses to undergo review by a Small Business Advocacy Review Panel to be established by the Small Business Administration. The review panel will consist of members from OSHA, the OMB, and a number of small business representatives. The panel will then review the applicable parts of the rulemaking record and will submit a report to OSHA, to which OSHA must respond in the rulemaking.

5.0 Delays in Rulemaking

The rulemaking process is often long and cumbersome. It is not uncommon for the period of time between the gathering of information and the publication of the Notice of Proposed Rulemaking to the publication of a Final Rule to span a period of many years. The APA directs an agency "to conclude [within a reasonable time] a matter presented to it."[44] Reviewing courts are to "hold unlawful and set aside agency action, findings, and conclusions found to be . . . arbitrary, capricious, an abuse of discretion, or otherwise not in accordance with law."[45]

In *Public Citizen Health Research Group v. Chao*,[46] the Public Citizen Health Research Group (PCHRG) petitioned the Third Circuit Court of Appeals to review the inaction of OSHA in promulgating a rule that would lower the permissible exposure limit for hexavalent chromium. Hexavalent chromium, a compound used for chrome plating, stainless steel welding, alloy production, and wood preservation, is classified as a carcinogen.[47] For decades, NIOSH had recommended that OSHA adopt a standard to minimize worker exposure to hexavalent chromium.

In 1993, PCHRG petitioned OSHA to issue an emergency temporary standard to limit worker exposure to this compound. OSHA declined to issue an emergency standard but announced that it was initiating the rulemaking process to develop a rule that would limit worker exposure to this compound. OSHA estimated that the Notice of Proposed Rulemaking would be published in the Federal Register no later than March 1995. OSHA then delayed the issuance of the proposed standard to May 1995, and then again to December 1995. This pattern of delay continued—the November 1995 agenda pushed the date to July 1996, then to June 1997 and again to September 1997.

In 1997, PCHRG petitioned the Third Circuit for review of OSHA's inaction alleging unreasonable delay. At that time, the Third Circuit denied PCHRG's petition stating that PCHRG did not "demonstrate that inaction is . . . unduly transgressive of the agency's own tentative deadlines."[48] However, OSHA's pattern of delay continued.

On December 3, 2001, OSHA issued a rulemaking agenda, which stated that the hexavalent chromium rulemaking was a "long-term action," with the timetable for the proposed rule "to be determined."[49] PCHRG again petitioned the court arguing that "deference to an agency's priorities and timetables only goes so far . . . at some point, a court must tell an agency that enough is enough." The Third Circuit did just that.

In holding that OSHA's inaction and continuous delays were unreasonable, the court noted that "OSHA has missed all ten of its self-imposed deadlines, including the September 1999 target it offered to this Court in *Oil Workers*. Far from drawing closer to a rulemaking, all evidence suggests that ground is being lost."[50] The court quoted the D.C. Circuit Court of Appeals in stating that "where the Secretary deems a problem significant enough to warrant initiation of the standard setting process, the Act requires that he have a plan to sheperd through the development of the standard . . . to ensure that the standard is not inadvertently lost in the process."[51] The court granted PCHRG's petition and ordered OSHA to begin the rulemaking process. It further ordered the parties to submit to mediation to work towards a "realistic timetable" to promulgate a rule. That timetable would be enforced by the court.[52]

6.0 Emergency Temporary Standards

The Secretary may issue an emergency temporary standard if a determination is made "(A) that employees are exposed to grave danger from exposure to substances or agents determined to be toxic or physically harmful or from new hazards, and (B) that such emergency standard is necessary to protect employees from such danger."[53] The emergency temporary standard is effective as soon as it is published in the *Federal Register* and stays in effect until it is replaced by a standard promulgated according to the informal rulemaking procedures under the APA. The published emergency temporary standard serves as the proposed rule for the promulgation of the final rule. A final standard must be promulgated no later than six months after publication of the emergency temporary standard.[54]

OSHA rarely issues emergency temporary standards. This is due in part to the fact that many emergency temporary standards have been successfully challenged in court. Additionally, the final rule must be promulgated within six months after an emergency temporary standard is issued, a nearly impossible task given the length of time OSHA typically takes to promulgate a final rule.

7.0 Variances

Under certain conditions, variances may be available for employers that are unable to comply with an OSHA standard. To obtain a variance, the employer must submit an application to OSHA requesting either a temporary variance if the employer can prove by a preponderance of the evidence that it is currently unable to comply with the standard but will be able to do so in the future, or for a permanent variance if the employer can demonstrate that there are no feasible means to comply with the standard and that the employer has another method or program in place to protect its workers from the hazard covered by the standard.

7.1 Temporary Variance

The OSH Act contains provisions for an employer to apply for a temporary order granting a variance from an OSHA standard or regulation. To receive a temporary variance, the employer must establish that it is "unable to comply with a standard by its effective date because of unavailability of professional or technical personnel or of materials and equipment needed to come into compliance with the standard or because necessary construction or alteration of facilities cannot be completed by the effective date."[55] The employer must also demonstrate that it is taking all possible precautions to safeguard their employees from any potential health or safety hazard covered by the standard.[56] In addition, the employer must demonstrate that it has developed a plan for coming into compliance with the standard at a future date.

Before OSHA grants a temporary variance, the employer must notify its employees and provide an opportunity for a hearing and comments. The temporary standard expires after the time needed by the employer to come into compliance with the standard or within one year, whichever period is shorter. An employer may apply for an extension but that extension cannot be granted more than twice. The application for extension must be filed at least 90 days prior to the expiration of the order granting the temporary variance.

An employer may apply for a temporary variance by submitting a temporary variance application or a letter to OSHA with the following information:

1. the specific standard or portion thereof from which the employer seeks a variance;

2. an explicit request for a variance with detailed explanation of the reasons why it is unable to comply with the standard;

3. the steps the employer has taken or will take (with specific dates) to protect employees against the hazard covered by the standard;

4. a statement of when the employer expects to come into compliance with the standard and what steps the employer has taken or will take along with specific dates to come into compliance; and

5. a certification that the employer has notified its employees of the application by giving a copy of the statement to the employees' authorized representative or posting the statement in a place where it is available for viewing by all employees.

7.2 Permanent Variance

The OSH Act also permits an employer may to apply for a permanent variance from an OSHA standard. When applying for a permanent variance, employees must be notified and given the opportunity to participate in a public hearing.[57] To apply for a variance, the employer must submit a permanent variance application or letter to OSHA containing similar information set forth in an application for a temporary variance. The Secretary will only issue a permanent variance if it is determined that the employer has demonstrated by a preponderance of the evidence that "the conditions, practices, means, methods, operations, or processes used or proposed to be used by an employer will provide employment and places of employment to his employees which are as safe and healthful as those which would prevail if he complied with the standard."[58] The order for a permanent variance may be revoked upon application by the employer, employees, or by the Secretary on her own motion six months after the permanent variance is issued. Very few permanent variances have ever been issued by OSHA.

7.3 Interim Order

When the employer applies for a permanent variance, OSHA can, upon request, grant an interim order allowing the employer to use the proposed alternative means of compliance while OSHA is reviewing their variance application.[59] OSHA will only grant an interim order after it determines that the interim order will protect workers at least as effectively as the standard from which the employer is seeking the variance. The employer may include in its application for an interim order a statement of facts and reasons why the order should be granted. OSHA may rule ex parte on a request for an interim order.[60] Interim orders are published in the *Federal Register* and the employer must give notice to its employees of the order in the same manner that notice of the application for the variance was made.[61]

8.0 State Law Standards/Jurisdiction

The OSH Act permits a state to regulate an occupational safety or health issue that is not governed by any standard, regulation, or rule under the OSH Act.[62] For example, in *Empire State Rest. & Tavern Assn. v. New York*,[63] at issue was the Clean Indoor Air Act amended by New York State, to prohibit smoking in various public establishments including bars and food service establishments. A number of owners of New York taverns and bars brought suit to permanently enjoin New York from enforcing the law, arguing that the Clean Indoor Air Act was preempted by the OSH Act, and more specifically that 29 C.F.R. § 1910.1000, which adopts standards relating to permissible safe exposure levels for employees exposed to "toxic and hazardous substances," preempted state legislation of occupational tobacco smoke. The court held that the OSH Act did not preempt New York's Clean Indoor Air Act because OSHA had made clear in various written statements that it declined to regulate environmental tobacco smoke. The court further stated that "many state and local governments already began to address this problem by curtailing smoking in public places and workplaces . . . [t]hus, formal OSHA policy indicates not only the compatibility of state and local smoking legislation and the OSH Act and regulations, but also the acknowledgement and approval of OSHA with such state and local action."[64]

Even if a state or local safety standard governs the same issues or area as the OSH Act, the state standard may not necessarily be preempted by the OSH Act. In *Gade v. National Solid Wastes Management Ass'n*, 550 U.S. 88 (1992), the Supreme Court held that a state law that "constitutes, in a direct, clear and substantial way, regulation of worker health and safety" is preempted under the OSH Act.[65] This includes laws that have dual impact – protecting both workers and the general public. However, *Gade* exempts from preemption "state laws of general applicability (such as laws regarding traffic safety or fire safety) that do not conflict with OSHA standards and that regulate the conduct of workers and nonworkers alike." Applying this principle in *Gade*, courts have found that certain state and local laws are not preempted by the OSH Act. [66]

9.0 Judicial Review

Any person who is adversely affected by a standard issued pursuant to the OSH Act may seek review at any time prior to the 60th day after such a standard is promulgated in the United States Court of Appeals for the circuit in which such person resides or has her principal place of business.[67] If the challenge to the standard is based on the failure of OSHA to adhere to a procedural requirement, the issue must be raised in pre-enforcement proceedings.[68] In pre-enforcement proceedings, OSHA has the affirmative burden to demonstrate the reasonableness of the standard based on "substantial evidence in the record considered as a whole."[69] OSHA must demonstrate by substantial evidence that the standard is feasible, although OSHA does not have the burden of proving feasibility with scientific certainty.[70]

Alternatively, an employer may challenge the validity of a standard after enforcement proceedings have commenced. If enforcement proceedings have already commenced, the standard is presumed to be valid and the burden of proof is on the petitioning employer.

Notes

1.   29 U.S.C. § 651(b).

2.   29 U.S.C. §§ 652(8), 655.

3. 29 C.F.R. Part 1910.

4. 29 C.F.R. Parts 1926.

5. 29 C.F.R. Part 1928.

6. 29 C.F.R. Parts 1915, 1917, and 1918.

7. 29 U.S.C. § 654(a)(1).

8. 29 U.S.C. § 657.

9. 29 U.S.C. § 655(a).

10. 29 U.S.C. § 655(b).

11. 29 U.S.C. § 655(c).

12. 29 U.S.C. § 655(b)(1); 29 C.F.R. § 1911.3.

13. UAW v. Chao, 361 F.3d 249 (3rd Cir. 2004).

14. Id. at 250.

15. Id.

16. 29 U.S.C. § 671.

17. 29 U.S.C. §§ 655(b)(1), 656.

18. 29 C.F.R. § 1912.2.

19. 29 C.F.R. § 1912.5.

20. 29 U.S.C. § 656(a).

21. Executive Order 11612 and continued most recently by Executive Order 13652, 78 Fed. Reg. 61817 (October 4, 2013).

22. FACOSH Charter, https://www.osha.gov/dep/facosh/facosh_charter.html.

23. 40 U.S.C. § 333.

24. See also 29 C.F.R. § 1911.10(a); 29 C.F.R. § 1912.3(a).

25. 29 C.F.R. § 1912.3(b); ACCSH Charter, https://www.osha.gov/doc/accsh/accshcharter.html.

26. 78 Fed. Reg. 21977.

27. https://www.osha.gov/dts/maritime/macosh/index.html.

28. 29 U.S.C. § 655(b)(2).

29. 29 U.S.C. § 655(b)(3).

30. 29 U.S.C. § 655(b)(3); 29 C.F.R. § 1911.15(a)(2).

31. 29 C.F.R. § 1911.15(a)(3) and (b); 29 C.F.R. § 1911.17..

32. 29 C.F.R. § 1911.15(a)(3).

33. 29 C.F.R. § 1911.16(e).

34. 29 C.F.R. § 1911.15(a)(3).

35. 29 U.S.C. § 655(b)(4); 29 C.F.R. § 1911.18(a)(1).

36. 29 C.F.R. § 1911.18(a)(1).

37. 29 C.F..R. § 1911.18(a)(2).

38. Id.

39. 29 U.S.C. § 655(e).

40. Dry Color Mfrs.' Assoc. v. Department of Labor, 486 F.2d 98, 105 (3rd Cir. 1973).

41. Id. at 105.

42. 29 C.F.R. § 1911.18(b).

43. 44 U.S.C. §§ 3501 – 3520.

44. 5 U.S.C. § 555(b).

45. 5 U.S.C. § 706(2).

46. 314 F.3d 143 (3rd Cir. 2002).

47. Id. at 146.

48. Oil, Chem., & Atomic Workers Union v. OSHA, 145 F.3d 120, 123 (3rd Cir. 1998).

49. PCHRG, 314 F.3d at 149.

50. Id. at 151.

51. Id. at 157 (quoting National Congress of Hispanic American Citizens v. Marshall, 626 F.2d 882, 890-91 (D.C.Cir.1979)).

52. See also United Steelworkers of America v. Pendergrass, 819 F.2d 1263 (3rd Circ. 1987) (holding the Secretary was unreasonably delayed in issuing hazard communication standard applicable to all employers covered under the OSH Act); In re International Chemical Workers Union, 958 F.2d 1144 (D.C. Cir. 1992) (imposing deadline for the Secretary to promulgate a standard to regulate cadmium exposure); Public Citizen Health Research Group v. Brock, 823 F.2d 626 (D.C. Cir. 1987) (concluding that the Secretary was unreasonably delayed in issuing a standard that limits workers' exposure to ethylene oxide)..

53. 29 U.S.C. § 655(c).

54. 29 U.S.C. § 655(b)(6)(c).

55. 29 U.S.C. § 655(b)(6)(A)(i); 29 C.F.R. § 1905.10.

56. 29 U.S.C. § 655(b)(6)(A)(ii); 29 C.F.R. § 1905.11.

57. 29 U.S.C. § 655(d).

58. Id.

59. 29 C.F.R. § 1905.10(c).

60. 29 C.F.R. § 1905.11(c)(1).

61. 29 C.F.R. § 1905.11(c)(3).

62. 29 U.S.C. § 667(a).

63. Empire State Rest. & Tavern Assn. v. New York, 360 F.Supp. 2d 454, 460 (D.N.Y. 2005).

64. Empire State Rest. & Tavern Assn. v. New York, 360 F.Supp. 2d at 459.

65. 550 U.S. 88 (1992).

66. See, e.g., Steel Institute of New York v. City of New York, 716 F.3d 31 (2nd Circuit 2013) (concluding that the city's crane regulations met the general applicability exception).

67. 29 U.S.C. § 655(6)(f). The petition must be filed by the 59th day after promulgation. American Federation of Labor and Congress of Industrial Organizations v. Occupational Safety and Health Admin., 905 F.2d 1568 (D.C. Cir. 1990).

68. National Industries Constructors, Inc. v. Occupational Safety & Health Review Comm., 583 F.2d 1048 (8th Cir. 1978).

69. Id.

70. AFL-CIO v. OSHA, 965 F.2d 962 (11th Cir. 1992).

Chapter 3

The Duty to Comply with Standards

Arthur G. Sapper, Esq.
McDermott Will & Emery LLP
Washington, D.C.

1.0 Overview

Section 5(a)(2) of the Occupational Safety and Health (OSH) Act, 29 U.S.C. § 654(a)(2), states that "each employer . . . shall comply with occupational safety and health standards promulgated under this Act." Other sections of the Act impose an implicit duty to comply with the Occupational Health and Safety Administration's (OSHA) regulations.[1] Although the duty to comply with standards and regulations seems unqualified, the courts and the Occupational Safety and Health Review Commission (OSHRC or Commission) have held that the duty is qualified in various ways.

2.0 Applicability of OSHA Standards

2.1 The General Principle of Preemption

The OSHA standards themselves state a general principle—the more specific standard prevails over the more general.[2] For this reason, decisions speak of the defense of preemption—that is, a citation will be vacated if the cited condition is regulated by a more specifically applicable standard.[3] While many factors are relevant to such an inquiry,[4] the basic question is whether application of the more generally applicable standard would defeat a rulemaking decision implicit in the more specifically applicable standard.[5]

In accordance with this principle, an employer must first determine whether his industry is specially regulated by one of the several industry-specific "parts" in Title 29 of the Code of Federal Regulations (C.F.R.). These industry-specific parts are Part 1913, which applies to shipyards; Part 1917, which applies to marine terminals; Part 1918, which applies to longshoring; Part 1926, which applies to construction; and Part 1928, which applies to agriculture.

If no industry-specific part applies, then an employer must look to Part 1910, which is entitled "General Industry Standards" and which applies to all employers engaged in businesses affecting commerce. The employer must then determine whether a special, industry-specific section within Subpart R of Part 1910 or an industry-specific subpart within Part 1910 regulates both his industry and the particular condition cited. For example, § 1910.261, the first section in Subpart R, regulates the paper industry, while Subpart T of Part 1910 covers commercial diving. If no industry-specific standard in Subpart R of Part 1910 applies, then the employer is regulated by the many generally applicable sub-parts in Part 1910. For example, Subpart O regulates the guarding of machinery generally.

The preemption principle—that is, the principle that the specific standard prevails over the general standard—applies even to standards within an industry-specific part (for example, within Part 1926, the construction part). Thus, the general provisions in Part 1926 governing the use of nonsparking electrical equipment in flammable gas concentrations were held to have been preempted by the specific provisions on the use of such equipment in flammable gas concentrations in tunnels under construction.[6]

2.2 Special Applicability Problems

May a standard in Part 1910 be applied to work regulated by an industry-specific part (e.g., construction work)? It has been held that, if there is no applicable construction standard in Part 1926, OSHA may cite an employer for a violation of a Part 1910 standard.[7] In addition, some Part 1910 standards expressly state that they apply to construction work.[8]

Nevertheless, some Part 1910 standards do not apply to construction work. Some expressly state they do not so apply,[9] and the preamble to at least one other set of standards states that those standards were not intended to so apply.[10] In at least one case, OSHA has refrained from applying a standard to construction work because its proposed version did not give that industry notice and opportunity to comment on its applicability.[11]

3.0 General Principles of the Duty to Comply

Although OSHA must show that a condition violative of a standard existed, OSHA need not always show that the cited employer *himself* violated the standard, that is, that the cited employer created the violative condition.[12]

Unless a standard explicitly or implicitly incorporates hazardousness as an element of a violation, OSHA need not show that a failure to comply with a standard creates a hazard.[13] Whether a hazard exists depends on whether there is a "significant risk," which in turn depends on the severity and probability of harm.[14]

3.1 The Exposure Rule

With respect to many standards,[15] employers must comply only if there is, or reasonably predictably will be, exposure of employees to the violative condition. This principle is reflected in the allocation to OSHA of the burden of proving either of the following:

1. Employees are or were in the zone of danger created by a violative condition; or

2. It is reasonably predictable that employees, by "operational necessity" or otherwise (including inadvertence) in the course of their work or associated activities (e.g., going to rest rooms) will be in the zone of danger created by the cited condition.[16]

The term "zone of danger" refers to "that area surrounding the violative condition that presents the danger to employees which the standard is intended to prevent."[17]

The Commission adopted this "reasonably predictable exposure" test after the courts rejected or suggested disapproval of the Commission's early requirement that OSHA prove actual exposure—that is, that an employee had actually been endangered by a violation.[18] Nevertheless, the mere possibility of exposure is insufficient.[19] That the employer is expected in the future to create a violative condition, but has not yet done so, is insufficient.[20] On the other hand, exposure of just a single employee is sufficient to trigger the employer's duty and to satisfy OSHA's burden of proof.[21]

OSHA need not show that a compliance officer personally witnessed facts supporting an exposure finding.[22] Nor need OSHA show that an employee is, for example, teetering on the edge of an unguarded floor.[23] Brevity of exposure is immaterial.[24] For some standards and regulations—particularly those requiring recordkeeping (e.g., an injury log)—no showing of exposure need be made.[25]

3.2 To Whose Employee Does the Duty Run?

This question was most vexing in the early years of the Act and, in some respects, the answer is only somewhat clearer today. The question first arose on multi-employer worksites, such as construction sites, where employees of Employer A (usually, a subcontractor) may be exposed to a violative condition created or controlled by Employer B (usually, the general contractor or another subcontractor). In its early days, the Review Commission followed a simple rule: The employer of the employees exposed to a violative condition (Employer A) could be cited, regardless of whether the condition had been created by another employer (Employer B) and regardless of whether abatement of the condition was controlled by that employer.[26] Moreover, *only* the employer of the exposed employee (Employer A) could be cited;[27] the employer who created or controlled the violative condition (Employer B) could not be cited unless one of his own employees was exposed to it as well. Complicating the matter was that Employer A may lack the expertise to recognize that the condition is violative or even hazardous.

Subcontractors complained that this policy was highly unfair to them, and they and OSHA complained that it allowed the most guilty to escape liability. Eventually, beginning with a decision by the Second Circuit,[28] the rules of liability changed in two ways: A new, expanded liability rule was developed (see § 3.2.1 below); and a new series of affirmative defenses was established (see § 3.2.2 below). OSHA has issued a directive to its enforcement personnel attempting to explain these rules in detail.[29]

3.2.1 The Multi-Employer Worksite Liability Rules

Although OSHA may satisfy its burden of proving exposure by proving an employment relationship between the exposed employees and the cited employer (i.e., showing that the exposed employees are those of the cited employer),[30] this is not necessary unless a standard or regulation provides otherwise.[31] OSHA may instead—again, unless a standard or regulation provides otherwise—show exposure of an employee of *some* other employer *and* that the cited employer controlled or created the violative condition.[32] The current general statutory principle on which multi-employer liability is based is that "an employer who either creates or controls the cited hazard has a duty . . . to protect not only its own employees, but those of other employers 'engaged in the common undertaking.'"[33] An allegedly controlling or creating employer is not liable if its employees are not present on the construction site and engaged in the construction work.[34]

3.2.1.1 General Construction Contractors

Hence, unless a standard or regulation provides otherwise, OSHA may cite general contractors for violations to which employees of subcontractors are exposed or which subcontractors created.[35] The general contractor is deemed to "have sufficient control over its subcontractors to require them to comply with the safety standards and to abate violations."[36] General contractors must take whatever measures are "commensurate with its degree of supervisory capacity,"[37] which includes some oversight over the work of subcontractors.[38] Hence, a general contractor was expected to detect a problem with a ground fault circuit interrupter installed by a subcontractor even though the condition was by nature latent and hidden from view.[39] On the other hand, a general contractor is responsible for only those violations that "it could reasonably be expected to prevent or detect."[40]

3.2.1.2 Legal Status of the Multi-Employer Liability Rules

Several courts have, to one degree or another, held that the wording of the OSH Act permitted this extra-employment liability to be imposed in the construction context.[41] The Fifth Circuit disagreed with the idea of extra-employment liability in a maritime industry case, holding that such liability was foreclosed by the statute and by an OSHA regulation governing maritime work.[42] The D.C. Circuit has twice reserved ruling on whether the imposition of extra-employment liability in the construction context is inconsistent with 29 C.F.R. § 1910.12, which regulates the application of the construction standards in Part 1926.[43] That provision requires a construction employer to protect "*his* employees" (emphasis added) by complying with the construction standards. Although the Occupational Safety and Health Review Commission held in 2007 that § 1910.12(a) foreclosed controlling-employer liability in construction work,[44] it later reversed its position after the Eighth Circuit disapproved of that view.[45]

3.2.1.3 Nonconstruction Applications of the Multi-Employer Liability Rules

Outside the construction industry, OSHA has occasionally attempted to hold boat owners[46] and factory owners[47] liable for violations committed by their independent contractors and to which only the contractors' employees are exposed. In a recent attempt, the D.C. Circuit held that a factory owner is not liable for a contractor's lockout violation that affected only its own employees when it had no authority over the contractor's employees and its only "control" stemmed from its ownership of the property or its contract with the contractor.[48] See the additional discussion of this point in § 5.3.2 of chapter 11.

3.2.2 Multi-Employer Worksite Defense Rules

As noted above, the liability rule followed in the early days of the Act was that exposure of one's own employee to a violative condition meant that one was liable, even if the cited employer did not create or control the violative condition. This rule has been partially reversed by the creation of a series of affirmative defenses by the Commission,[49] which has been accepted by several courts of appeals.[50] Today, a citation will be vacated if the cited employer on a multiemployer worksite—

a. Did not create or control the allegedly violative condition (such that he could not realistically correct the condition); and

b. Either

1. Took reasonable alternative protective measures; or

2. Did not know, nor with the exercise of reasonable diligence, could have known of the *hazardousness* of the cited condition.[51]

Element (a) may be established by, for example, showing that the employer was prevented from abating a hazardous working condition due to union jurisdictional rules.[52]

Although these defenses originally arose in the context of construction sites, where there are frequently a number of different employers working at the same time, the Commission later applied them to all multi-employer worksites.[53]

4.0 Actual or Constructive Knowledge

OSHA must prove that the cited employer actually knew, or could have known with the exercise of reasonable diligence, of the physical circumstances that comprise the violative condition.[54] This element must also be proved in General Duty Clause cases.[55] The element pertains to the physical circumstances that comprise the violative condition, not the violativeness or hazardousness of the condition.[56] Knowledge of a violative condition by a supervisory employee will be imputed to the employer.[57] That the violative condition was created by the supervisor's own conduct, however, will not alone suffice to prove employer knowledge.[58]

An employer is not reasonably diligent if he neither makes an attempt to become aware of the physical conditions facing his employees, nor trains his employees to recognize hazards arising from them.[59] Reasonable diligence includes "the obligation to inspect the work area, to anticipate hazards to which employees may be exposed, and to take measures to prevent the occurrence."[60] A citation will be vacated on this ground if an employer reasonably[61] relies on the expertise of an independent contractor who is himself subject to the cited standard[62] and who created the condition to which the cited employer's employees were exposed.[63]

In general, if a compliance officer can see a physical condition during a normal inspection, it will be inferred that the employer could, with reasonable diligence, have seen it too.[64] However, OSHA must show that the cited condition was present for a sufficient amount of time such that, with the exercise of reasonable diligence, the employer could have discovered its existence.[65]

5.0 Additional Elements That OSHA Must Sometimes Prove

Sometimes a standard is so vague that it deprives employers of fair notice of its requirements, contrary to the Due Process Clause of the Fifth Amendment to the Constitution.[66] To cure this vagueness, OSHA may be required to prove additional elements. For example, in *Granite City Terminals Corp.*,[67] the Commission held that, if a standard is vague, OSHA must prove that a reasonable person would have recognized a hazard warranting protective measures, and that the sought measures are feasible. A showing that a "reasonable person" would have recognized the hazardousness[68] or violativeness of the cited condition has been required, and held or implied to be sufficient, in a number of circuits and by the Commission; they have not expressly required, or have suggested to be unnecessary, a showing that the employer or its industry follow the practice that OSHA seeks to impose.[69] At least for the generally worded personal protection standards at 29 C.F.R. 1910.132(a) and 1926.28(a), however, the Fifth Circuit has determined that the employer's own conduct or "industry custom and practice will generally establish the conduct of the reasonably prudent employer."[70]

6.0 The Employer's Substantive Affirmative Defenses

This section discusses defenses that an employer may raise to a citation. The burden of pleading and proving these defenses is on the employer. Additional discussion of some of these defenses is in § 5.3.2 of chapter 11.

6.1 Infeasibility

The Commission has created a limited affirmative defense for the employer who finds that compliance is infeasible. A citation may be vacated if the employer proves that

1. [The Infeasibility Element:] the means of compliance prescribed by the applicable standard would have been infeasible under the circumstances in that either (a) its implementation would have been technologically or economically infeasible or (b) necessary work operations would have been technologically or economically infeasible after its implementation; and

2. [The Alternative Measures Element:] either (a) an alternative method of protection was used or (b) there was no feasible alternative means of protection."[71]

Element two effectively compels an employer to show that, although strict compliance was necessary, he took whatever steps were feasible. See § 6.1.2 below for more detail on this point. An employer need not show that a variance application was inappropriate,[72] which is an element of the defense of greater hazard. See § 6.2 below.

6.1.1 The Infeasibility Element of the Defense

In the early days of the act, this defense was known as "impossibility."[73] In 1986, in *Dun-Par Engd. Form*,[74] the Commission changed the name of the defense and its first element to "infeasibility" in part because "[s]trict application of an 'impossibility' defense does not accommodate considerations of reasonableness or common sense, or reflect the strong sense of the practical implicit in the standards adopted under § 6(a)" and the feasibility element in § 6(b)(5) of the act.[75]

The defense has often proved difficult to establish. An employer must at least attempt to adapt existing technology and use some creativity to solve the infeasibility problem.[76] An inability to comply because the appropriate equipment was not onsite is insufficient, for "it is the duty of an employer to use equipment that permits him to comply with the Secretary's standard."[77] The defense may also be rejected if it was feasible to preclude employee access to the zone of danger.[78]

Courts that have considered the infeasibility defense have concluded that it encompasses both technological and economic factors.[79] At one time, the Commission took the position that the economic effect of compliance was irrelevant.[80] However, in *State Sheet Metal Co.*,[81] the Commission stated that "evidence as to the unreasonable economic impact of compliance with a standard may be relevant to the infeasibility defense."[82] In *Peterson Bros. Steel Erec. Co.*,[83] the Commission stated that it would look to the effect that compliance would have on the company's "financial position as a whole" to determine whether the company would be "adversely affected." It is not sufficient that an employer who has failed to use safety measures would be at a competitive disadvantage with others that did not use the measures, for an "employer cannot be excused from compliance with the Act on the basis that everyone else will ignore the law."[84]

6.1.2 The Alternative Measures Element of the Infeasibility Defense

This element—which also appears in the greater hazard and multi-employer defenses—reflects the view that, even if full compliance is not feasible, "an employer [must] comply to the extent feasible."[85] "[B]efore an employer will be excused from ignoring a standard's requirements and leaving its employees unprotected, it must show that it has explored all possible alternate forms of protection."[86] At one time, the Commission in *Dun-Par Engd. Form*[87] shifted the burden of persuasion on this element to OSHA, but the Eighth Circuit held otherwise,[88] and the Commission later followed that holding.[89]

6.2 The Greater Hazard Defense

The Commission has also held that employers need not strictly comply with a standard to the extent that compliance would create greater hazards than noncompliance would. It created an affirmative defense based on the idea that "industry is so diverse that any rule is bound to be counterproductive now and again."[90] The defense has three elements:

1. Compliance with the standard would create greater hazards than noncompliance;

2. Alternative protective measures were taken or were not available; and

3. A variance application is inappropriate.[91]

The defense does not apply to the General Duty Clause because the usefulness of a proposed abatement method is part of the Secretary's burden in General Duty Clause cases.[92]

The first element of the defense requires a showing that compliance would create *greater* hazards than noncompliance—not new or different hazards.[93] It also requires a showing that *all* alternative ways of protection are more dangerous than noncompliance, not just the means of protection mentioned in the standard.[94]

The reason for the third element of the defense—the inappropriateness of a variance—is that "some employers will believe *incorrectly* that their working conditions are safer than those prescribed in the standards. . . . [R]emoving this incentive to seek variances [by eliminating the element] . . . would be allowing an employer to take chances not only with his money, but with the lives and limbs of his employees."[95] The third element does not apply to regulations because a variance cannot be sought from a regulation.[96]

6.3 Unpreventable Employee Misconduct

This defense has been stated in various ways, but it basically requires an employer to show that he required his employees to take protective measures that comply with the standard and that he enforced that requirement.[97] The Commission has distilled its decisions as requiring four elements of proof:

1. The employer has established work rules designed to prevent the violation;

2. It has adequately communicated those rules to its employees;

3. It has taken steps to discover violations; and

4. It has effectively enforced the rules when violations have been discovered.[98]

Effective enforcement generally must be progressive, that is, it must become increasingly severe as an employee commits additional infractions. Thus, an employer was held to have failed to establish the defense when an employer who broke a safety rule for the second time was given only an oral warning instead of a written reprimand.[99]

Although there is a similar doctrine of supervisory misconduct,[100] some cases characterize it as not an affirmative defense but as a rebuttal of the imputation to the employer of the supervisor's knowledge.[101] The Commission has stated that involvement by a supervisor in a violation is "strong evidence that the employer's safety program was lax."[102] "Where a supervisory employee is involved, the proof of unpreventable employee misconduct is more rigorous and the defense is more difficult to establish since it is the supervisor's duty to protect the safety of employees under his supervision."[103]

The objection has been made that the overlap of this defense with the knowledge element of OSHA's case is confusing, for OSHA must prove knowledge while the employer must prove the defense.[104]

6.4 Invalidity of the Standard

6.4.1 Violation of Statutory Procedural Requirements

A standard is invalid if it was not adopted in accordance with a statutory procedural requirement. *See generally* chapter 2. Two examples of invalidity resulting from violations of such requirements are

1. Making a substantive change in a national consensus or established federal standard adopted under § 6(a) of the OSH Act, 29 U.S.C. § 655(a).[105] This is a special case of failing to give the public notice or an opportunity for comment on the adoption or amendment of a standard.[106]

2. Failing to consult the Advisory Committee on Construction Safety and Health before proposing a standard regulating construction work.[107]

6.4.2 Violation of Constitutional Requirement of Fair Notice of Standard's Requirements

The Due Process Clause of the Fifth Amendment to the U.S. Constitution requires that persons subject to penalties be given fair notice of the law's requirements.[108] Hence, OSHA standards may not be construed or applied in a way that deprives employers of fair notice of their requirements.[109] This may occur if a standard's words would not apprise an employer of a requirement, OSHA's view was not announced in advance, and industry practice was uniformly to the contrary.[110] This requirement can also be violated when an OSHA interpretation letter available on the Internet is inconsistent with the interpretation implicit in a citation,[111] or when a course of conduct by OSHA induces confusion in the mind of an employer as to the requirements of a standard.[112] To avoid a violation of this requirement, several OSHA standards have been construed narrowly.[113]

6.5 De Minimis

Section 9(a) of the OSH Act, 29 U.S.C. § 658(a), states: "The Secretary may prescribe procedures for the issuance of a notice in lieu of a citation with respect to *de minimis* violations which have no direct or immediate relationship to safety or health." The consequence of characterizing a violation as *de minimis* is that the violation carries neither an abatement requirement nor a monetary penalty.[114] The Commission has long asserted that it may characterize a violation as *de minimis*.[115] There is a split in the circuits as to whether the Commission has this authority. The First, Third, Fifth, and Ninth Circuits have held that it does,[116] while the Seventh Circuit disagrees.[117]

As to what a *de minimis* violation is, the Commission has formulated the test in various ways, including asking whether the violation is "trifling."[118] In another case, it stated: "A *de minimis* violation is one in which there is technical noncompliance with a standard but the departure from the standard bears such a negligible relationship to employee safety and health as to render inappropriate the assessment of a penalty or the entry of an abatement order."[119] One circuit has held that a violation is *de minimis* if the employer's safety measures are as safe as those required by a standard.[120] The Commission has in effect held that the employer bears the burden of proof on the *de minimis* issue.[121]

Notes

1.   Section 9(a), 29 U.S.C. § 658(a) (permitting citation to be issued for violation of regulation); Sections 17(a)–(c) and (e), 666(a)–(c) and (e) (permitting civil and criminal penalties to be imposed for violating regulations).

2.   29 C.F.R. § 1910.5(c).

3. For example, McNally Constr. & Tunneling Co., 16 BNA OSHC 1886 (Rev. Comm'n 1994), aff'd and approved, 71 F.3d 208, 17 BNA OSHC 1412 (6th Cir. 1995).

4. See Trinity Indus., Inc., 20 BNA OSHC 1051, 1057 (Rev. Comm'n 2003) ("Where a standard provides meaningful protection to employees beyond the protection afforded by another standard, there is no preemption"), aff'd without pub. op., 20 BNA OSHC 1873 (5th Cir. July 23, 2004) (No. 03-60511); Bratton Corp., 14 BNA OSHC 1893 (Rev. Comm'n 1990) (steel erection standards do not preempt general fall protection standards) (acceding to view of several circuits); New England Tel. & Teleg. Co., 8 BNA OSHC 1478 (Rev. Comm'n 1980).

5. Lowe Constr. Co., 13 BNA OSHC 2182 (Rev. Comm'n 1989).

6. McNally Constr. & Tunneling Co., 16 BNA OSHC 1886 (Rev. Comm'n 1994), aff'd and approved, 71 F.3d 208, 17 BNA OSHC 1412 (6th Cir. 1995).

7. Western Waterproofing Inc., 7 BNA OSHC 1499, 1501-02 (Rev. Comm'n 1979). One administrative law judge has held, however, that a Part 1910 standard adopted under section 6(a) without notice-and-comment rulemaking from an established federal standard not applicable to construction work (such as a Walsh-Healy standard) may not be validly applied to construction work. Kiewit Power Construcs. Co., No. 11-2395 (Dec. 11, 2012) (ALJ), review directed, Jan. 24, 2013.

8. For example, § 1910.134 (introductory provision).

9. For example, § 1910.147(a)(1)(ii)(A) (lockout standard).

10. For example, the electrical standards in Subpart S of Part 1910. See 46 Fed. Reg. 4034, 4039 ("the electrical standards of Part 1910 do not apply to construction activities").

11. Thus, OSHA has stated that the Bloodborne Pathogens Standard, § 1910.1030, does not apply to construction work because the construction industry did not receive public notice of, and an advisory committee was not consulted about, such an application when the standard was proposed. Letter from Sec'y of Labor Lynn Martin to Robert Georgine, "Construction Activities and Operations and the Bloodborne Pathogens Standard" (Dec. 23, 1992), at http://www.osha.gov/pls/oshaweb/owadisp.show_document?p_table=INTERPRETATIONS&p_id=20968>.

12. See § 3.2 of this chapter.

13. Harry C. Crooker & Sons v. OSHRC, 537 F.3d 79, 22 BNA OSHC 1298 (1st Cir. 2008) (citing cases); Bunge Corp. v. Sec'y of Labor, 638 F.2d 831, 834 (5th Cir. 1981); Kaspar Electroplating Corp., 16 BNA OSHC 1517 (Rev. Comm'n 1993).

14. Weirton Steel Corp., 20 BNA OSHC 1255, 1259 (Rev. Comm'n 2003).

15. See text accompanying note 25 below.

16. Fabricated Metal Prods., 18 BNA OSHC 1072, 1074 (Rev. Comm'n 1997) (surveying cases); Gilles & Cotting, Inc., 3 BNA OSHC 2002 (Rev. Comm'n 1976).

17. RGM Constr. Co., 17 BNA OSHC 1229, 1234 (Rev. Comm'n 1995).

18. Brennan v. Gilles & Cotting, Inc., 504 F.2d 1255, 1263-66, 2 BNA OSHC 1243 (4th Cir. 1974) (remanding for reconsideration of actual exposure test); Brennan v. OSHRC (Underhill Constr. Corp.), 513 F.2d 1032, 2 BNA OSHC 1641 (2nd Cir. 1975) (rejecting actual exposure test). See also Adams Steel Erec., Inc., 766 F.2d 804, 812, 12 BNA OSHC 1393, 1398 (3rd Cir. 1985) (same).

19. Fabricated Metal Prods., 18 BNA OSHC 1072, 1074 & n.8 (Rev. Comm'n 1997) (rejecting "physically possible" test); Rockwell Intl. Corp., 9 BNA OSHC 1092 (Rev. Comm'n 1980).

20. Sharon Steel Corp., 12 BNA OSHC 1539 (Rev. Comm'n 1985).

21. For example, Mineral Industries v. OSHRC, 639 F.2d 1289, 1294-95, 9 BNA OSHC 1387 (5th Cir. 1981).

22. For example, Brennan v. OSHRC (Underhill Constr. Corp.), 513 F.2d 1032, 1038, 2 BNA OSHC 1641 (2d Cir. 1975); see North Berry Concrete Corp., 13 BNA OSHC 2055, 2055-56 (Rev. Comm'n 1989).

23. For example, Underhill Constr., 513 F.2d at 1038.

24. For example, Morgan & Culpepper, Inc. v. OSHRC, 676 F.2d 1065, 1069, 10 BNA OSHC 1629, 1632 (5th Cir. 1982) (short duration of exposure no defense); Brock v. L.R. Wilson & Sons, 773 F.2d 1377, 1386, 12 BNA OSHC 1499 (D.C. Cir. 1985); Walker Towing Corp., 14 BNA OSHC 2072, 2074 (Rev. Comm'n 1991); Frank Swidzinski Co., 9 BNA OSHC 1230, 1232 (Rev. Comm'n 1981); see Flint Engineering & Constr. Co., 15 BNA OSHC 2052, 2056 (Rev. Comm'n 1992).

25. Thermal Reduction Corp, 12 BNA OSHC 1264, 1268 (Rev. Comm'n 1985). One court of appeals suggested, albeit unclearly and perhaps due to a misapprehension, that exposure need not be shown with respect to training violations. Compass Environ. Inc. v. OSHRC, 663 F.3d 1164, 23 BNA OSHC 1772, 1774 (10th Cir. 2011) (discussing § 1926.21(b)(2)).

26. R.H. Bishop Co., 1 BNA OSHC 1767, 1769 (Rev. Comm'n 1974). Liability on the part of such a so-called "exposing" employee was recently re-affirmed by the Commission in S. Pan Servs. Co., 25 BNA OSHC 1081, 1085-87 (Rev. Comm'n 2014).

27. Martin Iron Works, Inc., 2 BNA OSHC 1063 (Rev. Comm'n 1974). See also Hawkins Constr. Co., 1 BNA OSHC 1761 (Rev. Comm'n 1974); Gilles & Cotting, Inc., 1 BNA OSHC 1388 (Rev. Comm'n 1973), aff'd in relevant part, 504 F.2d 1255 (4th Cir. 1974).

28. Brennan v. OSHRC (Underhill Constr. Corp.), 513 F.2d 1032, 2 BNA OSHC 1641 (2d Cir. 1975).

29. CPL 2-0.124, "Multi-Employer Citation Policy" (Dec. 10, 1999), at http://www.osha.gov/pls/oshaweb/owadisp.show_document?p_table=DIRECTIVES&p_id=2024.

30. For example, Van Buren-Madawaska Corp., 13 BNA OSHC 2157 (Rev. Comm'n 1989); MLB Industries, Inc., 12 BNA OSHC 1525, 1528 (Rev. Comm'n 1985).

31. See Summit Contracs., Inc., 21 BNA OSHC 2020 (Rev. Comm'n, 2007) (construing § 1910.12), rev'd on another ground, 558 F.3d 815, 22 BNA OSHC 1496 (8th Cir. 2009), in § 3.2.1.2 below.

32. See notes 33–41 below.

33. McDevitt Street Bovis, Inc., 19 BNA OSHC 1108 (Rev. Comm'n 2000).

34. United States v. MYR Group, Inc., 361 F.3d 364, 20 BNA OSHC 1614 (7th Cir. 2004); cf. Reich v. Simpson, Gumpertz & Heger, Inc., 3 F.3d 1, 16 BNA OSHC 1313 (1st Cir. 1993) (same holding based on 29 C.F.R. § 1910.12).

35. For example, Huber, Hunt & Nichols, Inc., 4 BNA OSHC 1406, 1407-08 (Rev. Comm'n 1976).

36. Gil Haugan d/b/a Haugan Constr. Co., 7 BNA OSHC 2004, 2006 (Rev. Comm'n 1979); see also Lewis & Lambert Metal Contract., Inc., 12 BNA OSHC 1026, 1030 (Rev. Comm'n 1984).

37. Marshall v. Knutson, 566 F.2d 596 (8th Cir. 1977).

38. McDevitt Street Bovis, Inc., 19 BNA OSHC 1108 (Rev. Comm'n 2000); Centex-Rooney Constr. Co., 16 BNA OSHC 2127 (Rev. Comm'n 1994).

39. Blount Int'l Ltd., 15 BNA OSHC 1897, 1899–1900 (Rev. Comm'n 1992). But see Knutson Constr. Co., 4 BNA OSHC 1759, 1761 (Rev. Comm'n 1976), aff'd 566 F.2d 596 (8th Cir. 1977) (general contractor not liable for failing to detect a one-inch crack on the underside of a scaffolding platform; unreasonable to expect general contractor to detect such a crack).

40. David Weekley Homes, 19 BNA OSHC 1116, 1119 (Rev. Comm'n 2000); Centex-Rooney Constr. Co., 16 BNA OSHC 2127, 2130 (Rev. Comm'n 1994); Blount Intl. Ltd., 15 BNA OSHC 1897, 1899 (Rev. Comm'n 1992).

41. Universal Constr. Corp., 182 F.3d 726, 728-31, 18 BNA OSHC 1769 (10th Cir. 1999); United States v. Pitt-Des Moines, Inc., 168 F.3d 976, 18 BNA OSHC 1609 (7th Cir. 1999); R.P. Carbone Constr. Co. v. OSHRC, 166 F.3d 815, 18 BNA OSHC 1551 (6th Cir. 1998); Beatty Equip. Leasing, Inc. v. Sec'y of Labor, 577 F.2d 534, 6 BNA OSHC 1699 (9th Cir. 1978); Marshall v. Knutson Constr. Co., 566 F.2d 596, 6 BNA OSHC 1077 (8th Cir. 1977); Brennan v. OSHRC (Underhill Constr. Corp.), 513 F.2d 1032, 1038, 2 BNA OSHC 1641 (2d Cir. 1975).

42. Melerine v. Avondale Shipyards, Inc., 659 F.2d 706, 10 BNA OSHC 1075 (5th Cir. 1981).

43. IBP, Inc. v. Herman, 144 F.3d 861, 865 and n.3, 18 BNA OSHC 1353 (D.C. Cir. 1998); Anthony Crane Rental, Inc. v. Reich, 70 F.3d 1298, 1306-07, 17 BNA OSHC 1447 (D.C. Cir. 1995) (noting "tension" between wording of § 1910.12 and liability doctrine).

44. Summit Contracs., Inc., 21 BNA OSHC 2020 (Rev. Comm'n, 2007), rev'd, 558 F.3d 815, 22 BNA OSHC 1496 (8th Cir. 2009). The same issue had been litigated in the courts of several states with OSHA-approved state plans, with mixed results. Compare Davenport v. Summit Contracs., Inc., 45 Va. App. 526, 612 S.E.2d 239, 21 BNA OSHC 1392 (Va. App. 2005) (rejecting liability), rev. denied, 21 BNA OSHC 1184 (Va. 2005), with Commissioner of Labor v. Weekley Homes, L.P., 609 S.E.2d 407, 21 BNA OSHC 1049 (N.C. App. 2005) (imposing liability), rev. denied, 21 BNA OSHC 1184 (N.C. 2005).

45. Summit Contracs., Inc., 23 BNA OSHC 1196 (No. 05-0839, 2010), acceding to Solis v. Summit Contracs., Inc., 558 F.3d 815, 22 BNA OSHC 1496 (8th Cir. 2009).

46. Harvey Workover, Inc., 7 BNA OSHC 1687, 1689 (Rev. Comm'n 1979); Camden Drilling Co., 6 BNA OSHC 1560, 1561 (Rev. Comm'n 1978) (barge owner responsible for compelling subcontractor to have its employees stop using their own defective fan and either repair it or remove it).

47. For example, IBP, Inc. v. Herman, 144 F.3d 861, 18 BNA OSHC 1353 (D.C. Cir. 1998), rev'g 17 BNA OSHC 2073 (Rev. Comm'n 1997).

48. IBP, Inc. v. Herman, 144 F.3d 861, 18 BNA OSHC 1353 (D.C. Cir. 1998), rev'g 17 BNA OSHC 2073 (Rev. Comm'n 1997).

49. See generally Anning-Johnson Co., 4 BNA OSHC 1193 (Rev. Comm'n 1976); Grossman Steel & Aluminum Corp., 4 BNA OSHC 1185 (Rev. Comm'n 1976).

50. Dun-Par Engd. Form Co. v. Marshall, 676 F.2d 1333, 10 BNA OSHC 1561 (10th Cir. 1982); Electric Smith, Inc. v. Sec'y of Labor, 666 F.2d 1267, 10 BNA OSHC 1329 (9th Cir. 1982); DeTrae Enters., Inc. v. Sec'y of Labor, 645 F.2d 103, 9 BNA OSHC 1425 (2d Cir. 1980); Bratton Corp. v. OSHRC, 590 F.2d 273, 7 BNA OSHC 1004 (8th Cir. 1979).

51. For example, LeeRoy Westbrook Constr. Co., 13 BNA OSHC 2101, 2103 (Rev. Comm'n 1989); Lewis & Lambert Metal Contractors, 12 BNA OSHC 1026 (Rev. Comm'n 1984).

52. See McLean-Behm Steel Erectors, Inc., 6 BNA OSHC 1712, 1715 (Rev. Comm'n 1978).

53. Rockwell International Corp., 17 BNA OSHC 1801 n. 11 (Rev. Comm'n 1996).

54. For example, Ragnar Benson, Inc., 18 BNA OSHC 1937, 1939 (Rev. Comm'n 1999); Continental Electric Co., 13 BNA OSHC 2153, 2154 (Rev. Comm'n 1989); Prestressed Systems, Inc., 9 BNA OSHC 1864, 1870 (Rev. Comm'n 1981).

55. See U.S. Steel Corp., 12 BNA OSHC 1692, 1699 (Rev. Comm'n 1986).

56. Ormet Corp., 14 BNA OSHC 2134, 2138 (Rev. Comm'n 1991); Southwestern Acoustics & Specialty, Inc., 5 BNA OSHC 1091 (Rev. Comm'n 1977) (employer need be shown only to have had knowledge of "physical conditions which constitute a violation," not that condition was prohibited by law).

57. E.g., Danis-Shook Joint Venture XXV v. Sec'y of Labor, 319 F.3d 805, 812 (6th Cir. 2003); Access Equip. Sys., Inc., 18 BNA OSHC 1718, 1726 (Rev. Comm'n 1999).

58. ComTran Grp, Inc. v. Dep't of Labor, 722 F.3d 1304, 24 BNA OSHC 1292 (11th Cir. 2013) (following four other circuits).

59. Beaver Plant Operations, Inc., 18 BNA OSHC 1972, 1976 (Rev. Comm'n 1999).

60. Frank Swidzinski Co., 9 BNA OSHC 1230, 1233 (Rev. Comm'n 1981).

61. Fabi Constr. Co. v. Sec'y of Labor, 508 F.3d 1077, 22 BNA OSHC 1001 (D.C. Cir. 2007) (doctrine does not apply when cited employer "has reason, by way of expertise, control, and time, to foresee a danger to its employees").

62. Summit Contractors, Inc., 23 BNA OSHC 1196, 1207 (No. 05-0839, 2010) (requiring that the contractor be subject to the construction standards).

63. E.g., Sasser Elec. & Mfg. Co., 11 BNA OSHC 2133 (Rev. Comm'n 1984), aff'd, 12 BNA OSHC 1445 (4th Cir. 1985) (not officially published).

64. See Green Constr. Co., 4 BNA OSHC 1808, 1810 (Rev. Comm'n 1976) (Barnako, Chairman, concurring).

65. David Weekley Homes, 19 BNA OSHC 1116, 1120-21 (Rev. Comm'n 2000).

66. For example, Kropp Forge Co. v. Sec'y of Labor, 657 F.2d 119, 9 BNA OSHC 2133 (7th Cir. 1981); Georgia Pacific Corp. v. OSHRC, 25 F.3d 399, 16 BNA OSHC 1895 (11th Cir. 1994) (standard vague as interpreted by OSHA). See generally § 6.4.1 below.

67. 12 BNA OSHC 1741 (Rev. Comm'n 1986).

68. Weirton Steel Corp., 20 BNA OSHC 1255, 1264 (Rev. Comm'n 2003).

69. For example, Voegele Company, Inc. v. OSHRC, 625 F.2d 1075, 1078, 8 BNA OSHC 1631 (3rd Cir. 1980); American Airlines, Inc. v. Sec'y of Labor, 578 F.2d 38, 41, 6 BNA OSHC 1691, 1692–1693 (2d Cir. 1978); Ray Evers Welding Co. v. OSHRC, 625 F.2d 726, 731-32, 8 BNA OSHC 1271 (6th Cir. 1980); Bristol Steel & Iron Works, Inc. v. OSHRC, 601 F.2d 717, 722-23, 7 BNA OSHC 1462 (4th Cir. 1979).

70. Cotter & Co. v. OSHRC, 598 F.2d 911, 913, 7 BNA OSHC 1510 (5th Cir. 1979); S & H Riggers & Erectors, Inc. v. OSHRC, 659 F.2d 1273, 1285, 10 BNA OSHC 1057 (5th Cir. 1981); Owens-Corning Fiberglas Corp. v. Donovan, 659 F.2d 1285, 1288, 10 BNA OSHC 1070 (5th Cir. 1981); B & B Insulation, Inc. v. OSHRC, 583 F.2d 1364, 1370, 6 BNA OSHC 2067 (5th Cir. 1978); Power Plant Div., Brown & Root, Inc. v. OSHRC, 590 F.2d 1363, 1365, 7 BNA OSHC 1137 (5th Cir. 1979).

71. Beaver Plant Operations, Inc., 18 BNA OSHC 1972, 1977 (Rev. Comm'n 1999), citing Gregory & Cook, Inc., 17 BNA OSHC 1189, 1190 (Rev. Comm'n 1995); Seibel Modern Manufacturing & Welding Corp., 15 BNA OSHC 1219, 1228 (Rev. Comm'n 1991); Mosser Constr. Co., 15 BNA OSHC 1408, 1416 (Rev. Comm'n 1991); Dun-Par Engineered Form Co., 12 BNA OSHC 1949 (Rev. Comm'n 1986), rev'd on another ground, 843 F.2d 1135, 13 BNA OSHC 1652 (8th Cir. 1988).

72. Dun-Par Engineered Form, 12 BNA OSHC at 1956.

73. For example, M.J. Lee Constr. Co., 7 BNA OSHC 1140, 1144 (Rev. Comm'n 1979).

74. 12 BNA OSHC at 1953–1956.

75. 12 BNA OSHC at 1955.

76. See Pitt-Des Moines Inc., 16 BNA OSHC 1429, 1433–1434 (Rev. Comm'n 1993); Gregory & Cook Inc., 17 BNA OSHC 1189, 1191 (Rev. Comm'n 1995) (employer should attempt to acquire more suitable guard; commission "expect[s] employers to exercise some creativity in seeking to achieve compliance").

77. Williams Enters. Inc., 13 BNA OSHC 1249 (Rev. Comm'n 1987).

78. Walker Towing Corp., 14 BNA OSHC 2072, 2075–2076 (Rev. Comm'n 1991).

79. Quality Stamping Products v. OSHRC, 709 F.2d 1093, 1099, 11 BNA OSHC 1550 (6th Cir. 1983) (employer must show not "economically practicable" because "prohibitively expensive"); Donovan v. Williams Enters., Inc., 744 F.2d 170, 178, 11 BNA OSHC 2241 (D.C. Cir. 1984); Faultless Div., Bliss & Laughlin Indus. v. Secretary, 674 F.2d 1177, 1189, 10 BNA OSHC 1481 (7th Cir. 1982); Southern Colo. Prestress Co. v. OSHRC, 586 F.2d 1342, 1351, 6 BNA OSHC 2032 (10th Cir. 1978); Atlantic & Gulf Stevedores, Inc. v. OSHRC, 534 F.2d 541, 4 BNA OSHC 1061 (3rd Cir. 1976)

80. See, for example, Stan-Best, Inc., 11 BNA OSHC 1222, 1231 (Rev. Comm'n 1983); Research Cottrell, Inc., 9 BNA OSHC 1489, 1498 (Rev. Comm'n 1981).

81. 16 BNA OSHC 1155 (Rev. Comm'n 1993).

82. In State Sheet Metal, the Commission stated that in Dun Par Engd. Form Co., 12 BNA OSHC 1962, 1966 (Rev. Comm'n 1986), it first implied that an infeasibility defense may include economic factors. There, it found that the employer had not demonstrated that the costs were unreasonable in light of the protection afforded and had not shown what effect, if any, the added costs would have on its contract or on its business as a whole. See also Walker Towing Corp., 14 BNA OSHC 2072, 2077 (Rev. Comm'n 1991).

83. 16 BNA OSHC 1196, 1203 (Rev. Comm'n 1993), aff'd, 26 F.3d 573 (5th Cir.1994).

84. Gregory & Cook, Inc., 17 BNA OSHC 1189, 1192 (Rev. Comm'n 1995). Accord, Peterson Bros. Steel Erection Co. v. Sec'y of Labor, 26 F.3d 573, 16 BNA OSHC 1900 (5th Cir. 1994) (employer's claim that it would be disadvantaged as against competitors that did not comply is not relevant because "an employer cannot be excused from non-compliance on the assumption that everyone else will ignore the law"). See also Peterson Bros., 16 BNA OSHC 1196, 1203 (Rev. Comm'n 1993) (evidence that costs, while substantial, could be absorbed on the project negated the employer's claim of economic infeasibility); State Sheet Metal, 16 BNA OSHC 1155, 1159, 1160–1161 (Rev. Comm'n 1993).

85. Donley's Inc., 17 BNA OSHC 1227 (Rev. Comm'n 1995). But see Spancrete Northeast Inc. v. OSHRC, 905 F.2d 589, 14 BNA OSHC 1585 (2d Cir. 1990), which suggests that the defense does not have a second element. The court held that if compliance with the cited standard is infeasible, the Secretary must plead in the alternative and prove a failure to comply with another standard.

86. State Sheet Metal Co., 16 BNA OSHC 1155 (Rev. Comm'n 1993).

87. 12 BNA OSHC at 1956–1959.

88. Brock v. Dun-Par Engineered Form Co., 843 F.2d 1135 (8th Cir. 1988), rev'g 12 BNA OSHC 1949 (Rev. Comm'n 1986).

89. Seibel, 15 BNA OSHC at 1227–1228.

90. Caterpillar Inc. v. Herman, 131 F.3d 666, 18 BNA OSHC 1104 (7th Cir. 1997).

91. Russ Kaller Inc., 4 BNA OSHC 1758 (Rev. Comm'n 1976). See also PBR, Inc. v. Sec'y of Labor, 643 F.2d 890, 9 BNA OSHC 1357 (1st Cir. 1981); John H. Quinlan, 17 BNA OSHC 1194 (Rev. Comm'n 1995).

92. Royal Logging Co., 7 BNA OSHC 1744, 1751 (Rev. Comm'n 1979), aff'd, 645 F.2d 822, 9 BNA OSHC 1755 (9th Cir. 1981).

93. See Dun-Par, 12 BNA OSHC at 1967; Williams Enters. Inc., 13 BNA OSHC 1249 (Rev. Comm'n 1987).

94. John H. Quinlan, 17 BNA OSHC 1194 (Rev. Comm'n 1995).

95. General Elec. Co. v. Sec'y, 576 F.2d 558, 561, 6 BNA OSHC 1541 (3d Cir. 1978) (emphasis in the original). See also Reich v. Trinity Indus., Inc., 16 F.3d 1149, 1154, 16 BNA OSHC 1670 (11th Cir. 1994); Modern Drop Forge Co. v. Sec'y of Labor, 683 F.2d 1105, 1116, 10 BNA OSHC 1852 (7th Cir. 1982); Dole v. Williams Enters., Inc., 876 F.2d 186, 190 n.7, 14 BNA OSHC 1001 (D.C. Cir. 1989).

96. Caterpillar Inc. v. Herman, 131 F.3d 666, 18 BNA OSHC 1104, 1106 (7th Cir. 1997).

97. E.g., Sec'y of Labor v. L.E. Myers Co., 818 F.2d 1270, 13 BNA OSHC 1289 (6th Cir.), cert. denied, 484 U.S. 989 (1987); Texland Drilling Corp., 9 BNA OSHC 1023 (Rev. Comm'n 1980).

98. E.g., Capform, Inc., 16 BNA OSHC 2040, 2043 (Rev. Comm'n 1994).

99. E.g., Gem Industrial, Inc., 17 BNA OSHC 1861 (Rev. Comm'n 1996), and particularly note 8 of the lead opinion and note 12 of the dissenting opinion there. See generally Arthur G. Sapper, "The Oft-Missed Step: Documentation of Safety Discipline," Occupational Hazards (January 2006).

100. Daniel Constr. Co., 10 BNA OSHC 1549, 1552 (Rev. Comm'n 1982).

101. E.g., Consolidated Freightways Corp., 15 BNA OSHC 1317, 1321 (Rev. Comm'n 1991).

102. Daniel Constr., 10 BNA OSHC at 1552.

103. Seyforth Roofing Co., 16 BNA OSHC 2031 (Rev. Comm'n 1994).

104. See, e.g., New York State Gas & Elec. v. Sec'y of Labor, 88 F.3d 98, 17 BNA OSHC 1650, 1655-1657 (2d Cir. 1996) (reviewing confusing state of law and criticizing Commission for inconsistency); L.E. Myers v. Brock, 484 U.S. 989 (1987) (White, J., dissenting from denial of certiorari) ("confusing patchwork of conflicting approaches").

105. E.g., Usery v. Kennecott Copper Corp., 577 F.2d 1113, 6 BNA OSHC 1197 (10th Cir. 1977). See Kiewit Power Construcs. Co., No. 11-2395 (Dec. 11, 2012) (ALJ), review directed, Jan. 24, 2013, summarized in n. 7 above.

106. See 5 U.S.C. § 553(b); OSH Act § 6(b)(2); 29 U.S.C. § 655(b)(2); Kooritzky v. Reich, 17 F.3d 1509 (D.C. Cir. 1994) (non-OSHA case).

107. National Constrs. Ass'n v. Marshall, 581 F.2d 960, 970-71, 6 BNA OSHC 1721 (D.C. Cir. 1978).

108. E.g., Gen. Elec. Co. v. EPA, 53 F.3d 1324, 1328 (D.C. Cir. 1995) (non-OSHA case).

109. Beaver Plant Opers., Inc. v. Herman, 223 F.3d 25, 19 BNA OSHC 1053 (1st Cir. 2000) ("The burden is on the Secretary to establish that [the employer] had actual or constructive notice" of the standard's requirement); Diebold, Inc. v. Marshall, 585 F.2d 1327, 6 BNA OSHC 2002 (6th Cir. 1978). See also the cases cited in note 61 above.

110. Fabi Constr. Co. v. Sec'y of Labor, 508 F.3d 1077, 22 BNA OSHC 1001 (D.C. Cir. 2007).

111. Oberdorfer Indus. Inc., 20 BNA OSHC 1321, 1329 (Rev. Comm'n 2003).

112. Latite Roofing & Sheet Metal Co., 21 BNA OSHC 1282 (Rev. Comm'n 2005); Trinity Marine Nashville Inc. v. OSHRC, 275 F.3d 423, 430-31, 19 BNA OSHC 1673, 1676-77 (5th Cir. 2001).

113. See the cases cited in note 62-65 above.

114. For example, Keco Indus., Inc., 11 BNA OSHC 1832, 1834 (Rev. Comm'n 1984).

115. For example, General Electric Co., 3 BNA OSHC 1031, 1040 (Rev. Comm'n 1975).

116. Donovan v. Daniel Constr. Co., 692 F.2d 818, 10 BNA OSHC 2188 (1st Cir. 1982); Sec'y of Labor v. OSHRC (Erie Coke Corp.), 998 F.2d 134, 16 BNA OSHC 1241 (3d Cir. 1993); Phoenix Roofing, Inc. v. Dole, 874 F.2d 1027, 14 BNA OSHC 1036 (5th Cir. 1989); Chao v. Symms Fruit Ranch, Inc., 242 F.3d 894, 19 BNA OSHC 1337 (9th Cir. 2001).

117. Caterpillar Inc. v. Herman, 131 F.3d 666, 18 BNA OSHC 1104 (5th Cir. 1989).

118. El Paso Crane & Rigging Co., 16 BNA OSHC 1419, 1429 (Rev. Comm'n 1993) (failure to attest and sign OSHA injury log "trifling").

119. Keco Indus., Inc., 11 BNA OSHC 1832, 1834 (Rev. Comm'n 1984).

120. Phoenix Roofing, Inc. v. Dole, 874 F.2d 1027, 14 BNA OSHC 1036 (5th Cir. 1989).

121. See Holly Springs Brick & Tile Co., 16 BNA OSHC 1856 (Rev. Comm'n 1994) (rejecting de minimis argument for lack of evidence).

Chapter 4

The General Duty Clause

William K. Doran, Esq.
Matthew C. Cooper, Esq.
Ogletree, Deakins, Nash, Smoak & Stewart, P.C.
Washington, D.C.

1.0 Overview

Section 5 of the Occupational Safety and Health Act (OSH Act), commonly referred to as the "General Duty Clause," was designed by Congress to be "an enforcement tool of last resort."[1] It places nonspecific, broad safety requirements on employers when more specific standards or regulations are not applicable. The general duty clause, as it pertains to employers, states that:

(a) Each employer—

(1) shall furnish to each of his employees employment and a place of employment which are free from recognized hazards that are causing or are likely to cause death or serious physical harm to his employees[2]

In recent years the General Duty Clause has been utilized as a sometimes controversial mechanism for enforcement of safety guidelines that have not yet been specifically addressed by statute or regulation. The most notable example of this was the Occupational Safety and Health Review Commission's (Review Commission, OSHRC) application of the general duty clause to ergonomic hazards in the *Pepperidge Farm* case.[3] Similarly, the clause is relied upon by the Occupational Safety and Health Administration (OSHA) to address issues such as bloodborne pathogen hazards in industries not covered by the general industry regulation at 29 C.F.R. § 1910.1030 and reactive chemical process safety.[4] OSHA has also used the general duty clause to address issues related to risks of workplace violence, such as when "security measures were not taken to minimize or eliminate employee exposure to assault and battery by tenants of [an] apartment complex [owned by the employer]."[1] It has relied on the clause to regulate illnesses related to employee heat exposure.[2] And, most recently, OSHA even utilized the general duty clause to cite SeaWorld for exposing its trainers to recognized hazards when working in close contact with killer whales during performances.[3]

As noted above, the general duty clause is not applicable "if a standard specifically addresses the hazard cited."[5] This specificity can extend to the type of industry for which the standard was promulgated.[6] As noted in the legislative history of the general duty clause: "The general duty clause in this bill would not be a general substitute for reliance on standards, but would simply enable the Secretary to insure the protection of employees who are working under special circumstances for which no standard has yet been adopted."[7]

The courts and the Review Commission have precisely laid out the Secretary of Labor's (Secretary) burden of proof under the general duty clause. In order to demonstrate a violation, the Secretary must show that:

(1) a workplace condition presented a hazard, (2) the employer or its industry recognized the hazard, (3) the hazard was likely to cause serious physical harm, and (4) there was a feasible and useful means of abatement that would eliminate or materially reduce the hazard.[8]

These criteria were originally set down in 1973 by the D.C. Circuit in *National Realty*.[9] In *National Realty*, a case involving the workplace death of an employee riding on the running board of a loader, the D.C. Circuit reversed a decision of the Review Commission that a company had committed a "serious violation" of the general duty clause. The appellate court held that Congress did not intend for the general duty clause to impose strict liability on employers. Instead, the court held that the employer is obligated only to protect its employees from "preventable hazards." The court found that a specific instance of equipment riding does not qualify as "recognized" or "preventable" under the general duty clause.

Although the general duty clause is cited against employers for workplace hazards, the Act does not absolve employees of all responsibility. In fact, Section 105(b), 29 U.S.C. § 654(b), places a duty upon employees to "comply with occupational safety and health standards and all rules, regulations and orders issued pursuant to this Act which are applicable to his own actions and conduct." Yet, OSHA only has enforcement authority over employers.[4]

This employee responsibility to "comply" with all safety and health standards and rules is significant, but it does not release the employer from its obligations to provide a safe workplace.[10] The legislative history of the Act specifically states that "final responsibility for compliance with the requirements of this act remains with the employer."[11] The Review Commission has reinforced this proposition.[12]

However, in demonstrating the elements of a general duty clause violation, "the Secretary must define the alleged recognized hazard in a manner that gives the employer fair notice of its obligations under the Act by specifying conditions or practices which are within the employer's control."[13] For instance, in a fatal accident case in which an employee was struck by an unloaded truck that was backing up, OSHA issued the following general duty clause citation:

> Section 5(a)(1) of the Occupational Safety and Health Act of 1970: The employer did not furnish employment and a place of employment which were free from recognized hazards that were causing or likely to cause death or serious physical harm to employees in that employees were exposed to a struck by and run over by a winch truck hazard:
>
> At the rig site, the employee(s) toolpusher was exposed to a struck by hazard from a winch truck which was not equipped with a reverse audible warning device.
>
> * * *
>
> Some feasible and acceptable means of abatement, among others, are:
>
> a. Install reverse audible warning devices (back-up alarms) on all winch trucks (vehicles) which are operated in reverse.
>
> b. Have the [flagman] guide the trucks traveling in reverse at the rig site.
>
> c. Have all trucks travel in a forward motion if at all possible.[14]

Only one of the proposed abatement measures need ultimately be proved to be feasible.

2.0 Who Is Protected by the General Duty Clause?

Unlike under specific OSHA safety and health standards, the employer's obligations under the general duty clause extend only to its own employees.[15] OSHA will not cite an employer for a Section 5(a)(1) violation unless the employer's own employees are exposed to a hazard created or controlled by the employer.[16] A common employer defense in a general duty clause case is that it is not an employer of the employees exposed to a particular hazard, and thus owes them no general duty under Section 105(a). The Sixth Circuit in *Stein, Inc.* explained that "the most important factor in determining whether an entity is an employer is 'who has control over the work environment such that abatement of the hazards can be obtained.'"[17]

The court in *Stein, Inc.* also cited the usefulness of the Review Commission's "economic realities test."[18] This test evaluates whether an employment relationship exists by considering the following factors:

1. Whom do the workers consider their employer?

2. Who pays the workers' wages?

3. Who has the responsibility to control the workers?

4. Does the alleged employer have the power to control the workers?

5. Does the alleged employer have the power to fire, hire, or modify the employment condition of the workers?

6. Does the workers' ability to increase their income depend on efficiency rather than initiative, judgment, and foresight?

7. How are the workers' wages established?[19]

Thus, in *Stein, Inc.*, a company was deemed to be an employer because it paid the specific workers, had the power to hire, fire, or modify their employment, made their job assignments, and had "sufficient control" over the job.[20]

In a multi-employer setting, OSHA will only cite the "exposing employer" for a general duty clause violation.[21] An exposing employer is "an employer whose own employees were exposed to a hazard."[22]

If the employer-employee relationship is demonstrated, that employer is obligated to protect each of his employees from hazards. With that said, the courts have clarified that "the phrase 'each of his employees' in the general duty clause is an inclusive expression which 'means that an employer's duty extends to all employees regardless of their individual susceptibilities.'"[23]

3.0 The Existence of a Hazard

The threshold inquiry in evaluating whether a general duty clause obligation exists is whether there is a hazard. A hazard "is not defined in terms of the absence of a particular abatement method."[24] Rather, it is defined "in terms of the physical agents that could injure employees."[25] Matters of "human comfort," as opposed to hazards that could cause death or serious physical harm, are not covered by the general duty clause.[26]

An employer may be cited for a general duty clause violation despite the lack of an accident or incident causing injury.[27] Further, hazardous conduct need not actually have occurred.[28] If the hazardous condition was preventable, the fact that no injuries occurred will not insulate the employer from enforcement action.[29] Conversely, an accident in the workplace does not conclusively prove a violation. If the hazard was not preventable, then regardless of injury, the general duty clause has not been violated.[30]

The Review Commission and the courts have not provided a firm standard for determining the existence of a hazard. In large part, this is due to disagreements over the applicability of a "significant risk" standard to "harm that has already occurred," as opposed to "prospective harm."[31] In *Kastalon, Inc.*, for example, the Review Commission determined that the Secretary had the burden of establishing a significant risk of harm in a case involving a possible human carcinogen in the workplace that was suspected on the basis of extrapolation from animal tests.[32]

In circumstances involving harm that has already occurred, such as repetitive motion and lifting injuries, the Review Commission has not required the Secretary to demonstrate significant risk. The Review Commission explained that "under section 5(a)(1), the Secretary need only show that the alleged hazards [in the workplace] were causing or likely to cause serious physical harm to employees there."[33]

A significant number of decisions have rejected outright the application of the "significant risk" test in a general duty clause case. In *Kelly Springfield Tire Co. v. Donovan*, the Fifth Circuit argued that the "extension of the significant risk standard to enforcements of the general duty clause would constitute an abandonment of the *National Realty* standard."[34]

4.0 Recognized Hazard

In defining the second element underlying the general duty clause, that there be a "recognized hazard," the D.C. Circuit in *National Realty* focused on a floor speech by Representative Dominick V. Daniels.

A recognized hazard is a condition that is known to be hazardous, and is known not necessarily by each and every individual employer but is known taking into account the standard of knowledge in the industry. In other words, whether or not a hazard is "recognized" is a matter for objective determination; it does not depend on whether the particular employer is aware of it.[35]

In order to meet its burden of proof, OSHA must demonstrate that a hazard was recognized either by the individual employer or the employer's industry.[36] An employer need not be specifically aware of a hazard to be cited under the general duty clause.[37] Individual employer knowledge of the hazard will, however, be sufficient to prove that the hazard was recognizable.[38] (The obvious problem with this approach, though, is that it potentially penalizes proactive employers and incentivizes employers to be uninformed about, not actively looking for, potential but unknown hazards.) In the absence of individual employer knowledge, OSHA will evaluate the hazard to determine if it is generally "recognized" within that employer's industry or if common sense would make the hazard recognizable.[39]

It is important to note that in evaluating employer or industry recognition, knowledge of the condition presenting the hazard is not sufficient. Rather, "it is the dangerous potential of the condition or activity being scrutinized that must be known specifically by the employer or known generally in the industry."[40]

4.1 Industry Recognition

If a particular hazard is well known or documented within an industry, an employer in that industry has an obligation under the general duty clause to keep its workplace free of that hazard.[41] In determining industry recognition of a hazard, the Review Commission and the courts have relied on a variety of indices. For instance, it has been "consistently held that voluntary industry codes and guidelines are evidence of industry recognition."[2] (Note, however, that an advisory code, i.e. "companies *should* do this," is different from a mandatory obligation in code. OSHA cannot take an advisory industry guideline and make it mandatory under the general duty clause.) Similarly, the opinion of "safety experts familiar with the workplace conditions or the hazard in question" is also pertinent.[43] It is not necessary that the experts be employed in the industry.[44]

Additional bases for finding industry recognition of a hazard include "evidence of industry safety practices,"[45] proposed OSHA regulations,[46] and warnings provided by manufacturers.[47] Thus, in *Cormier Well Serv.*, the Review Commission relied on an OSHA compliance officer's testimony regarding oil derrick fall protection practices to conclude that the hazard of standing on the platform without being protected by a safety belt secured to a lifeline was recognized in the industry.[48] The decision further noted that the compliance officer had "a degree in petroleum engineering and five years of oil industry experience."[49]

This list is not exhaustive. OSHA will use any available evidence to determine whether the hazard is generally recognized.

4.2 Employer Recognition

As set out above, an employer's actual knowledge of a hazard in the workplace is not required to prove that a hazard was recognizable. However, if it can be demonstrated that the employer *was* aware of the hazard, the hazard will be deemed recognizable despite a lack of industry recognition or common sense recognition.[50] The Review Commission has expressed a reluctance to rely solely on voluntary safety efforts by an employer in determining recognition of a hazard.[51]

OSHA has demonstrated employer recognition of hazards in a number of ways. For instance, recognition has been imputed from the knowledge and expertise of safety personnel.[52] In *Pegasus Tower*, 21 OSHC BNA 1190, 1191 (2005), the Review Commission found that the potential 400-foot-fall hazard produced by employees riding a base-mounted dual-drum hoist line at a tower erection construction site was recognized by the employer. The Review Commission based its finding on the testimony of the employer's experienced safety instructor, which indicated recognition of "the dangers associated with raising and lowering employees on the hoist line."[53]

Similarly, in *Arcadian Corporation*, the Review Commission found that a urea manufacturer recognized the hazard of an explosion caused by leaks that could erode the lining of the reactor.[54] Specifically, the Review Commission pointed to testimony regarding the statement of a supervisor regarding the potential for explosion:

The only thing you have to worry about is if that reactor ever leaks or if it ever blows up. You won't be here to tell about it.[55]

Other circumstances demonstrating employer knowledge include the existence of voluntary employer safety rules and procedures (in combination with other factors),[56] specific warnings given by company management,[57] manufacturers' warnings,[58] complaints by employees,[59] and past accidents and injuries.[60]

It is important to note that the fact that an employer's work practices are consistent with those in the industry does not necessarily demonstrate that the employer could not recognize that the practices are hazardous.[61] In *Beverly Industries,* for instance, the hazards related to patient lifting practices that were standard in the nursing home industry were deemed by the Review Commission to be recognized by the employer because, among other things, management implemented a back belt program to address the hazard, specifically warned personnel of the hazard, and distributed a manufacturer's warning to personnel.[62]

4.3 Obvious Hazard Recognition

Even if industry or specific employer recognition cannot be established, recognition can nonetheless be demonstrated if the hazard is deemed to be "obvious."[63] OSHA refers to such hazard recognition as "common sense recognition."[64] OSHA enforcement policy restricts the application of this theory of recognition to "flagrant cases."[65]

In *Safeway, for example,* the Tenth Circuit found that the recognized hazard element of the general duty clause criteria was met where a bread-baking plant employer utilized a forty-pound propane tank on an outdoor gas grill designed for twenty-pound tanks. The court rejected the employer's argument that the bread-baking industry had never recognized this type of hazard, explaining that "Safeway cannot ignore the presence of an obviously hazardous condition by asserting that its industry is ignorant of such hazards."[66] Likewise, in *Tri-State Roofing & Sheet Metal,* the Fourth Circuit found that work on an unguarded platform in excess of 40 feet above a concrete floor without protective equipment was "patently dangerous," and therefore, upheld the administrative law judge's finding that the hazard was obvious and thus recognized by the industry.[5]

The courts and the Review Commission sometimes apply a reasonable person standard in evaluating the obviousness of a hazard. If a reasonable person would have recognized the existence of a hazard in the workplace, that hazard will be deemed recognizable.[67] This analysis is not always so simple, however. In *Kelly Springfield Tire Co.*, for example, Fifth Circuit judges were split over whether an explosion hazard related to the employer's dust collection system was "obvious."[6] In that case, it was undisputed that the employer did not know of the explosion hazard, and there was no evidence that anyone within the tire industry knew of the hazard.[7] Nevertheless, the majority concluded that when a hazard is "obvious and glaring" or where a practice is plainly recognized as hazardous in one industry, the Commission may determine that a hazard was recognized without reference to specific knowledge within the specific industry at hand.[8] The majority concluded that employer recognition centers on "the common knowledge of safety experts" and relied solely on the testimony of one Secretary of Labor expert.[9] The dissenting judge, however, pointed out that simply because an "expert can postulate the theoretical possibility of a hazardous condition does not mean that an employer should have recognized it as such."[10] The dissent points out that in doing so the majority mistakes "recognition of hazards" with "foreseeability of hazards."[11]

The bottom line is that whether a hazard can be recognized as obvious is not always so obvious.

5.0 Causing or Likely to Cause Death or Serious Physical Harm

Not all workplace hazards are prohibited by the general duty clause. Employers are liable only for preventing hazards that are "causing or are likely to cause death or serious physical harm."[68] This language has been interpreted as applying the same standard as that applied to a "serious" violation.[69] Specifically, there must be a "substantial probability that death or serious physical harm could result from a condition which exists."[70] The occurrence of a death "constitutes at least prima facie evidence of likelihood."[71]

In evaluating likelihood, the Review Commission has determined that the general duty clause does not require the Secretary of Labor to prove that an accident is likely.[72] Rather, the Secretary need only establish that, *if an accident occurs*, death or serious physical harm would be the likely result.[73] Similarly, in evaluating likelihood with respect to an occupational illness caused by a substance in the workplace, the Secretary must only prove that death or serious harm is likely if an employee contracts the illness.[74]

Injuries and illnesses that the courts and the Review Commission have indicated provide a sufficient basis for a finding that death or serious physical harm could result include "physical disorders that so adversely affect employees that they are disabled from doing their jobs"[75] or from performing their normal activities.[76] The Review Commission has also noted that serious physical harm can be demonstrated even in circumstances where there is no evidence of "pathological anatomic change or injury"[77]—a condition "which cannot be linked to any detectable tissue or body damage or injury."[78]

Further, this is true "even if the disability is not permanent."[79] In *Consolidated Freightways Corp.*, for instance, the Review Commission found that temporary nausea, vomiting, light sensitivity, and the potential for temporary loss of vision, which caused employees to miss up to seven and a half weeks of work, constituted serious physical harm.[80] According to the Review Commission, these conditions, caused by exposure to toxic dye, were indicative of substantial impairment to functions of the body.[81]

6.0 Feasible Measures to Correct the Hazard

The general duty clause has not been interpreted as imposing an impossible responsibility on employers. The *National Realty* court explained that "Congress quite clearly did not intend the general duty clause to impose strict liability: The duty was to be an achievable one."[82] Further, the D.C. Circuit acknowledged that a "demented, suicidal, or willfully reckless employee may on occasion circumvent the best conceived and most vigorously enforced safety regime."[83] Each employer, therefore, is required by the general duty clause to keep its workplace free of all *preventable* hazards.[84]

Determining whether a hazard is preventable is essentially dictated by an evaluation of the feasibility of abating it. Feasibility "means economically and technologically capable of being done."[85] The Secretary of Labor must set out the proposed abatement measures and "demonstrate both that the measures are capable of being put into effect and that they would be effective in materially reducing the incidence of the hazard."[86] The Secretary must also establish that a proposed measure of abatement is not cost prohibitive.[87]

Feasibility can, in some cases, be determined by common sense.[88] OSHA can also point to employer or general industry practices to prove that a feasible method of abatement exists.[89] Alternatively, an employer may use evidence of its own abatement efforts or standard industry abatement methods in demonstrating that "it took all necessary precautions to prevent the occurrence of the violation."[90] However, the courts and the Review Commission have made it clear that a safety precaution does not have to have "general usage" in an industry in order to be deemed feasible.[91] As explained by the D.C. Circuit in *National Realty*, "the question is whether a precaution is recognized by safety experts as feasible, not whether the precaution's use has become customary."[92]

In essence, the basic focus in evaluating feasibility is whether the measure is "reasonable and practical."[93] A key component of this inquiry is whether the measure will "materially reduce the hazard,"[94] or, in other words, whether it will "eliminate or substantially reduce" the hazard. In *Waldon Healthcare Center,* the Review Commission found that the Secretary's proposed pre-exposure vaccination measure for Hepatitis provided "virtually the same effectiveness rate" as the postexposure treatment measure offered by the employer.[95] Consequently, the Review Commission determined that the Secretary had failed to demonstrate that the abatement measure would result in a material reduction of the hazard.[96]

In *Arcadian Corp.,* the Review Commission found that three specific abatement methods proposed by the Secretary would materially reduce the reactor explosion hazard caused by the leakage of a corrosive chemical (urea).[97] Based on the abatement practices in the fertilizer industry and the employer's own policy for dealing with urea leaks, the Review Commission specifically identified the following abatement methods: (1) shutting down the urea reactor upon detection of the leak; (2) inspection of the leak detection system; and (3) regular inspections of the reactor.[98]

An equally important component of the feasibility determination is whether the employer can absorb the costs of the proposed abatement measures without threatening its "economic viability."[99] In this regard, two issues that courts and the Review Commission generally consider are: (1) whether the costs of a proposed abatement method will "jeopardize a company's long-term profitability and competitiveness," and/or (2) whether those costs can be passed on to the customer.[100] In *Sun Ship, Inc.,* the Review Commission found that the Secretary had made a *prima facie* case of economic feasibility because the proposed $2,500 noise abatement control would not adversely impact an employer with annual sales of $100 million.[101] In *SeaWorld,* the court similarly found that SeaWorld's voluntary abatement measures with regard to one killer whale without any suggestion of harm to its profits demonstrated that close trainer contact with other killer whales was not integral to SeaWorld's workplace.[12]

7.0 Practical Enforcement of the General Duty Clause

The mechanics of general duty clause enforcement derive from the central tenet that the clause can only be used when there is no safety standard applicable to the particular hazard involved.[102] With this said, agency policy, consistent with court and Review Commission decisions, allows enforcement personnel to cite the general duty clause, along with a standard, when there is doubt as to the application of the standard.[103] OSHA's *Field Inspection Reference Manual* provides the OSHA area manager and assistant area manager discretion to conduct a precitation review when section 105(a)(1) is contemplated.[104]

Upon a determination that no standard applies to an identified hazard, OSHA policy directs its enforcement personnel to evaluate whether the four elements of a general duty clause violation, laid out by the courts and the Review Commission, are met (hazard in the workplace, recognition of hazard, hazard causing or likely to cause death or serious physical harm, and feasible abatement).[105] Under this policy, if no standard exists and all four elements cannot be demonstrated, the area office must forward a letter "to the employer and the employee representative describing the hazard and suggesting corrective action."[106]

The agency also acknowledges a number of limitations on its ability to cite the general duty clause. The policy specifically states that Section 5(a)(1):

1. Shall not group violations together (can be grouped with related violation of a specific standard);

2. Shall not normally be used to impose a stricter requirement than that required by the standard (unless it can be documented the employer knows that the standard is ineffective in protecting the employees from the specific hazard);

3. Shall normally not require an abatement method not set forth in a standard;

4. Shall not be used to enforce 'should' standards;

5. Shall not normally be used to cover categories of hazards exempted by a standard (unless exemption was based on something other than lack of a hazard).[107]

Despite this guidance, recent activity by OSHA indicates the agency is moving in another direction. In September 2013, OSHA issued a general-duty citation for alleged exposure to excessive amounts of styrene, which were beyond the recommended occupational exposure limits (OELs) set by NIOSH of 50 ppm.[13] (OSHA has posted these numbers on its website as a recommendation to employers, acknowledging that many of the agency's PELs are outdated.)[14] However, OSHA already has a standard for permissible exposure limits (PELs) for styrene set forth in 29 CFR 1910.1000, which sets the PEL at 100 ppm, noticeably above the 65.2 ppm for which the employer was cited. Whether this attempt to circumvent the rulemaking process will be upheld is yet to be conclusively decided. Regardless, this attempt shows OSHA will attempt to impute employer and industry knowledge for general-duty citations based on "recommended" OELs based on the theory that the hazard is "recognized." This effort is merely an extension of OSHA's stated policy in a related 2003 interpretation letter, in which OSHA stated it will issue general-duty citations when chemical hazards are not presently covered by an OSHA PEL if an employer fails "to control or eliminate a 'recognized hazard.'"[15] While currently limited to PELs, this approach could obviously spread to other areas of enforcement that are monitored by occupational safety and health organizations such as NIOSH.

In proposing civil penalties for general duty clause violations, OSHA has, in the past, attempted to issue penalties based on the number of employees affected by the hazardous condition rather than the condition itself. This, of course, had the effect of greatly increasing the overall civil penalty directed to the employer. In *Arcadian Corp.*, the Fifth Circuit affirmed the Review Commission's rejection of this position. The court noted that the Act focuses civil penalties on conditions, practices, and violations, and not on employees.[108] With this stated, the court acknowledged that the number of employees affected by a hazardous condition can be considered by the Review Commission in carrying out its "exclusive role" of assessing a civil penalty that has been proposed by the Secretary and contested by an employer.[109]

8.0 Conclusion

The primary purpose of the general duty clause is to offer an extra measure of protection to employees in the workplace. Most standards implemented under OSHA are targeted at a specific hazard. The general duty clause, however, allows inspectors to cite employers for exposing its employees to a recognized hazard that has not been specifically addressed in the regulations.

Notes

1. See Sec'y of Labor v. Megawest Financial, Inc., 17 OSH Cas. (BNA) 1337 (OSHRC 1995); see also OSHA Interpretation Letter, OSHA Policy Regarding Violent Employee Behavior (December 10, 1992) ('where the risk of violence and serious personal injury are significant enough to be 'recognized hazards,' the general duty clause would require the employer to take feasible steps to minimize those risks."); OSHA Recommendation, Recommendations for Workplace Violence Prevention Programs, (OSHA 3153-12R, 2009) ('workplace violence has emerged as a major occupational safety and health issue in many industries, especially the retail trade.").

2. See OSHA Campaign to Prevent Heat Illness in Outdoor Workers, at https://www.osha.gov/SLTC/heatillness/index.html; see also General Query on OSHA citation database for "heat" citations.

3. See SeaWorld of Florida, LLC v. Sec'y of Labor, 748 F.3d 1202 (D.C. Cir. 2014).

4. 29 U.S.C. §§ 658, 659.

5. Tri-State Roofing & Sheet Metal, Inc. v. Occupational Safety & Health Review Comm'n, 685 F.2d 878, 880 (4th Cir. 1982).

6. See Kelly Springfield Tire Co., Inc. v. Donovan, 729 F.2d 317 (5th Cir. 1984)

7. Id. (dissenting opinion) at 325.

8. Id. at 321.

9. Id. at 322.

10. Id. (dissenting opinion) at 325.

11. Id. (dissenting opinion) at 326 ('there is a vast difference between what is known or recognized and what is reasonably foreseeable.") (quoting Pratt & Whitney Aircraft v. Secretary of Labor, 649 F.2d 96, 100 (2d. Cir. 1981)).

12. SeaWorld, 748 F.3d at 1213-1215.

13. See EHS Today, Exposure Limits: Is OSHA Circumventing the Rulemaking Process?, Mark S. Dreux, Aaron S. Brand, and Matt Throne (May 12, 2014).

14. See Permissible Exposure Limits – Annotated Tables, at https://www.osha.gov/dsg/annotated-pels/ ('OSHA's mandatory PELs in the Z-Tables remain in effect. However, OSHA recommends that employers consider using the alternative occupational exposure limits because the Agency believes that exposures above some of these alternative occupational exposure limits may be hazardous to workers, even when the exposure levels are in compliance with the relevant PELs").

15. OSHA Interpretation Letter, Standard 1910.1000 (January 24, 2003, corrected 9/29/2008) ("An OEL is sometimes used as a reference point to support a finding of a 'recognized hazard,' but it should be only to buttress other evidence that a hazard exists...[e]xceeding an OEL recommendation, therefore, is not the hazardous or violative condition, but rather that employees were exposed to harmful levels of a chemical." See also OSHA Memorandum for Regional Administrators, Enforcement Policy for Respiratory Hazards not Covered by OSHA Permissible Exposure Limits, January 24, 2003.

Chapter 5

Recordkeeping

Melissa A. Bailey
Shontell Powell
Ogletree, Deakins, Nash, Smoak & Stewart, P. C.
Washington, D. C.

1.0 Overview

The Occupational Safety and Health Administration (OSHA) requires employers to maintain several different types of records. First, the Recording and Reporting Occupational Injuries and Illnesses regulation requires employers to record work-related injuries and illnesses that meet certain criteria on standardized OSHA forms or equivalent forms.[1] The recordkeeping regulation also requires employers to report a fatality within eight hours and an in-patient hospitalization of one or more employees; amputation; or loss of an eye within twenty-four hours to the local OSHA area office.

Second, OSHA safety standards, which are generally intended to prevent physical hazards in the workplace, require a variety of different types of records. For example, OSHA safety standards may require documentation of employee training, written compliance programs, inspection reports and hazard assessments.

Third, OSHA health standards, which are generally intended to limit exposure to substances that may be hazardous to employee health, typically require the employer to prepare and maintain records of employee exposure levels as well as employee medical records. For example, certain substance-specific standards, such as the lead,[2] cadmium,[3] and benzene[4] standards, require employers to perform monitoring to determine employee exposure levels and to provide medical surveillance for certain employees. OSHA's Access to Employee and Medical Records standard mandates that employers maintain the records generated pursuant to these types of health standards.[5]

Finally, employers often maintain additional records that are not explicitly required by an OSHA safety or health standard to prove compliance with a standard if OSHA takes enforcement action. For example, Subpart L contains requirements for fire protection, including maintenance and testing provisions for fire suppression systems such as portable fire extinguishers, standpipe and hose systems, and sprinkler systems.[6] Although the Subpart L standards do not require employers to maintain records of tests and inspections, the employer may, as a practical matter, be unable to show compliance without the testing and inspection records. Many other OSHA standards essentially require employers to keep records to prove compliance in the event of an enforcement action.

2.0 Statutory Authority

When the Occupational Safety and Health Act (OSH Act) was enacted, Congress recognized that data about injuries, illnesses, and hazards would assist OSHA in developing standards to address hazards and in concentrating enforcement resources on workplaces with employees experiencing injuries and illnesses. As such, Congress sought to "assure so far as possible every working man and woman in the National safe and healthful working conditions."[7] To achieve this goal, Congress authorized "research in the field of occupational safety and health," the "development and promulgation of occupational safety and health standards," and "appropriate reporting procedures with respect to occupational safety and health" to "accurately describe the nature of the occupational safety and health problem."[8]

Section 8 of the OSH Act includes several recordkeeping provisions. Section 8(c)(1) requires each employer to "make, keep and preserve" any records OSHA "may prescribe by regulation as necessary or appropriate for the enforcement of this Act or for developing information regarding the causes and prevention of occupational accidents and illnesses."[9] Section 8(c)(2) authorizes OSHA to "prescribe regulations requiring employers to maintain accurate records of, and make periodic reports on, work-related deaths, injuries and illnesses other than minor injuries requiring only first aid treatment and which do not involve medical treatment, loss of consciousness, restriction of work or motion, or transfer to another job."[10] Section 8(c)(3) requires OSHA to "issue regulations requiring employers to maintain accurate records of employee exposures to potentially toxic materials or harmful physical agents."[11]

The remainder of this chapter describes the recordkeeping and documentation provisions OSHA has instituted pursuant to these congressional directives.

3.0 Injury and Illness Recordkeeping

As discussed further below, the Recording and Reporting Occupational Injuries and Illnesses regulation, commonly known as the recordkeeping regulation, requires employers to record certain work-related injuries and illnesses on the OSHA 300 Log, to keep other data concerning these injuries, and to post a summary of injuries and illnesses each February. In addition, employers must report a fatality within eight hours and an in-patient hospitalization of one or more employees; an amputation; or an loss of an eye within twenty-four hours to the local OSHA area office. The regulation also requires employers to respond to requests from OSHA or the Bureau of Labor Statistics (BLS) for injury and illness data.

3.1 History of the Recordkeeping Requirements

OSHA has required employers to keep injury and illness records since shortly after the OSH Act was enacted in 1971. From 1971 to 1990, OSHA and the Bureau of Labor Statistics jointly administered the injury and illness recordkeeping system. In a Memorandum of Understanding executed in 1990, BLS agreed to conduct annual surveys of occupational injuries and illnesses and compile the data, and OSHA agreed to administer the enforcement and rulemaking aspects of recordkeeping.

On January 19, 2001, OSHA issued a revised recordkeeping regulation.[12] Prior to these revisions, OSHA's recordkeeping regulation was considered by many employers to be confusing, and employers often had difficulty determining when an injury or illness had to be recorded. Most provisions of the revised recordkeeping regulation became effective on January 19, 2002.[13]

3.2 OSHA's Authority to Require Employers to Keep Records

OSHA's authority to require employers to keep injury and illness records derives from Section 8(c) of the OSH Act. Also, Section 24(a) requires the Secretary to develop and implement a program to "compile accurate statistics on work injuries and illnesses which shall include all disabling, serious, or significant injuries and illnesses, whether or not involving loss of time from work, other than minor injuries requiring only first aid treatment and which do not involve medical treatment, loss of consciousness, restriction of work or motion, or transfer to another job."

3.3 Identifying Injuries and Illnesses that Must be Recorded

OSHA's recordkeeping regulation generally requires employers to record each "non-minor" work-related injury or illness on the establishment's OSHA 300 Log. The regulation provides a series of criteria that must be met for an injury or illness to be recorded. These criteria are:

- Has an "injury" or "illness" occurred?

- Is the injury or illness "work-related"?

- Is the work-related injury or illness a "new case"?

- Does the injury or illness meet the general recording criteria, such as days away from work, restricted work or job transfer, or the provision of medical treatment?

- Is the injury or illness in a special category, such as an occupational hearing loss or tuberculosis, a "significant" injury or illness, or a sharps/needlestick injury that requires recording?[14]

Each recordable injury and illness must also be described on an OSHA 301 Incident Report. The OSHA 301 report requires information about the employee, where the employee was treated, how the injury or illness occurred, and the extent of the injury or illness.[15] The employer is also required to execute and post an annual summary of illnesses and injuries on the Form 300-A, which describes the injuries and illnesses from the previous year, and must be posted each February.[16]

3.3.1 Determining Whether an Injury or Illness Has Occurred

In the majority of cases, it is obvious whether an injury or illness has occurred. For example, an employee who trips and breaks his ankle has clearly experienced an injury, and an employee who contracts tuberculosis clearly has an illness.[17]

Section 1904.46 defines an injury or illness as an "abnormal condition or disorder," including but not limited to a "cut, fracture, sprain, or amputation," or "a skin disease, respiratory disorder, or poisoning."[18] OSHA intentionally defined "injury or illness" very broadly because the "series of screening mechanisms for recording," such as the work-related requirement and the various recording criteria, are intended to weed out what OSHA considers to be minor injuries and illnesses. Nevertheless, an injury or illness has to constitute an abnormal condition and includes only "those changes that reflect an adverse change in the employee's condition that is of some significance." Although injury and illness are broadly defined, a "mere change in mood or experiencing normal end-of-the-day tiredness" do not meet the definition.[19]

3.3.2 Defining "Work-Related": The Geographic Presumption

Employers are required to record only those injuries and illnesses that are work-related. While this determination may seem simple on its face, the question of whether an injury or illness is work-related often presents difficult and complex issues. The recordkeeping regulation is not intended to require recording of only those injuries that result from a hazard in the workplace. Instead, the regulation is designed to capture the vast majority of significant injuries or illnesses that occur or surface at the workplace on a "no fault" basis. The OSHA 300 Log itself states that recording "does not mean that the employer or a worker was at fault or that an OSHA standard was violated."[20]

The regulation also requires employers to presume that any injury or illness that occurs or surfaces in the workplace is "work-related." The only exceptions to this geographic presumption are set out in the regulation. This geographic presumption may lead to counterintuitive results. For example, an employee may slip in the company parking lot and suffer an injury. This injury would be considered work-related even though it did not occur because of a hazard in the workplace; the employee was simply clumsy.[21]

Section 1904.5(a) of the regulation states that an injury or illness is "work-related if an event or exposure in the work environment either caused or contributed to the resulting condition or significantly aggravated a pre-existing injury or illness." Section 1904.5(b) defines work environment as "the establishment and other locations where one or more employees are working or are present as a condition of their employment," and the term "includes not only physical locations, but also the equipment or materials used by the employee during the course of his or her work."

The "geographic presumption" is described as follows: "Work-relatedness is presumed for injuries and illnesses resulting from events or exposures occurring in the work environment, unless an exception in § 1904.5(b)(2) specifically applies."[22] As such, an injury or illness is work-related if the injury or illness would not have happened but for the presence of the employee in the workplace. The fact that the employer could not have prevented the injury or illness through measures such as a better safety program or machine guards is irrelevant.

OSHA illustrated this concept during the rulemaking by giving several examples, including the following:

> Injuries and illnesses also occur at work that do not have a clear connection to specific work activity, condition or substance that is peculiar to the employment environment. For example, an employee may trip for no apparent reason while walking across a level factory floor; be sexually assaulted by a co-worker; or be injured accidentally as a result of violence perpetrated by one co-worker against a third party. In these and similar cases, the employee's job-related tasks or exposures did not create or contribute to the risk that such an injury would occur. *Instead, a casual connection is established by the fact that the injury would not have occurred but for the conditions and obligations of employment that placed the employee in the position in which he or she was injured or made ill.*[23]

Similarly, OSHA stated:

> If an event, such as a fall, an awkward motion or lift, an assault or an instance of horseplay, occurs at work, the geographic presumption applies and the case is work-related unless it otherwise falls within an exception. Thus, if an employee trips while walking across a level factory floor, the resulting injury is considered work-related under the geographic presumption because the precipitating event—the tripping accident— occurred in the workplace. The case is work-related even if the employer cannot determine why the employee tripped, or whether any particular workplace hazard caused the accident to occur.[24]

OSHA provides additional guidance about determining whether an injury or illness is work-related in the Recordkeeping Policies and Procedures ("Compliance Directive").[25] The guidance states that "a case is presumed work-related if, and only if, an event or exposure in the work environment is the discernable cause of the injury or illness or of a significant aggravation to a pre-existing condition," and notes that the "work event or exposure need only be one of the discernable causes; it need not be the sole or predominant cause."

3.3.3 Preexisting Conditions

An injury or illness is also considered work-related if an "event or exposure . . . significantly aggravated a pre-existing injury or illness."[26] A condition is preexisting if "it resulted solely from a non-work-related event or exposure that occurred outside the work environment."[27] A preexisting condition is considered aggravated, and therefore work-related, only if the workplace event or exposure worsens the preexisting condition such that one of the following conditions occurs:

- Death, provided that the preexisting injury or illness would likely not have resulted in death but for the occupational event or exposure;

- Loss of consciousness, provided that the preexisting injury or illness would likely not have resulted in death but for the occupational event or exposure;

- One or more days away from work, or days or restricted work, or days of job transfer that otherwise would not have occurred but for the occupational event or exposure;

- Medical treatment in a case where no medical treatment was needed for the injury or illness before the workplace event or exposure, or a change in medical treatment was necessitated by the workplace event or exposure.[28]

For example, an employee may have preexisting asthma that is not work-related. If the employee inhales dust while at work, and the dust aggravates his asthma such that he loses consciousness, the incident is considered work-related.

3.3.4 The Employer's Obligation to Determine Work-Relatedness

Although the employer may rely upon the opinion of a physician in determining whether an injury or illness is work-related, the employer is ultimately responsible for making the decision. OSHA commented during the rulemaking that "the employer is in the best position to obtain the information, both from the employee and the workplace, that is necessary to make the determination" of whether an injury or illness is work-related.[29] Further, "[a]lthough expert advice may occasionally be sought by employers in particularly complex cases, the final rule provides that the determination of work-relatedness ultimately rests with the employer."[30]

Accordingly, Section 1904.5(b)(3) instructs that "if it is not obvious whether the precipitating event or exposure occurred in the work environment or occurred away from work," the employer must "evaluate the employee's work duties and environment and decide whether or not one or more events or exposures in the work environment either caused or contributed to the resulting condition or significantly aggravated a pre-existing condition."

3.3.5 Exceptions to Work-Relatedness

Section 19054.5(b)(2) lists nine situations in which an injury or illness will not be considered work-related even though it occurs in the work environment. These nine exceptions essentially invalidate the geographic presumption for injuries and illnesses that occur in the workplace.

The broadest and most complex exception applies to injuries and illnesses that "involve signs or symptoms that surface at work, but result solely from a non-work-related event or exposure that occurs outside the work environment." As stated, Section 1904.5(b)(3) requires the employer to evaluate the employee's work duties and determine whether an event or exposure in the workplace caused the injury or illness.

In the rulemaking, OSHA explained this exception as follows:

> This exception is consistent with the position followed by OSHA for many years and reiterated in the final rule: that any job-related contribution to the injury or illness makes the incident work-related, and its corollary—that any injury or illness in which work makes no actual contribution is not work-related.[31]

OSHA uses the following example in the Compliance Directive to illustrate this exception:

> Question 5-8: If an employee's pre-existing medical condition causes an incident which results in a subsequent injury, is the case work-related? For example, if an employee suffers an epileptic seizure, falls, and breaks his arm, is the case covered by the exception in section 1904.5(b)(2)(ii)?
>
> Neither the seizures nor the broken arm are recordable. Injuries and illnesses that result solely from non-work-related events or exposures are not recordable under the exception in section 1904.5(b)(2)(ii). Epileptic seizures are a symptom of a disease of non-occupational origin, and the fact that they occur at work does not make them work-related. Because epileptic seizures are not work-related, injuries resulting solely from the

seizures, such as the broken arm in the case in question, are not recordable.

The other exceptions to the geographic presumption are much more narrow, and are strictly interpreted by OSHA.

3.3.5.1 General Public

An injury or illness that occurs while the employee is at the workplace as a member of the general public are not considered work-related.[32] For example, a worker employed at a restaurant who stays after his shift to eat with friends has not suffered a work-related injury if he slips on the floor.

3.3.5.2 Voluntary Participation in Wellness Program

An employee who is injured during voluntary participation in a "wellness program or in a medical, fitness or recreational activity such as blood donation, physical examination, flu shot, exercise class, racquetball, or baseball" has not suffered a work-related injury or illness.[33] For example, no work-related injury occurs when an employee voluntarily gives blood and faints. The participation must be voluntary. For example, some employers are required to have employees trained to provide first aid in the event of a medical emergency.[34] Employees who provide first aid are required to receive hepatitis vaccinations pursuant to OSHA's Bloodborne Pathogens standard.[35] An employee who suffers a reaction to the vaccination has experienced a work-related injury or illness because his participation on the emergency response team is not voluntary. Similarly, an employee who is required to wear a respirator is required to undergo a medical examination that includes a pulmonary function test, and any injury experienced during the test would also be work-related.[36]

3.3.5.3 Injury from Eating or Drinking

An injury or illness that is "solely the result of employee eating, drinking or preparing food or drink for personal consumption" is not work-related.[37] For example, an employee who is drinking a can of soda has not experienced a work-related illness when a bee comes out of the can and stings him on the lip. If, however, an employee becomes ill from eating food from the company cafeteria, then the illness is work-related. For example, an employee who eats food prepared by the employer that is contaminated with lead has experienced a work-related illness.

3.3.5.4 Injury from Performing Personal Tasks Outside of Working Hours

This exception applies to an injury or illness that occurs when an employee comes to the workplace outside of his or her working hours to perform personal tasks.[38] For example, an injury that occurs when an employee comes to the establishment over the weekend to photocopy personal materials is not work-related. The injury must, however, occur outside of normal working hours. As such, an injury that occurs while an employee is photocopying personal materials during normal working hours is work-related. Similarly, an employee who falls on a sidewalk during a smoking break has experienced a work-related injury.

3.3.5.5 Personal Grooming, Self-Medication for Non-Work-Related Condition, or Self-Inflicted Injury

This exception covers situations such as an injury that occurs while an employee is brushing her hair, putting in contact lenses, or applying make-up.[39] Similarly, an illness that occurs when an employee reacts to two aspirin he took to alleviate a headache is not work-related.

3.3.5.6 Common Cold or Flu

Cold or flu, regardless of severity or where the employee contracted it, are not work-related illnesses.[40] However, an employee who contracts a contagious disease, such as tuberculosis, brucellosis, or hepatitis at work has experienced a work-related illness. For example, an employee who works with animals and contracts brucellosis, a flu-like illness that is transmitted by infected animals, has experienced a work-related illness.

3.3.5.7 Motor Vehicle Accidents

An injury caused by a motor vehicle accident that "occurs on a company parking lot or company access road while the employee is commuting to or from work" is not work related.[41] If, however, an employee on a public road on his way to the bank to make a deposit on behalf of the employer is hit by a car, any injury is work-related.

3.3.5.8 Mental Illness

Mental illnesses are not considered work-related unless the "employee voluntarily provides the employer with an opinion from a physician or other licensed health care professional with appropriate training and experience (psychiatrist, psychologist, psychiatric nurse practitioner, etc.) stating that the employee has a mental illness that is work-related."[42]

3.3.6 Injuries or Illnesses that Occur While Traveling

Injuries or illnesses that occur while an employee is on travel status are generally considered work-related if the employee is acting in the interest of the employer. Section 1904.5(b)(6) states:

Injuries or illnesses that occur while an employee is on travel status are work-related if, at the time of the injury or illness, the employee was engaged in work activities "in the interest of the employer." Examples of such activities include travel to and from customer contacts, conducting job tasks, and entertaining or being entertained to transact, discuss or promote business (work-related entertainment includes only entertainment activities being engaged in at the direction of the employer).

The definition of "work environment" also establishes that injuries or illnesses that occur during business travel are "work-related." The work environment includes any location "where one or more employees are working or are present as a condition of their employment."[43] As such, the work environment of a traveling employee is essentially wherever he or she is traveling.

An injury that occurs while an employee is driving to the airport to return from a business trip is work-related. If the employee takes a side trip to visit relatives and is in a car accident during that trip, any injury is not work-related. Similarly, a bad reaction to food eaten at a sporting event attended by an employee and business associates would be considered work-related, while the same situation at a sporting event the employee attends with personal friends during a business trip would not be work-related.

3.3.7 Injuries and Illnesses Resulting from Work at Home

Section 1904.5(b)(7) states that "injuries and illnesses that occur while an employee is working from home, including work in a home office, will be considered work-related if the injury or illness occurs while the employee is performing work for pay or compensation in the home, and the injury or illness is directly related to the performance of work rather than to the general home environment or setting." Section 1904.5(b)(7) also offers the following examples:

For example, if an employee drops a box of work documents and injures his or her foot, the case is considered work-related. If an employee's fingernail is punctured by a needle from a sewing machine used to perform garment work at home, becomes infected and requires medical treatment, the injury is considered work-related. If the employee is injured because he or she trips on the family dog while rushing to answer a work phone call, the case is not considered work-related. If an employee working at home is electrocuted because of faulty home wiring, the injury is not considered work-related.

3.3.8 New Cases

Employers are only required to record work-related injuries and illnesses that are new cases on the OSHA 300 Log. Section 1904.6 states that an injury or illness is a new case if: "the employee has not previously experienced an injury or illness of the same type that affects the same part of the body," or "the employee previously experienced a recorded injury or illness of the same type that affected the same part of the body but had recovered completely (all signs and symptoms had disappeared) from the previous injury or illness and an event of exposure in the work environment caused the signs or symptoms to reappear." OSHA only requires employers to record new cases to prevent double counting when a previously recorded injury or illness fails to heal.

Section 1904.6(b) distinguishes between symptoms of chronic illnesses that surface at work regardless of workplace exposures or events, and chronic illnesses that are aggravated by workplace exposures. Symptoms of some chronic work-related illnesses like occupational cancer may surface at work, but are not considered new cases because no new work-related exposure has occurred. These types of chronic illnesses are only recorded once.[44] Employees with other types of chronic work-related illnesses, like occupational asthma, may experience symptoms when a new workplace exposure occurs.[45] If these symptoms are the result of a workplace event or exposure, a new injury or illness has occurred. Employers are not required to consult a physician or other licensed healthcare professional in determining whether a new case has occurred. If the employer does rely upon a medical professional, then it must follow the advice given.[46]

During the rulemaking, OSHA observed that employers "may occasionally have difficulty in determining whether new signs or symptoms are due to a new event or exposure in the workplace or whether they are the continuation of an existing workplace injury or illness."[47] In most cases, it will be obvious whether a new case has occurred because a workplace exposure will be the cause of the injury or illness. For example, a "worker may suffer a cut, bruise or rash from a clearly recognized event in the workplace, receive treatment, and recover fully within a few weeks." Then, "[a]t some future time, the worker may suffer another cut, bruise or rash from another workplace event." These two events are unrelated and both are considered "new cases."[48]

The difficulty in determining whether a new case has occurred typically arises with chronic illnesses. According to OSHA, the key distinction is between chronic illnesses that continue regardless of workplace exposures and those that recur because of workplace exposures. Occupational cancer, asbestosis, and similar illnesses are diseases that are never cured or completely resolved. Such cases are never closed under the OSHA recordkeeping system, even though the signs and symptoms of the condition may alternate between remission and active disease."[49]

Other chronic illnesses like reactive airways dysfunction syndrome (RADS) or sensitization to certain workplace chemicals "recur if the ill individual is exposed to the agent . . . that triggers the illness again." Each time symptoms of these types of diseases appear, a separate recordable incident has occurred.[50]

The Compliance Directive provides some additional guidance on how to determine whether an employee has recovered completely from a previous injury such that an injury or illness would be a new case. The directive states:

An employee has "recovered completely" from a previous injury or illness . . . when he or she is fully healed or cured. The employer must use his best judgment based on factors such as the passage of time since the symptoms last occurred and the physical appearance of affected part of the body. If the signs and symptoms of a previous injury disappear for a day only to reappear the following day, that is strong evidence that the injury has not properly healed.

3.3.9 Recording Criteria

Even if an injury or illness is work-related and a new case, the employer is only required to record those meeting one of the recording criteria. Specifically, the injury or illness must result in one of the following to be recordable:

- Death

- Days away from work

- Restricted work or transfer to another job

- Medical treatment beyond first aid

- Loss of consciousness

- A "significant injury or illness diagnosed by a physician or other licensed health care provider"[51]

The OSHA 300 Log contains columns that are marked by the employer to designate which of the recording criteria are met.

3.3.9.1 Death

The requirement to record an injury or illness resulting in death is typically fairly clear. In the Compliance Guide, however, OSHA does discuss one unusual example. If an employee dies during surgery necessitated by the workplace injury or illness, then the injury or illness is recordable. For example, if an employee dies as a result of work-related knee surgery because he contracts a Staphylococcus infection or the physician commits gross medical malpractice, then the employer must record the original injury as resulting in death.

3.3.9.2 Days Away from Work

If an injury or illness requires the employee to miss work, the employer must record it on the OSHA 300 Log. The box on the OSHA 300 log for "days away from work" is marked, and the employer must also record the exact number of days the employee misses. The day that the injury or illness actually occurs is not counted.[52] For example, if an employee hurts his ankle on Monday and misses the remainder of that day, but comes to work on Tuesday, then no recordable injury has occurred.

The employer is required to record days away from work even if the employee ignores the recommendations of a physician or other licensed health care professional and comes to work. If, for example, a physician recommends that an employee take five days off, and the employee comes to work on the fourth day and says he is feeling better and is able to work, the employer must nevertheless record each day the physician found that the employee should not work.[53] Similarly, if the employee does not return to work even though the physician states that he is capable of working, the employer is not required to record the extra days the employee does not work.[54] OSHA stated during the rulemaking that the "employer is the ultimate recordkeeping decision-maker and must resolve the differences in opinion" that may arise between the employer's physician and the employee's physician.[55]

Employers are required to record calendar days away from work. Specifically, each day the employee would not have been able to work but for the injury or illness, including holidays, weekends, and vacation days, must be recorded.[56] The regulation contains two exceptions to this general rule. First, if an employee is injured or becomes ill on a Friday, reports to work on Monday, and was not scheduled to work on Saturday and Sunday, then the employer is not required to record the case on the OSHA 300 Log, unless the employer receives information from the physician or other licensed health care professional that the employee should not have worked on Saturday and Sunday.[57]

Second, the employer is not required to record an injury or illness that occurs on the day before scheduled time off, such as a vacation or temporary plant closing, unless the employer receives information from the physician or other licensed health care professional that the employee should not have worked for some period of time during the scheduled time off.[58]

The maximum number of days away from work that must be recorded is 180.[59] If, for example, an employee is away from work for 200 days because of a work-related injury or illness, the employer is only required to record 180 days away from work. Also, the employer is permitted to stop counting days away from work if the employee retires or leaves the company for a reason unrelated to the injury or illness. If, however, the employee leaves the company because of his injury or illness, then the employer must record the number of days the employee would not have been able to work.[60] For example, if an employee breaks his arm and is projected to be away from work for 30 days, but retires ten days later, the employer would be required to record only ten days away from work. If the employee cannot afford financially to be away from work drawing workers' compensation pay for 30 days and takes a new job ten days after the injury, the employer would have to record 30 days away from work because the employee left the company because of his injury. Similarly, if an employee is terminated because of drug use revealed during a post-accident investigation, the employer must record all of the days the employee would not have been able to work, even after the termination, because the termination is related to the injury.

Finally, Section 1904.7(b)(3)(ix) addresses cases that span more than one year. For example, an employee may be injured in December 2006 and unable to return to work until January 2007. The employer must record all of the days away from work on the 2006 log.[61]

3.3.9.3 Restricted Work or Job Transfer

The employer must record each day the employee is unable to perform the routine functions of his normal job and performs restricted work or transfers to a different job.[62]

A work-related injury or illness results in "restricted work" when the employer keeps "the employee from performing one or more of the routine functions of his or her job, or from working the full workday that he or she would otherwise have been scheduled to work"; or "a physician or other licensed health care professional recommends that the employee not perform one or more of the routine job functions of his or her job, or not work the full workday that he or she would otherwise have been scheduled to work."[63] A "routine function" means "those work activities the employee regularly performs at least once per week."[64]

The regulation addresses vague restrictions from physicians, such as instructions to perform only "light duty" or to "take it easy" for some period of time. Section 1904.7(4)(vii) states that if a physician's recommendation is not clear, the employer should ask "whether the employee can do all of his or her routine job functions and work all of his or her normally assigned work shift." If the physician concludes that the employee can perform routine job functions for a normal work shift, then the employee's work is not restricted, and the injury or illness is not recordable. The regulation also states that if the employer is "unable to obtain this additional information from the physician or other licensed health care professional who recommended the restriction," the injury or illnesses should be recorded "as a case involving restricted work."[65]

A job restriction ordered by a physician does not necessarily result in a recordable injury or illness. For example, a physician may recommend that an employee not lift over 20 pounds for one week. If the employee does not lift 20 pounds or more at least once each week as part of his job, then the employee's routine functions are not restricted, and the employer is not required to record the injury or illness. Similarly, OSHA states in the

Compliance Directive that if a physician instructs the employee not to use his left arm for one week, and "the employee is able to perform all of his or her routine job functions using only the right arm (though at a slower pace)," the injury has not resulted in restricted work because "loss of productivity is not considered restricted work."[66]

The "employer has the ultimate authority to restrict an employee's work, so the definition is clear that, although a health care professional may recommend the restriction, the employer makes the final determination of whether or not the health care professional's recommended restriction involves the employee's routine functions."[67] As such, "[r]estricted work assignments may involve several steps: an HCP's recommendation, or employer's determination to restrict the employee's work, the employer[']s analysis of jobs to determine whether a suitable job is available, and assignment of the employee to that job."[68]

If the employee refuses to follow the work restriction, the employer must nevertheless record the injury or illness. Section 1904.7(b)(4)(viii) states that employers must "ensure that the employee complies with [the] restriction."[69] The employer may also "receive recommendations from two or more physicians or other licensed health care professionals" and decide "which recommendation is more authoritative and record the case based upon that recommendation."[70]

As with days away from work, the employer is not required to record a day of restricted work or job transfer if it only applies to the day the injury or illness occurred.[71]

3.3.9.4 Medical Treatment beyond First Aid

Any injury or illness that requires medical treatment must be recorded. Medical treatment is defined as "the management and care of a patient to combat disease and disorder," and does not include: "visits to a physician or other licensed health care professional solely for observation or counseling," or "the conduct of diagnostic procedures, such as x-rays and blood tests, including the administration of prescription medications used solely for diagnostic purposes (e.g. eye drops to dilate pupils)."[72]

The definition of medical treatment also explicitly excludes any treatment meeting the definition of first aid. Section 1904.7(b)(5)(i) provides the following comprehensive list of treatments qualifying as first aid:

- Using a non-prescription medication at nonprescription strength (for medications available in both prescription and non-prescription form, a recommendation by a physician or other licensed health care professional to use a non-prescription medication at prescription strength is considered medical treatment for recordkeeping purposes);

- Administering tetanus immunizations (other immunizations, such as Hepatitis B vaccine or rabies vaccine, are considered medical treatment);

- Cleaning, flushing or soaking wounds on the surface of the skin;

- Using wound coverings such as bandages, Band-Aids™, gauze pads, etc.; or using butterfly bandages or Steri-Strips™ (other wound closing devices such as sutures, staples, etc., are considered medical treatment);

- Using hot or cold therapy;

- Using any non-rigid means of support, such as elastic bandages, wraps, non-rigid back belts, etc. (devices with rigid stays or other systems designed to immobilize parts of the body are considered medical treatment for recordkeeping purposes);

- Using temporary immobilization devices while transporting an accident victim (e.g., splints, slings, neck collars, back boards, etc.);

- Drilling of a fingernail or toenail to relieve pressure, or draining fluid from a blister;

- Using eye patches;

- Removing foreign bodies from the eye using only irrigation or a cotton swab;

- Removing splinters or foreign material from areas other than the eye by irrigation, tweezers, cotton swabs, or other simple means;

- Using finger guards;

- Using massages (physical therapy or chiropractic treatment are considered medical treatment for recordkeeping purposes); or

- Drinking fluids for relief of heat stress.[73]

Any treatment not listed in the definition of first aid is considered medical treatment that triggers a recordable incident. The listed treatments are also considered first aid regardless of whether a physician, licensed health care professional, or another person, like a member of the employer's first aid team or a "Good Samaritan," provides them.[74]

The Compliance Directive provides some additional information about first aid and medical treatment. The use of surgical glue is considered medical treatment because, "all wound closing devices except for butterfly and steri strip are by definition 'medical treatment' because they are not included on the first aid list."[75] Although "drinking fluids for relief from heat stress" is first aid, "intravenous administration of fluids to treat work-related heat stress is medical treatment."[76] Finally, the administration of oxygen as a "purely precautionary measure" when an employee is exposed to "chlorine or some other substance" is first aid, but the administration of oxygen to an exposed employee who "exhibits symptoms of an injury or illness" is medical treatment.[77]

3.3.9.5 Loss of Consciousness

Every work-related injury or illness that results in a loss of consciousness must be recorded, "regardless of the length of time the employee remains unconscious."[78] The loss of consciousness must result from a workplace exposure rather than a condition that is unrelated to work, such as pregnancy or epilepsy. In addition, the employee must actually be unconscious for this provision to apply. Feelings of wooziness or dizziness are not recordable unless one of the other recording criteria is met.[79]

3.3.9.6 "Significant" Diagnosed Injury or Illness

Section 1904.7(b)(7) states that "[w]ork-related cases involving cancer, chronic irreversible disease, a fractured or cracked bone, or a punctured eardrum must always be recorded under the general criteria at the time of diagnosis by a physician or other licensed health care professional."[80] OSHA considers this a "catch-all" provision to record those rare significant injuries or illnesses that are not required to be recorded under the other criteria. "There are some significant injuries, such as a punctured eardrum or a fractured toe or rib, for which neither medical treatment nor work restrictions may be recommended," and "significant progressive diseases, such as byssinosis, silicosis, and some types of cancers" may also not meet the other recording criteria.[81]

Section 1904.7(b)(7) is designed to capture two categories of injuries and illnesses. First, the provision requires recording of "significant injuries and illnesses" that "are not amenable to medical treatment" at the time of initial diagnosis, such as "a fractured rib, a broken toe, or a punctured eardrum," which are often "left to heal on their own," or "untreatable occupational cancer." Second, "chronic irreversible diseases" are "cases that would clearly become recordable at some point in the future (unless the employee leaves employment before medical treatment is provided), when the employee's condition worsens to a point where medical treatment, time away from work, or restricted work are needed." Injuries and illnesses in the second category are "expected to progressively worsen and become serious over time (chronic irreversible diseases)."[82]

3.4 Special Cases

The recordkeeping regulation has special recording criteria for certain types of injuries or illnesses.

3.4.1 Hearing Losses

Section 1904.10 contains a two-part test for determining whether a hearing loss must be recorded. First, the employer must determine whether an employee has experienced a Standard Threshold Shift (STS) in one or both ears since his last audiogram. Pursuant to OSHA's Occupational Noise Exposure standard, 29 C.F.R. Section 1910.95, an STS is defined as a change in hearing threshold of an average of 10 decibels (dB). The employer determines whether an STS has occurred by comparing the annual audiogram to the employee's baseline audiogram, which is performed before or soon after the employee begins work. If the employee has previously experienced a recordable hearing loss, then the employer must compare the new audiogram with the revised baseline audiogram, which will reflect the employee's previous recordable hearing loss.

Second, the employer must determine whether the STS represents an overall hearing level of 25 dB or more. As explained in Section 1904.10(b)(2)(ii), "[a]udiometric test results reflect the employee's overall hearing ability in comparison to audiometric zero." Using the current audiogram, the employer "must use the average hearing level at 2000, 3000 and 4000 Hz [hertz]to determine whether or not the employee's total hearing level is 25 dB or more."

If both of these conditions are met, then the hearing loss is recorded in the hearing loss column on the OSHA 300 Log.

In determining whether an STS has occurred, Section 1904.10(b)(3) allows the employer to adjust for age using Table F-1 or F-2 in the Occupational Noise Exposure standard. In addition, the employer may retest the employee's hearing within 30 days of the first test, and if the retest does not confirm the recordable hearing loss, it does not have to be recorded.[83]

3.4.2 Needlestick Injuries

Section 1904.8 requires "all work-related needlestick injuries and cuts from sharp objects that are contaminated with another person's blood or other potentially infectious material" to be recorded. The term other potentially infectious material is defined in OSHA's Bloodborne Pathogens standard as including human body fluids, such as semen or spinal fluid, "any body fluid that is visibly contaminated with blood," and "all body fluid in situations where it is difficult or impossible to differentiate between body fluids," and "any unfixed tissue or organ (other than intact skin) from a human (living or dead)."[84] The incident is recorded on the OSHA 300 Log as an injury regardless of whether one of the other recording criteria, such as medical treatment, days away from work, or job transfer, is met.

Employers are only required to record needlestick or sharps injuries that "bring an employee into contact with another person's blood or other potentially infectious materials."[85] If an employee has contact with a "clean object" or "a contaminant other than blood or other potentially infectious material," the injury is recordable only if it meets one of the other recording criteria, such as medical treatment or restricted work.

Employers are also required to update the OSHA 300 Log if the employee is later diagnosed with an infectious disease, such as hepatitis or human immunodeficiency virus (HIV), resulting from the sharps or needlestick injury. In addition, the classification of the case must be updated if the case results in death, days away from work, restricted work, or job transfer.[86]

The Bloodborne Pathogens standard applies to all employers with employees who have "occupational exposure to blood or other potentially infectious material."[87] These employees typically include health care workers and employees who are designated to provide first aid or to clean up blood or other infectious material after an accident. The Bloodborne Pathogens standard requires all employers who are required to keep injury and illness records under the recordkeeping regulations to keep a sharps injury log that includes: "the type and brand of device involved," "the department or work area where the exposure incident occurred," and "an explanation of how the incident occurred."[88] According to the Compliance Directive, an employer may use the OSHA 300 Log to meet the requirements for a sharps injury log as long as "the type and brand of device" is listed, if applicable, and the records are maintained such that sharps injuries are segregated from other types of injuries.[89]

3.4.3 Medical Removal

Many OSHA standards that address exposure to hazardous substances, such as lead, cadmium, benzene, methylene chloride, and formaldehyde, require employers to remove employees from the work area if certain criteria are met.[90] For example, OSHA's Lead standard requires employees to be removed from work resulting in lead exposure if the required periodic blood tests show a certain level of lead.[91] These types of provisions, known as "medical removals", typically require the employer to move an employee to an alternate job without exposure or, if there is no alternate job, provide paid time off.

Section 1904.9 requires employers to record medical removal cases as days away from work, if the employee is not offered a different job without exposure, or days of restricted work, if the employee is moved to a different job.[92] If the medical removal is required by a standard addressing chemical exposure, the case is recorded by checking the "poisoning" column on the OSHA 300 Log.[93]

Employers may voluntarily remove an employee from a particular job before the criteria for medical removal are applicable. For example, an employer may move an employee to another job even though his blood lead levels are below the threshold. The recordkeeping regulation states that if "the case involves voluntary medical removal before the medical removal levels required by an OSHA standard," the employer is not required to record the case unless or until one of the other recording criteria is met.[94]

Medical removal may be applicable even if no specific standard applies. For example, an employer may temporarily transfer an employee from a particular job for preventive purposes to prevent a work-related injury from occurring. The rulemaking record states:

Transfers or restrictions taken *before* the employee has experienced an injury or illness do not meet the first recording requirement of the recordkeeping rule, i.e. that a work-related injury or illness must have occurred for recording to be considered at all. . . . However, transfers or restrictions whose purpose is to allow an employee to recover from an injury or illness as well as to keep the injury or illness from becoming worse are recordable because they involve restriction or work transfer caused by the injury or illness. . . . A work restriction that is made for another reason, such as to meet reduced production demands, is not a recordable restricted work case. For example, an employer might "restrict" employees from entering the area in which a toxic chemical spill has occurred or make an accommodation for an employee who is disabled as a result of a non-work-related injury or illness. These cases would not be recordable as restricted work cases because they are not associated with a work-related injury or illness.[95]

3.4.4 Tuberculosis

If an employee is occupationally exposed to a "known case of active tuberculosis," and the employee "subsequently develops a tuberculosis infection, as evidenced by a positive skin test or diagnosis by a physician or other licensed health care professional," then a recordable "respiratory condition" has occurred.[96]

Tuberculosis (TB) may be work-related if it is contracted by a health care worker or if an employee infects other employees. OSHA specifically addressed employee-to-employee transmission of tuberculosis during the rulemaking. First, OSHA pointed out that tuberculosis is "clearly a non-minor" illness that should be recorded if it is work-related even though it is likely impossible for an employer to prevent an employee from contracting TB from a coworker. Under the geographic presumption OSHA set out for work-relatedness, an employee-to-employee transmission is work-related because the employee would not have contracted TB but for his presence in the workplace. Second, for the case to be recordable, the employee must have been exposed to someone at work that has a known case of active TB. As such, the fact that an employee contracts TB and the source cannot be identified is not enough to result in a recordable illness.[97]

The regulation states that employers may "line out or erase a recorded TB case from the OSHA 300 Log under the following circumstances: (i) The worker is living in a household with a person who has been diagnosed with active TB; (ii) The Public Health Department has identified the worker as a contact of an individual with a case of active TB unrelated to the workplace; or (iii) A medical investigation shows that the employee's infection was caused by exposure to TB away from work, or proves that the case was not related to the workplace TB exposure."[98]

3.5 Recordkeeping Forms and Retention Periods

The regulation requires employers to use three forms: the OSHA 300 Log, the OSHA 301 report, and the OSHA 300-A form.[99] The OSHA 300 Log is the document on which each work-related injury or illness that meets one of the general recording criteria is recorded. An OSHA 301 report is executed for each recordable injury and illness, and describes the injury or illness in greater detail than the OSHA 300 Log. Each recordable injury or illness must be recorded on the OSHA 300 Log and an OSHA 301 form within seven calendar days of the employer "receiving information that a recordable injury or illness has occurred."[100]

The OSHA 300-A form is an annual summary of recordable injuries and illnesses. On February 1 of each year, employer must post the 300-A form and keep it posted until April 30.[101] To prepare the OSHA 300-A, the employer must "review the OSHA 300 Log to verify that the entries are complete and accurate, and correct any deficiencies identified," and then record the following information on the OSHA 300-A: (1) the total for each column on the OSHA 300 Log; (2) the year covered by the annual summary; (3) the name and address of the employer's establishment; (4) the average number of employees at the establishment during the covered year; and (5) the total hours worked by the employees during that year.[102] A "company executive must certify that he or she has examined the OSHA 300 Log and that he or she reasonably believes, based on his or her knowledge of the process by which the information was recorded, that the annual summary is correct and complete."[103] The company executive who certifies the OSHA 300-A must be an owner, an officer of the corporation, the "highest ranking company official working at the establishment," or the "immediate supervisor of the highest ranking company official working at the establishment."[104]

The employer must retain the OSHA 300 Log, OSHA 300-A, and OSHA 301 forms for five years.[105] During the five-year period, the employer must update the OSHA 300 Log to include "newly discovered recordable injuries and illnesses" and must update the "classification of previously recorded injuries and illnesses."[106] If, for example, an employee experiences a recordable injury in 2006 that steadily worsens until surgery is required in 2007, and the employee is completely unable to work after the surgery, then the employer must update the 2006 OSHA 300 Log to reflect this. Employers are not required to update the OSHA 301 forms or OSHA 300-As.[107]

3.6 Employee Involvement and Access to Records

Employers have several recordkeeping obligations to employees. First, employers must "inform each employee of how he or she is to report an injury or illness."[108] To meet this requirement, employers typically establish a work rule requiring employees to report all work-related injuries to supervisors immediately.

Second, employers must provide access to the OSHA 300 Log to employees, former employees, and certain employee representatives. The regulation defines two types of representatives. A "personal representative" is any person designated in writing by an employee or former employee as a representative, or the legal representative of a deceased or legally incapacitated employee or former employee.[109] An "authorized employee representative" is the "authorized collective bargaining agent of employees."[110] The employer must provide access to the current or retained OSHA 300 Log to employees, former employees, personal representatives, and authorized employee representatives the next business day after a request is made.[111]

Third, employers must provide access to certain OSHA 301 forms. An employee, former employee, or personal representative is entitled to the OSHA 301 form describing the injury or illness experienced by the employee or former employee. The form must be provided the next business day after the request.[112] An authorized employee representative is entitled to certain portions of OSHA 301 forms for employees at the establishment. The employer must only provide the information under the section on the OSHA 301 form entitled "Tell us about the case," which describes how the injury or illness occurred. All other information must be redacted.[113]

Employers are also required to provide records kept pursuant to a recordkeeping requirement to OSHA compliance officers conducting an inspection, representatives of the National Institute of Occupational Safety and Health who are conducting a health hazard evaluation, and state agencies administering an OSHA state plan.[114] The records must be provided within four hours of the request.[115]

3.7 Privacy Cases

Employers are not permitted to include the name of the employee on the OSHA 300 Log if the injury or illness is a "privacy concern case."[116] A privacy concern case is defined as an "injury or illness to an intimate body part or reproductive system," an "injury or illness resulting from a sexual assault," a "mental illness," "HIV infection, hepatitis, or tuberculosis," "[n]eedlestick injuries and cuts from sharp objects that are contaminated with another person's blood or other potentially infectious materials," and "other illnesses, if the employee voluntarily requests that his or her name not be entered on the log."[117] No other situations are considered "privacy concern cases," and the employer must therefore include the names of employees on the OSHA 300 Log.[118] The employer may also redact other information on the OSHA 300 Log if it would identify an employee who has experienced a privacy case injury or illness but must "enter enough information to identify the cause of the incident and the general severity of the injury or illness."[119]

The name of the employee and other identifying information in a privacy case must not be disclosed on the OSHA 300 Log that may be provided to employees, former employees, or representatives. If the employer decides to share the recordkeeping forms with third parties voluntarily, identifying information for privacy cases must be redacted unless the forms are disclosed to "an auditor or consultant hired by the employer to evaluate the safety and health program," a "public health authority or law enforcement agency," or "to the extent necessary for processing a claim for workers' compensation or other insurance benefits."[120]

The employer must keep a "separate, confidential list of the case numbers and employee names" for "privacy concern cases," and must update the list when changes occur. Government representatives are entitled to this information.[121]

3.8 Reporting Injuries and Fatalities

Section 1904.39 requires employers to report fatalities within eight hours and in-patient hospitalizations of one or more employees;[122] an amputation;[123] or an loss of an eye within twenty-four hours to the OSHA.[124] Employers may report the fatality, in-patient hospitalization, amputation, or loss of an eye by telephone or in person to the nearest local OSHA Area.[125] Employers may also call OSHA's central telephone number at 1-800-321-OSHA (1-800-321-6742) or make an electronic submission using the application on OSHA's website to report the fatality, in-patient hospitalization, amputation, or loss of an eye.[126] The information provided to OSHA must include: (1) the establishment name; (2) the location of the incident; (3) the time of the incident; (4) the number of employees who suffered a fatality, in-patient hospitalization, amputation, or loss of an eye;

(5) the names of any injured employees; (5) the contact person at the facility and his or her telephone number; and (6) a brief description of the incident.[127] If the local OSHA area office is closed when an injury must be reported, the employer must call the toll-free number or use the reporting application located on OSHA's website. Leaving a message or sending a facsimile or e-mail to the local OSHA office is insufficient.[128]

A fatality must be reported within 8 hours of its occurrence.[129] An in-patient hospitalization, amputation, or loss of an eye must be reported within 24 hours of occurrence. If the employer does not learn of these type of injuries right away, then the employer must report the fatality within 8 hours and an in-patient hospitalization, amputation, or loss of an eye within 24 hours of learning of the incident.[130]

Two types of fatalities, in-patient hospitalizations, amputations, and losses of an eye are exempt from the reporting requirements. First, employers are not required to report fatalities, in-patient hospitalizations, amputations, and losses of an eye that result from motor vehicle accidents on public streets or highways or on a public transportation system, such as a subway or commercial airplane.[131] However, fatalities, in-patient hospitalizations, amputations, or a loss of an eye resulting from motor vehicle accidents that occur in road construction work zones must be reported to OSHA.[132] Second, employers are not required to report fatalities that occur more than 30 days after the work-related incident.[133] For example, if an employee is in an accident, but dies 60 days after the accident, the employer is not required to report the fatality. Also, employers are not required to report in-patient hospitalizations, amputations, or a loss of an eye that occur more than 24 hours after the work-related incident.[134]

The recordkeeping regulation requires employers to report all heart attacks that occur at work regardless of whether it is work-related.[135] OSHA will investigate and determine whether the heart attack is work-related.

Some states with state OSHA plans have different reporting requirements. For example, employers in California are required to report any "serious injury or illness" or death.[136] A "serious injury or illness" is defined as requiring inpatient hospitalization for over 24 hours for treatment other than medical observation, loss of a "member of the body," or a "serious degree of permanent disfigurement."[137] Injuries or illnesses meeting the criteria or deaths must be reported "immediately," which is defined as "as soon as practically possible, but not longer than 8 hours after the employer knows or with diligent inquiry would have known" of the incident.[138]

3.9 Exemptions from Recordkeeping Requirements

The recordkeeping regulation contains two partial exemptions. First, Section 1904.1 exempts employers with ten or fewer employees during the last calendar year from the requirement to keep OSHA 300 logs and prepare OSHA 301 forms and the OSHA 300-A summary. These employers are required to respond to BLS surveys and must report workplace fatalities, in-patient hospitalizations, amputations, and a loss of an eye.[139]

Second, Non-Mandatory Appendix A to the recordkeeping regulation provides a partial recordkeeping exemption for certain industries, including retail stores (e.g., hardware stores, retail bakeries, car dealerships, service stations, apparel stores, drug stores), certain health care offices (e.g., offices of doctors and dentists, medical and dental laboratories, offices of other health care practitioners), recreational facilities (e.g., dance studies, orchestras or entertainers, museums and art galleries, bowing centers), and traditional offices (e.g., law, engineering, accounting, and research offices, computer and data processing services, mailing, production and stenographic services).[140] Facilities that are partially exempt pursuant to Appendix A must also respond to a BLS survey, and must report workplace fatalities, in-patient hospitalizations, amputations, and a loss of an eye.[141]

4.0 OSHA Standards Requiring Written Documents

OSHA issues two types of standards—health standards and safety standards—and the types of records required generally depend upon the type of standard at issue. Safety standards generally address physical conditions. For example, the standards in Subpart J of Part 1910 (Sections 1910.141 through 1910.147) contain requirements for addressing general work hazards such as hazardous conditions in confined spaces like tanks, sewers, storage bins, and silos (Section 1910.146) and hazards that may result from servicing and maintenance work on equipment (Section 1910.147). Similarly, the Subpart I of Part 1910 (Section 1910.132 through 1910.138) standards address personal protective equipment, the Subpart D of Part 1910 (Section 1910.21 through 1910.130) standards are intended to protect employees from hazards such as falls, and the Subpart E (Section 1910.33 through 1910.39) standards apply to emergency situations such as fires. The standards applicable to construction that are in Part 1926 also include many safety standards, such as requirements for materials handling, storage, use and disposal (Subpart H), fall protection systems (Subpart M), and crane and similar equipment (Subpart N).

Health standards typically address exposure to a substance or condition in the workplace. The standards in Subpart Z of Part 1910 (Section 1910.1000 through 1910.1450) contain requirements for controlling exposure to toxic and hazardous substances. Some of the Subpart Z standards have permissible exposure limits ("PEL") for exposure to various hazardous substances. Other Subpart Z standards address chemical exposure in broader terms. For example, the Hazard Communication standard (Section 1910.1200) requires employers to provide employees with information about all hazardous substances in the workplace.

4.1 Safety Standard Recordkeeping Requirements

OSHA's safety standards contain many recordkeeping requirements. The types of records that must be kept fall into three general categories. First, some standards require the employer to develop a written plan describing how a particular hazard will be addressed. For example, the Respiratory Protection standard (Section 1910.134) requires employers with employees that use respirators to develop a written program. Similarly, the Control of Hazardous Energy (Lockout/Tagout) (Section 1910.147) standard requires employers with employees who perform servicing or maintenance on equipment to develop a written lockout/tagout program.

Second, some standards require the employer to maintain written training records. For example, the Process Safety Management standard (Section 1910.119), which applies to facilities that manufacture or handle certain quantities of hazardous chemicals, requires employers to "prepare a record" that includes "the identity of the employee, the date of the training, and the means used to verify that the employee understood the training."[142]

Third, some standards require the employer to maintain written inspection or preventative maintenance programs. For example, the Powered Platforms for Building Maintenance standard (Section 1910.66) requires employers to perform periodic inspections and keep a certification record of each inspection or test.[143] The Portable Fire Extinguisher standard (Section 1910.157) requires an annual inspection that must be documented.[144]

The following table provides a list of safety standards that have recordkeeping requirements. The list is not exhaustive, and is intended to provide a sample of safety standard recordkeeping requirements.

Topic	Standard	Recordkeeping Required
General Industry		

Fire Prevention/Emergency Action	1910.38	Written Compliance Plans
Powered Platforms	1910.66	Log of Maintenance Inspections and
Manlifts	1910.68	Inspection Records
Noise	1910.95	Hearing Conservation Program
Respirators	1910.134	Respiratory Protection Program
Permit-Required Confined Spaces	1910.146	Permit System Training Records
Lockout/Tagout	1910.147	Lockout/Tagout Program
Fire Protection	1910.156	Fire Brigade Program
Fixed Fire Extinguishing Systems	1910.160	Records of Inspection and Maintenance
Portable Fire Extinguishers	1910.157	Records of Annual Inspections
Automatic Fire Sprinklers	1910.159	Records of Tests
Overhead and Gantry Cranes	1910.179	Certification and Inspection Records
Crawler, Locomotive and Truck Cranes	1910.180	Certification and Inspection Records
Derricks	1910.181	Preventive Maintenance Program
Industrial Slings	1910.184	Records of Inspections and Repairs
Mechanical Power Presses	1910.217	Certification, Maintenance, Injury Reports
Forging Machines	1910.218	Safety Checks and Certification
Welding, Cutting, and Brazing	1910.255	Certification
Telecommunications	1910.268	Training Certification Inspection Records
Grain-handling Facilities	1910.272	Emergency Action Plan; Permit for Hot Work; Confined Space Permit
Electrical Wiring Design and Protection	1910.304	Assured Equipment Grounding Conductor; Program Tests
Electrical Work	1910.333	Lockout/Tagout Procedure
Commercial Diving, Post-Dive Procedures	1910.423	Various Records of Dives
Maritime		
Shipyards	1915.7(a)(2)	Competent Person Designation
	1915.7(c)(1)-(2)	Record of Tests and Inspections
	1917.7(c)(3)	Hot Work and Fumigation Certificates
Explosive and Dangerous Atmospheres	1915.12(c)(4)	Tests, Inspections, Instructions
Shackles and Hooks	1915.113(b)(1)	Certificates and Tests
Portable Air Receivers and Other Unfired Pressure Vessels	1915.172(d)	Certification of Examinations and Tests
Hazardous Atmospheres and Substances	1917.23(b)	Tests

Vessel Cargo Gear Certification	1919.11(b)	Certification of Vessels
Operators or Offices of Vessels Construction	1919.12	Register and Certificate of Inspection, Testing, Heat Treatment of Derricks, Cranes, Etc.
Medical and First Aid	1926.50(c)	Certification and First Aid Testing
	1926.50(f)	Contact Information for Medical Response
Rigging Equipment and Material Handling	1926.251	Tests on Hooks
Ground Fault Protection	1926.404	Grounding Conductor Program, Tests
Cranes and Derricks	1926.550(a)	Inspections
Crawler, Locomotive and Truck Cranes	1926.550(b)	Certification
Personnel Hoists	1926.552	Certification
Excavations	1926.652(b)	Written Design
	1926.652(c)	Supporting Data for Deviations from Manufacturer Specifications for Support System
Lift-Slab Construction Operations	1926.705	Instructions

4.2 The Health Standards

As discussed, OSHA has promulgated standards designed to limit exposure to a variety of hazardous substances. These standards include requirements for performing monitoring to determine exposure levels as well as mandatory medical examinations. Many of the retention and other requirements for maintaining records required by health standards are set out in the Access to Employee Exposure and Medical Records standard.[145]

4.2.1 The Typical Health Standard

Section 6(b)(5) of the OSH Act contains the requirements OSHA must follow in "promulgating standards dealing with toxic materials or harmful physical agents," including setting "the standard which most adequately assures, to the extent feasible, on the basis of the best available evidence, that no employee will suffer material impairment of health or functional capacity even if such employee has regular exposure to the hazard dealt with by such standard for the period of his working life."[146]

As stated, OSHA has promulgated health standards addressing a variety of toxic substances.[147] While each standard is necessarily different because unique substances are addressed, each health standard generally follows a common pattern in that three distinct types of records are required. First, health standards generally require the employer to perform initial monitoring to determine whether the "action level"—the level at which the employer is obligated to take action—is met. Assuming the regulated substance is present at levels at or above the action level, the employer must then perform periodic monitoring to determine whether the control measures, such as work and engineering controls, are effectively reducing exposure to levels at or below the permissible exposure limit (PEL).

Second, health standards typically require the employer to provide a medical surveillance program. Under the program, employees undergo periodic medical examinations performed by a physician or other licensed health care professional. The PLHCP prepares a written opinion, and the employer must provide a copy to the employee and retain a copy. In addition, some health standards contain medical removal provisions that require an employer to remove an employee from a particular job when his or her medical examination or other biological monitoring, such as blood tests, show a certain level of the substance in the employee's body. If medical removal is required, then additional medical examinations are required, and the employer must maintain these records.

Third, some health standards require the employer to maintain "objective data" supporting any exemption from complying with the standard. For example, the Asbestos standard allows employers to rely upon "objective data that demonstrates that asbestos is not capable of being released in airborne concentrations at or above" the permissible exposure limit rather than performing initial monitoring.[148] Similarly, the Chromium (VI) standard allows the employer to rely upon "historical monitoring data" obtained before the effective date of the standard rather than performing initial monitoring.[149] These standards as well as other health standards require the employer to maintain records regarding this type of historical or objective data.

4.2.2 Health Standards Applicable to General Industry

The recordkeeping requirements for some typical health standards applicable to general industry are listed below:[150]

Topic	Standard	Recordkeeping Required
Asbestos	1910.1001	Exposure monitoring, medical surveillance, objective data
13 Carcinogens	1910.1003	Medical surveillance
Vinyl chloride	1910.1017	Exposure monitoring, medical surveillance

Inorganic arsenic	1910.1018	Exposure monitoring, medical surveillance
Lead	1910.1025	Exposure monitoring, medical surveillance, medical removal
Cadmium	1910.1027	Exposure monitoring, medical surveillance, medical removal, objective data
Benzene	1910.1028	Exposure monitoring, medical surveillance, medical removal
Coke oven	1910.1029	Exposure monitoring, medical surveillance, emissions
Cotton dust	1910.1043	Exposure monitoring, medical surveillance
1, 2-dibromo-	1910.1044	Exposure monitoring, medical surveillance
Acrylonitrile	1910.1045	Exposure monitoring, medical surveillance
Ethylene oxide	1910.1047	Exposure monitoring, medical surveillance
Formaldehyde	1910.1048	Exposure monitoring, medical surveillance, objective data
Methylenedianiline	1910.1050	Exposure monitoring, medical surveillance, medical removal, objective data
1,3 Butadiene	1910.1051	Exposure monitoring, medical surveillance, objective data
Methylene chloride	1910.1052	Exposure monitoring, medical surveillance, medical removal, objective data
Ionizing radiation	1910.1096	Exposure monitoring

4.3 Hazard Communication and Bloodborne Pathogens

Two health standards—Hazard Communication and Bloodborne Pathogens—merit special mention as they apply to many workplaces.

4.3.1 Hazard Communication

The Hazard Communication standard (HCS), Section 1910.1200, applies to general industry as well as construction, marine terminal, shipyard and longshoring industries.[151] Pursuant to the HCS, employers are required to maintain Safety Data Sheets (SDS) (formerly Material Safety Data Sheets (MSDS)) for each hazardous chemical at the worksite.[152] The SDS, which contains information regarding the physical and health hazards of substances, is generally prepared by chemical manufacturers or distributors and is sent to purchasers with the shipments.[153] Employers are required to maintain SDSs for hazardous substances that are present in the workplace and ensure that SDSs are "readily accessible" to employees.[154]

Employers are also required to ensure that labels on incoming shipments are not defaced and must also label each container of hazardous chemicals in the workplace with information about the identity of the chemicals and hazard warnings.[155] In addition, the standard requires employers to develop a written program describing how the employer will comply with the standard.[156]

Pursuant to OSHA's Medical Records Access standard (Section 1910.1020), employers are required to keep MSDSs for hazardous substances that are no longer used, or to keep records of when and where hazardous substances were used. The requirements of Section 1910.1020 are discussed later in this chapter.[157]

4.3.2 Bloodborne Pathogens

The Bloodborne Pathogens standard (Section 1910.1030) applies to general industry and shipyard employment, but does not apply to construction, agriculture, marine terminal, or longshoring operations.[158] Within general industry and shipyard employment, the standard applies to any employee with "occupational exposure" to blood or other potentially infectious material such as human body fluids, organs, or tissue cultures.[159] "Occupational exposure" is defined as "reasonably anticipated" contact with blood or other potentially infectious materials that result from the employee's job duties.[160] Based on these definitions, employees in many health care industries, laboratories that handle blood or other specimens, and similar facilities are covered by the standard. In addition, employees at non-health care facilities are covered to the extent that they provide first aid or other medical services to other employees.[161]

Employers with covered employees must develop and maintain several types of documents. First, the employer must offer a Hepatitis B vaccination to any employee with occupational exposure to blood or other potentially infectious material. If the employee declines the vaccination, he or she must sign a declination form that must be maintained by the employer.[162]

Second, in the event an employee is exposed to blood or other potentially infectious material, the employer must provide a "confidential medical evaluation and follow-up," and must obtain and maintain the resulting report from the physician or other health care professional.[163]

Third, the employer must provide training to employees with occupational exposure. The training records must be kept for three years from the date of the training.[164]

Fourth, the employer must maintain a "sharps injury log" with information about any "percutaneous injuries from contaminated sharps," such as needlesticks.[165]

Fifth, the employer must develop a written "exposure control plan" that describes how the employer determined which employees have occupational exposure as well as methods the employer will use to comply with the standard.[166]

Finally, required employee medical records, such as post-exposure medical examinations, must be kept in accordance with the Medical Records Access standard (Section 1910.1020), which is discussed in the next section.

4.4 The Access Standard

The Access to Employee Exposure and Medical Records standard (the "Access standard") contains requirements for providing access to specific types of records to employees and their representatives, and also governs record retention. The standard applies to employers in the general, construction and maritime industries.[167]

The Access standard requires employers to retain the types of records required by the health standards discussed in the previous section. Specifically, employers are required to retain "employee medical records" and "employee exposure records." An employee medical record concerns the "health status of an employee" and is "made or maintained by a physician, nurse, or other health care personnel, or technician." Medical records include: "medical and employment questionnaires or histories"; "results of medical examinations," such as pre-employment or periodic physical examinations, work-related laboratory tests or X-rays; "medical opinions, diagnoses, progress notes and recommendation"; "first aid records," "descriptions of treatments and prescriptions," and; "employee medical complaints."[168] Medical records do not include actual physical specimens, health insurance records that are maintained by the employer's health insurance company that are "not accessible to the employer by employee name or other direct personal identifier," records that are privileged under the attorney-client privilege or attorney work product doctrine, or records resulting from employee assistance programs, such as drug or alcohol counseling services.[169]

"Employee exposure records" are defined as having the following types of information: results from "environmental (workplace) monitoring or measuring of a toxic substance or harmful physical agent," such as personal or area sampling results; "biological monitoring results," such as blood or urine tests; material safety data sheets; and records like chemical inventories that show where and when chemicals were used.[170]

Employers are required to maintain employee medical records for the duration of employment plus thirty years.[171] Employers are not required to meet this retention period for "health insurance claims records maintained separately from the employer's medical program and its records" or "first aid records" for minor scratches, cuts or similar injuries that do not result in an entry on the employer's OSHA 300 Log.[172] Employers are not required to maintain medical records for employees who work at the facility for less than one year as long as the records are provided to the employee upon termination of employment.[173]

Employers are generally required to maintain employee exposure records for a total of thirty years regardless of how long the employee is employed at the facility. "Background data" like laboratory reports and worksheets from workplace monitoring must only be kept for one year as long as a report with the results is maintained. MSDSs for substances that are no longer at the facility do not have to be maintained for any specified period as long as the employer maintains for 30 years a record of when and where particular substances were used.[174]

The purpose of maintaining employee medical and exposure records is to provide access to employees, their designated representatives, and OSHA. Employees and their designated representatives must generally be given access to all employee exposure records upon request.[175] Employees must be given access to their own medical records, and designated employee representatives and OSHA are permitted to review personally identifiable medical records only if they have written consent from the employee.[176]

If an employer ceases to do business, all records covered by the standard must be transferred to the successor employer. If there is no successor employer, then current employees must be notified of their rights to access records at least three months before the business closes.

5.0 Using Records to Prove Compliance

Even if a particular OSHA standard does not require a written record, employers may, as a practical matter, need to prepare written records to prove compliance in the event of an enforcement action. For example, the Respiratory Protection standard requires employers to provide training to employees who wear respirators, but does not have any explicit requirements to document the training.[177] Similarly, the Occupational Noise Exposure standard requires employers to provide annual training on the "effects of noise on hearing," "the purpose of hearing protectors," and "the purpose of audiometric testing."[178] The standard requires employers to provide OSHA with the "materials related to the employer's training and education program" but contains no explicit recordkeeping provisions.[179] Even though no records are required by these and other standards, written records will be the most effective way for an employer to prove compliance with the standards in the event of an OSHA inspection.

In addition to proving compliance with certain training requirements, employers often rely on written records to prove that its employees have been informed of safety requirements. For example, the employer may have specific safety rules that prohibit employees from operating equipment when machine guarding has been removed. No OSHA standard requires this type of written rule. If, however, OSHA inspects the facility and observes employees working on unguarded equipment, the employer can show that it had written rules prohibiting this conduct and trained employees on the requirements of the rule. By presenting these documents to OSHA, the employer may be able to show that the OSHA violation resulted from unpreventable employee misconduct and the employer is therefore not liable.

Finally, employers often develop documents such as audits or inspection checklists that may be used to prove that a particular hazard was addressed. For example, OSHA's general housekeeping standard requires floors to be clean and dry and requires aisles to be clear.[180] Written safety rules and checklists for periodic inspections of the worksite may be used to show OSHA that the employer made reasonable efforts to ensure that the housekeeping requirements were met. Through these types of documents, the employer may be able to show that it could not have had knowledge of housekeeping violations that did not exist for extended periods of time.

Notes

1 .   9 C.F.R. Part 1904.

2.   29 C.F.R. § 1910.1025.

3.   29 C.F.R. § 1910.1027.

4.   29 C.F.R. § 1910.1028.

5.   29 C.F.R. § 1910.1020.

6. 29 C.F.R. Subpart L, 1910.155 et seq.

7. 29 U.S.C. § 651(b).

8. Id.

9. 29 U.S.C. § 657(c)(1).

10. 29 U.S.C. § 657(c)(2).

11. 29 U.S.C. § 657(c)(3).

12. Occupational Injury and Illness Recording and Reporting Requirements: Final Rule, 66 Fed. Reg. 5916 (January 19, 2001). Most parts of the revised regulation became effective on January 19, 2002.

13. Id.

14. The recordkeeping regulation contains a flowchart to assist employers in determining whether an injury or illness is recordable. See 29 C.F.R. Section 1904.4(b)(2).

15. 29 C.F.R. § 1904.29(b)(1).

16. 29 C.F.R. § 1904.32.

17. Prior to the revision of the recordkeeping regulation in 2001, the recording criteria for injuries and illnesses were different. In the revision, OSHA simplified the regulation by treating the recording of injuries and illnesses the same.

18. 29 C.F.R. § 1904.46.

19. 66 Fed. Reg. 5916, 6080 (January 19, 2001).

20. See also Note to 29 C.F.R. 1904.0.

21. See generally 66 Fed. Reg. 5916, 5928-29 (January 19, 2001).

22. 29 C.F.R. § 1904.5(a). The exceptions in Section 1904.5(b)(2) are discussed later in this section.

23. 66 Fed. Reg. 5916, 5946 (January 19, 2001) (emphasis added).

24. 66 Fed. Reg. 5916, 5959 (January 19, 2001).

25. Compliance Directive 02-00-135, December 30, 2004, available at http://www.osha.gov/pls/oshaweb/owadisp.show_document?p_table=DIRECTIVES &p_id=3205. OSHA also published a Recordkeeping Handbook in 2005, available at http://www.osha.gov/recordkeeping/handbook/index.html.

26. 29 C.F.R. § 1904.5(a).

27. 29 C.F.R. §1904.5(b)(5).

28. 29 C.F.R. § 1904.5(b)(4).

29. 66 Fed. Reg. 5916, 5965 (January 19, 2001).

30. 66 Fed. Reg. 5916, 5950 (January 19, 2001).

31. Id.

32. 29 C.F.R. § 1904.5(b)(2)(i).

33. 29 C.F.R. § 1904.5(b)(2)(iii).

34. OSHA's Medical Service and First Aid standard requires the employer to ensure the "ready availability" of medical services and states that if there is not an "infirmary, clinic or hospital in near proximity," the employer must train employees to provide first aid. 29 C.F.R. § 1910.151(b).

35. 29 C.F.R. § 1910.1030.

36. 29 C.F.R. § 1910.134.

37. 29 C.F.R. § 1904.5(b)(2)(iv).

38. 29 C.F.R. § 1904.5(b)(2)(v).

39. 29 C.F.R. § 1904.5(b)(2)(vi).

40. 29 C.F.R. § 1904.5(b)(2)(viii).

41. 29 C.F.R. § 1904.5(b)(2)(vii).

42. 29 C.F.R. § 1904.(b)(2)(ix).

43. 29 C.F.R. § 1904.5(b)(1).

44. 29 C.F.R. § 1904.6(b)(1).

45. 29 C.F.R. § 1904.6(b)(2).

46. 29 C.F.R. § 1904.6(b)(3).

47. 66 Fed. Reg. 5916, 5962 (January 19, 2001).

48. Id. at 5962-63.

49. Id. at 5962.

50. Id.

51. 29 C.F.R. § 1904.7(b).

52. 29 C.F.R. § 1904.7(b)(3)(i).

53. 29 C.F.R. § 1904.7(b)(3)(ii).

54. 29 C.F.R. § 1904.7(b)(3)(iii).

55. 66 Fed. Reg. 5916, 5969 (January 19, 2001).

56. 29 C.F.R. § 1904.7(b)(3)(iv).

57. 29 C.F.R. § 1904.7(b)(3)(v).

58. 29 C.F.R. § 1904.7(b)(3)(vi).

59. 29 C.F.R. § 1904.7(b)(3)(vii).

60. 29 C.F.R. § 1904.7(b)(3)(viii).

61. 29 C.F.R. § 1904.7(b)(3)(ix).

62. 29 C.F.R. § 19904.7(b)(4).

63. 29 C.F.R. § 19904.7(b)(4)(i).

64. 29 C.F.R. § 19904.7(b)(4)(ii).

65. 29 C.F.R. § 1904.7(b)(4)(vii).

66. Compliance Directive at Question 7-4.

67. 66 Fed. Reg. 5916, 5980 (January 19, 2001).

68. Id.

69. 29 C.F.R. § 1904.7(b)(4)(viii).

70. Id.

71. 29 C.F.R. § 1904.7(b)(4)(iii).

72. 29 C.F.R. § 19904.7(b)(5)(i).

73. 29 C.F.R. § 1904.7(b)(5)(ii).

74. 29 C.F.R. § 1904.7(b)(5)(iv).

75. Compliance Directive at Question 7-5.

76. Id. at Question 7-6.

77. Id. at Question 7-15.

78. 29 C.F.R. § 1904.7(b)(6).

79. 66 Fed. Reg. 5916, 5994 (January 19, 2001).

80. 29 C.F.R. § 1904.7(b)(7).

81. Note to 29 C.F.R. § 1904.7.

82. 66 Fed. Reg. 5916, 5998 (January 19, 2001).

83. 29 C.F.R. 1904.10(b)(4). The Occupational Noise Exposure standard allows the employer to substitute an annual audiogram for the baseline audiogram if a physician, otolaryngologist, or audiologist determines that the STS is persistent or that the "hearing threshold shown in the annual audiogram indicates significant improvement over the baseline audiogram." 29 C.F.R. § 1910.95(g)(9).

84. 29 C.F.R. § 1910.1030(b).

85. 29 C.F.R. § 1904.8(b)(2).

86. 29 C.F.R. § 1904.8(b)(3) and (b)(4).

87. 29 C.F.R. § 1910.1030.

88. 29 C.F.R. § 1910.1030(h)(5).

89. Compliance Directive at Question 8-2.

90. 29 C.F.R. § 1910.1025(k) (lead medical removal requirements); 29 C.F.R. § 1910.1027(i) (cadmium medical removal requirements); 29 C.F.R. § 1910.1028(i) (benzene medical removal requirements); 29 C.F.R. § 1052(j) (methylene chloride medical removal requirements); 29 C.F.R. 1910.1048(l) (formaldehyde medical removal requirements).

91. 29 C.F.R. § 1910.1025(k).

92. 29 C.F.R. § 1904.9.

93. 29 C.F.R. § 1904.9(b)(1).

94. 29 C.F.R. § 1904.9(b)(3).

95. 66 Fed. Reg. 5916, 5981 (January 19, 2001) (emphasis in original).

96. 29 C.F.R. § 1904.11(a).

97. 66 Fed. Reg. 5916, 6013 (January 19, 2001).

98. 29 C.F.R. § 1904.11(b)(2).

99. 29 C.F.R. 1904.29. The regulation allows employers to use "equivalent forms" that have the "same information" as the OSHA forms, are "readable and understandable," and are "completed using the same instructions as the OSHA form [they] replace." 29 C.F.R. 1904.29(b)(4).

100. 29 C.F.R. § 1904.29(b)(3). The D.C. Circuit has held that pursuant to the six month statute of limitation under the OSH Act, OSHA may not issue citations alleging a failure to comply with this requirement more than six months after the occurrence of the failure to record. AKM/Volks LLC v. Secretary of Labor, 675 F3d 752 (D.C. Cir. 2012). For example, "[i]f an injury is reported on May 1, OSHA can cite an employer for failure to create a record beginning May 8, a citation issued within the following six months, and only the following six months, would be valid." Id. at 756.

101. 29 C.F.R. § 1904.32(a) and (b)(6).

102. 29 C.F.R. § 1904.32(a) and (b)(2).

103. 29 C.F.R. § 1904.32(b)(3).

104. 29 C.F.R. § 1904.32(b)(4).

105. 29 C.F.R. § 1904.33(a).

106. 29 C.F.R. § 1904.33(b).

107. 29 C.F.R. § 1904.33(b)(2) and (b)(3).

108. 29 C.F.R. § 1904.35(a).

109. 29 C.F.R. § 1904.35(b)(2)(ii).

110. 29 C.F.R. § 1904.35(b)(2)(i).

111. 29 C.F.R. § 1904.35(b)(2).

112. 29 C.F.R. § 1904.35(b)(2)(v)(A).

113. 29 C.F.R. § 1904.35(b)(2)(v)(B).

114. 29 C.F.R. § 1904.40(b)(1).

115. 29 C.F.R. § 1904.40(b).

116. 29 C.F.R. § 1904.29(b)(6).

117. 29 C.F.R. § 1904.29(b)(7).

118. 29 C.F.R. § 1904.29(b)(8).

119. 29 C.F.R. § 1904.29(b)(9).

120. 29 C.F.R. § 1904.29(b)(10).

121. 29 C.F.R. § 1094.29(b)(6).

122. An in-patient hospitalization is "a formal admission to the in-patient service of a hospital or clinic for care or treatment." 29 C.F.R. § 1904.39(b)(9). Employers are not required to report in-patient hospitalizations that only involve observation or diagnostic testing. 29 C.F.R. § 1904.39(b)(10).

123. An amputation is "the traumatic loss of a limb or other external body part. Amputations include a part, such as a limb or appendage, that has been severed or cut off, amputated (either completely or partially); fingertip amputations with or without bone loss; medical amputations resulting from irreparable damage; amputations of body parts that have since been reattached." 29 C.F.R. § 1904.39(b)(11). But "[a]mputations do not include avulsions, enucleations, deglovings, scalpings, severed ears, or broken or chipped teeth."

124. 29 C.F.R. § 1904.39(a)(1) – (2).

125. 29 C.F.R. § 1904.39(a)(3)(i).

126. 29 C.F.R. § 1904.39(a)(3)(ii) – (iii).

127. 29 C.F.R. § 1904.39(b)(2).

128. 29 C.F.R. § 1904.30(b)(1).

129. 29 C.F.R. § 1904.39(a).

130. 29 C.F.R. § 1904.39(b)(7).

131. 29 C.F.R. § 1904.39(b)(3) and (b)(4). If these injuries or illnesses are work-related, they must be recorded on the OSHA 300 Log.

132. Id. Motor vehicle accidents that occur while an employee is commuting to or from work is not considered work-related, and therefore, do not need to be reported to OSHA or recorded on an OSHA 300 Log.

133. 29 C.F.R. § 1904.39(b)(6).

134. Id.

135. 29 C.F.R. § 1904.39(b)(5).

136. 8 CCR § 342(a).

137. 8 CCR § 330(h).

138. 8 CCR § 342(a).

139. See Section 3.8 of this chapter discussion regarding the requirement to report workplace fatalities, in-patient hospitalizations, amputations, and a loss of an eye.

140. 29 C.F.R. § 1904, Non-Mandatory Appendix to Subpart B.

141. See Section 3.8 of this chapter for discussion regarding the requirement to report workplace fatalities, in-patient hospitalizations, amputations, and a loss of an eye.

142. 29 C.F.R. § 1910.119(g)(3).

143. 29 C.F.R. §1910.66(h).

144. 29 C.F.R. § 1910.157(e). The standard also requires monthly inspection, but does not explicitly require these inspections to be documented.

145. 29 C.F.R. § 1910.1020.

146. 29 U.S.C. § 955(b)(5).

147. Health standards are contained in Subpart Z of the OSHA general industry standards. OSHA has also promulgated health standards for the construction industry (Subpart Z of Part 1926) and shipbuilding industry (Subpart Z of Part 1915).

148. 29 C.F.R. § 1910.1001(d)(2)(iii).

149. 29 C.F.R. § 1910.1026(b) and (d)(3).

150. Section 1910.1048(o)(2) requires the employer to "maintain a record of the objective data relied upon to support the determination that no employee is exposed to formaldehyde at or above the action level." Section 1910.1096(d) requires covered employers to "make such surveys as may be necessary to comply" with the standard, and to provide personal monitoring equipment to employees who enter "restricted areas."

151. See 29 C.F.R. Section 1926.59 (applying Section 1910.1200 to construction); 29 C.F.R. § 1918.90 (applying Section 1910.1200 to longshoring); 29 C.F.R. § 1915.1200 (applying Section 1910.1200 to shipyard employment); 29 C.F.R. § 1917.28 (applying Section 1910.1200 to marine terminal operations). See details in Chapter 8 Hazard Communication.

152. 29 C.F.R. § 1910.1200(g)(1). In 2012, OSHA revised the HCS, in part, to replace the MSDS with the SDS. The SDS and MSDS contain similar information, except the SDS has a standardized format. 77 Fed. Reg. 17574, 17593 (March 26, 2012). As of June 1, 2015, the HCS will require new SDSs to be in a16-section, uniform format. 29 C.F.R. § 1910.1200(j)(2).

153. 29 C.F.R. § 1910.1200(b)(1).

154. 29 C.F.R. § 1910.1200(g)(8).

155. 29 C.F.R. § 1910.1200(f)(6). The employer is also prohibited from removing or defacing existing labels on incoming shipments. 29 C.F.R. § 1910.1200(f)(9).

156. 29 C.F.R. § 1910.1200(e).

157. The Occupational Exposure to Hazardous Chemicals in Laboratories standard, 29 C.F.R. § 1910.1450, sets forth hazard communication requirements for laboratories.

158. See 29 C.F.R. § 1915.1030 (applying general industry standard at 20 C.F.R. Section 1910.1030 to shipyard operations).

159. See definitions at 29 C.F.R. § 1910.1030(b).

160. Id.

161. OSHA's Medical and First Aid standard requires employers at facilities without nearby hospitals or clinics to train employees to provide first aid treatment. 29 C.F.R. § 1910.151.

162. 29 C.F.R. § 1910.1030(f).

163. 29 C.F.R. § 1910.1030(f)(5).

164. 29 C.F.R. § 1910.1030(h)(2).

165. 29 C.F.R. § 1910.1030(h)(5).

166. 29 C.F.R. § 1910.1030(c)(1).

167. 29 C.F.R. § 1910.1020(b)(1).

168. 29 C.F.R. § 1910.1020(c)(6)(i).

169. 29 C.F.R. § 1910.1020(c)(6)(ii).

170. 29 C.F.R. § 1910.1020(c)(5).

171. 29 C.F.R. § 1910.1020(d)(1)(i).

172. Id.

173. Id.

174. 29 C.F.R. § 1910.1020(d)(1)(ii).

175. 29 C.F.R. § 1910.1020(e)(2)(i).

176. 29 C.F.R. § 1910.1020(e)(2)(ii). See also 29 C.F.R. § 1913.10 (rules governing OSHA access to medical records).

177. 29 C.F.R. § 1910.134(k).

178. 29 C.F.R. § 1910.95(k).

179. Id.

180. 29 C.F.R. § 1910.22(a) and (b).

Chapter 6

Employees' and Employers' Rights

Phillip B. Russell, Esq.
Ogletree, Deakins, Nash, Smoak & Stewart, P.C.
Tampa, FL

1.0 Introduction

The Occupational Safety and Health Act of 1970 was passed in large part on a basic political agreement that something had to be done about workplace safety and health and that the federal government should finally lead the way. The major legislative disputes were centered mostly on the "how" and the "who" of the new law. In other words, there was considerable debate and disagreement about *how* to best protect worker safety and *who* should have primary responsibility. What emerged is a law with a balance of mechanisms and responsibilities – although whether that balance is just, appropriate, or even effective, has been the subject of ongoing debate since 1970.

In this Chapter, we will take a closer look at the allocation of responsibilities under the Occupational Safety and Health Act of 1970 ("OSH Act" or "Act") – the "who" part of the legislative structure so that we have a better of understanding of employers' rights and employees' rights under the Act. It his helpful to everyone involved in workplace safety and health compliance and enforcement to have a better understanding of these allocated, but shared responsibilities.

No situation more starkly illustrates employees' and employers' rights under the OSH Act than fatality cases. When an employee loses his or her life at work, it is a crisis for everyone. Owners, executives, managers, supervisors, and employees all need to know employees' and employers' rights (and conversely, responsibilities) under very tragic and stressful circumstances. Unfortunately, all too often confusion and misunderstandings dominate practical and workable understandings of how the OSH Act operates when a crisis hits.

This Chapter sets forth a summary of employees' and employers' rights as illustrated by a fatality scenario. Other Chapters in this guide provide deeper explanations of several topics, such as Chapter 7's discussion of whistleblower protections for employees and Chapter 11's discussion of understanding and contesting OSHA citations.

This scenario is for illustration and discussion purposes in this chapter. Although it contains some actual events, they did not all occur in one particular case. Rather, it is a compilation of situations from various fatality cases.

Put yourself into the position of Safety Director for a utility construction contractor called WeDigHoles, Inc. WeDigHoles has 100 employees deployed to multiple job sites. Each job site has a job site foreman as its most senior management position. WeDigHoles contracts with government contracting agencies and general contractors, which both strictly scrutinize safety records during the bid process so having a clean safety record is very important to WeDigHoles.

Jones County awards a contract to WeDigHoles to build a 30 foot high berm between two properties—one is a residential development and the other is a wetland preservation. To build the berm, WeDigHoles must move substantial dirt from one side of the wetland preservation property to the other and build up the berm in layers. Each layer must be compacted with a roller before adding other layers.

John is the job site superintendent for the Jones County berm project. John's 10-person crew includes 5 operators and 5 laborers. The operators are all very experienced and have extensive training on their equipment, including back hoes, dump trucks, and rollers. As the Safety Director, you have provided all operators and laborers safety training on the specific equipment and tasks they will be performing at the berm project. In his role, John also performs daily tailgate talks at the job site trailer in which he discusses various safety issues, such as confined space entry (for the manholes on the job site), trenching, and wearing seat belts.

George has the longest tenure of any employee on the job site. Although he has no formal management responsibilities, the operators and laborers look up to him as a leader. When George operates a roller, he does not wear his seat belt because he believes it is safer for him to be able to jump from the roller rather than to stay strapped in if it were to roll over.

He knows OSHA standards and company policy and training all require him and others to wear their seat belts, but he chooses not to do so. John knows George does not regularly wear his seat belt on the rollers, but John only verbally counsels or coaches George from time to time. John does not formally discipline George for not wearing his seat belt.

Susan is an operator with over 5 years' experience operating a roller. She has extensive training in safety and is known and respected by her coworkers as a safe operator who carefully operates her roller. Susan actively participates in John's tailgate talks and frequently emphasizes the importance of safely operating equipment at the job site.

Frank is another operator who is concerned about how George has ridiculed him and others for wearing their seat belts. Frank has called you to complain that he believes seat belts are required and must be worn by everyone. You call John to discuss Frank's concerns. John tells you he will take care of it, but he does not report back to you and you move on to other projects and tasks without following-up.

Toward the end of the berm project, tragedy strikes. At 4:38 p.m. on Friday, just before her shift ends, Susan is operating her roller on top of the berm. Susan is moving her roller in reverse at the edge of the berm when it is about 20 feet high. The compaction for this layer is almost complete and she is finishing the edges before the end of the day. Susan's coworker, George, is focused on shaping the edge of the berm about 200 yards away using his backhoe. Susan is moving her roller toward George. George looks up and toward Susan's roller just in time to see it veer off the edge and flip over side to side. Susan attempts to jump out of her seat inside the rollers' ROPS or roll over protection structure. She was not wearing her seat belt. Tragically, the roller flips too quickly and the top of the ROPS catches Susan and pins her underneath the roller on the side of the berm. Although George runs to see if he can help, when he gets there, Susan has already lost her life.

2.0 What Are the Employees' Rights in This Scenario?

First, let's start with Susan's estate as represented through her family. Can they sue under the OSH Act or any other law for Susan's loss of life while working? The answer in most states is no. Susan's fatality will usually be covered by state workers' compensation laws, which typically provide the exclusive remedy for any workplace injury, even a fatality. The OSH Act provides a regulatory structure, but it does not provide Susan's estate or any employee a private right of action. Susan's estate will pursue a workers' compensation benefits claim and unless there are some exceptions or exclusions (e.g., intoxication), it is likely the estate will receive the death benefit.

2.1 Right to Complain

One of the most important rights employees do have under the OSH Act is the right to complain about unsafe workplace conditions or practices that they believe violate OSHA standards or pose an imminent danger to their safety and health.[1] In our illustrative case study above, Frank complained to you, the Safety Director, about John's failure to enforce the seat belt requirement. When Frank called you to complain, he admitted John talks about wearing seat belts during the tailgate talks and at other times. But, Frank tells you John never takes any action to enforce the requirement and George keeps making fun of anyone who is not "man enough" to operate equipment without being strapped in like a baby. If John ignored your admonitions – or worse – had took any action against Frank, such as giving him tougher assignments or cutting his hours, Frank could reach out to OSHA. In our scenario above, if Frank did not complain before the accident, he likely will tell the OSHA investigator about how he complained and what the company did or failed to do. Section 11(c) of the Act provides employees broad protection from retaliation for exercising their rights under the Act.[2] Chapter 7 provides a more detailed discussion. It is important to understand that the Act does not require employees to make a formal complaint to OSHA for protection under Section 11(c), even if unwarranted, as long as the employee's complaint is made in good faith.

2.2 Right to Refuse to Work

A corollary right for employees is the refusal to work when they believe they will be exposed to an imminent hazard.[3] In our scenario, if Frank, Susan, or any other employee refused to operate a roller because it had a broken seat belt that would not securely latch, could WeDigHoles discipline them? Not without exposing the company to a valid Section 11(c) retaliation charge with OSHA. If Frank, Susan, and other employees combined their efforts or sought to act on behalf of other employees, there would also be potential legal risk for WeDigHoles for such protected concerted activity under the National Labor Relations Act. For more details on employees' right to refuse to work, see Chapter 7. It is important to note that OSHA has statutory authority to seek a federal court injunction to require an employer to avoid, correct, or remove any imminent danger.[4]

2.3 Right to Information

Employees also have a right to certain occupational safety and health information so they can better fulfill their role in protecting their own safety and health. Section 8(c) of the OSH Act and OSHA's implementing regulations require employers to provide employees: (1) a poster provided by OSHA containing basic information about the OSHA Act and employees' rights;[5] (2) injury and illness logs containing information about all recordable workplace injuries or illnesses;[6] (3) exposure monitoring information and medical records;[7] and (4) information regarding chemicals to which they may be exposed in the workplace.[8] All of these are covered in more detail in Chapters 5 and 8.

In our scenario, when OSHA's compliance officer reports to the job site to investigate Susan's death, one of the first things he or she will request are the OSHA injury/illness logs. Steve, our OSHA compliance officer for the case study, requests the logs and notes there is no OSHA poster in the break room or anywhere else in WeDigHoles' work trailer at the Jones County berm construction site. Although failing to properly keep logs and post the OSHA poster are citable violations, perhaps a more practical negative implication is that Steve will form an early opinion that WeDigHoles does not take its responsibilities under the OSH Act seriously. During the opening conference of his investigation, Steve specifically tells you he is concerned that WeDigHoles did not provide employees with information they have a right to access. Although you may be able to illustrate otherwise as the investigation progresses, it is best to avoid this initial negative impression by making sure you comply with all recordkeeping and information requirements well in advance of tragedy striking.

2.4 Right to Participate

Employees have a statutory right to participate in OSHA inspections either directly or through their representatives in the opening and closing conferences, walk-around inspections, as well as subsequent enforcement proceedings.[9] Chapter 10 has much more detail on this right. As a practical matter, employees rarely participate in any manner unless they are represented by a labor union. Nonetheless, employers must still keep employees informed of their rights through various postings and notices, as discussed in Chapter 10.

Back to our scenario, Frank is distraught over Susan's death and what he perceives as the company's lack of responsiveness to his concerns about seat belts – especially after he complained to you and John. Although there is no labor union at WeDigHoles, Frank insists on participating in the entire investigation. He openly tells you he is taking detailed notes and will take pictures of everything so he can work with Susan's estate and family lawyers to get to the bottom of this. Although you will not be required to pay Frank for his time spent participating in the investigation, he will have great latitude in which to lawfully participate.

2.5 Right to a Workplace Free From Recognized Hazards

As has been covered in Chapter 4, the General Duty Clause of the Act includes a provision requiring employers to provide a workplace "free from recognized hazards that are causing or are likely to cause death or serious physical harm to [its] employees[.]"[10] It is significant that the final language was "recognized hazards," rather than "hazards which are readily apparent," as OSHA has continued to broadly apply the General Duty Clause under this much broader definition. Thus, employees' right to a workplace free from recognized hazards stems from employers' general duty under the Act. One could also say employees' rights to be free from recognized hazards flows from specific OSHA standards that regulate certain industries and activities. These are covered in more detail in other Chapters.

In our berm construction fatality scenario, the General Duty Clause likely comes into play here because there is considerable debate among OSHA practitioners whether there is a specific construction industry standard that requires employees to wear seat belts while operating heavy construction equipment. When OSHA attempts to cite an employer for employees not using their seat belts on construction equipment, it frequently relies on the General Duty Clause.

3.0 What Are the Employer's Rights in This Scenario?

The OSH Act and its regulations not only provide certain rights to employees, it gives employers certain rights and protections – especially from unreasonable conduct (e.g., searches and seizures) that could violate constitutional protections. Through our case study above, we will explore some of the most important rights of employers.

In most instances, OSHA arrives to inspect a facility unexpectedly or when a major disruptive event, such as a fatality, catastrophe, or serious employee injury, has recently occurred. Too often, employers are not prepared to handle the inspection so as to minimize liability as well as reduce interference with operations. The result may be significant OSHA citations, civil liability, and even criminal penalties where a fatality occurs. This is especially so in the current environment of aggressive OSHA enforcement. Also, the outcome of the OSHA investigation may have an adverse effect on related liability issues, such as damage or personal injury claims.

To reduce this risk, the best way to address OSHA inspections is to prepare in advance by considering the issues that arise, and deciding to the extent possible how they will be addressed. While every inspection is different and relationships with local OSHA Area Offices may influence the approach to an inspection, there are certain issues which may be anticipated.

Of course, the best way to prepare for an inspection is to comply with applicable OSHA standards and regulations, and to practice sound safety and health principles. Striving for a professional relationship with OSHA representatives is also beneficial. *The main point, however, is that with proper preparation, the OSHA inspection process can be managed.*

3.1 Right to a Reasonable Inspection

Under the Fourth Amendment to the United States Constitution, OSHA cannot enter an employer's property without either a warrant or the employer's consent, except under very limited circumstances. Section 8(a) of the OSH Act provides OSHA the authority to enter and inspect worksites during reasonable times.[11] Specifically, Section 8(a) authorizes OSHA "to inspect and investigate during regular working hours and at other reasonable times, and within reasonable limits and in a reasonable manner, any such place of employment and all pertinent conditions . . . ." See Chapters 10 and 14 for a detailed discussion.

Under this structure, an employer certainly has property rights it can and should exercise appropriately. One prevalent myth is that employers must allow OSHA inspectors unconditional and unlimited access to their property simply because someone says "I'm here from the government and I'm here to help." Tongue-in-cheek political references aside, it is conversely not a good idea in practice to make OSHA inspectors get a warrant every time they want to inspect an employer's property. In practice, there is a balance between an employer's privacy rights and OSHA's duty to use inspections as its primary means of enforcing the OSH Act. By understanding the reason for the inspection and its scope, an employer will be in a better position to have a cooperative, problem-solving approach with OSHA.

In our fatality scenario, Steve, our OSHA inspector, will be judged by a reasonableness standard. If he conducts the investigation in an unreasonable manner, OSHA runs a substantial risk of having evidence suppressed in subsequent litigation. In our scenario, if Steve immediately begins interviewing John although John has told him the company's attorney is on his way, Steve could seriously undermine the integrity of the investigation. If he takes a statement from John or any other management personnel, such as you, the Safety Director, without allowing your attorney to participate, those statements could be thrown out. Other ways Steve could be unreasonable would be to require witnesses to be available for interviews when they are not scheduled to work or to give no notice at all to the employer and just start showing up at the worksite. Productive and cooperative communication is the key to avoiding any issues with the process of the investigation. Employers can and should assert their right to have a reasonable investigation conducted by the government.

Employers also have a vindictive prosecution defense to citations in extremely narrow circumstances. However, to establish the defense, it is not enough for the employer to show OSHA or any of its personnel had a bias or adversarial attitude toward the employer. The employer must also show more than OSHA acting unreasonably. The reality is that even political or personal bias is probably not enough to use this defense, which is why it is only appropriate in the most extreme circumstances. In our scenario, even if Steve says during the investigation that he thinks the company is careless and should be "hammered" by OSHA to send a message to other bad employers, WeDigHoles should not rely on the vindictive prosecution defense. Perhaps it would be appropriate if the company, through legal counsel, elevated the issue to the local area office director, rather than wait and try to assert the defense in subsequent legal proceedings.

3.2 Right to Representation During Any Interview of a Management Employee

The employer has the right to have a representative present during any interview of a management employee. In general, a management employee is one who has any type of supervisory responsibility and is paid on a salaried basis. The Company's right to be present should not be given up.

Most compliance officers recognize this right, but even if resistance is encountered, it is not advisable to give up this right.

The reason is that statements made by management employees are nearly always binding on the company. They may be considered legally as admissions against the employer's interest.

As such, it is worth investing the time to prepare management employees for their interviews. They must, of course, be instructed to answer questions truthfully, but carefully. Management employees should be urged to answer only the question asked without volunteering information and to avoid admitting that a certain condition or practice violates an OSHA standard. As noted above, there is no requirement that an employer allow a manager to sign a statement or permit recording of the interview. If the company has a policy that forbids signing interview statements or the recordation of interviews, taking the burden off the manager-witness by having the employer inform the compliance officer of this policy in advance of the interview is highly recommended.

In our scenario, if Steve wants to interview John about the accident and any other issue that may bear on OSHA's investigation, he would have to make that request through the company's representative and coordinate a reasonable time. If Steve fails to do so and interviews John without the company's representative present, OSHA would have substantial risk of the interview and any statement by John being deemed inadmissible in any legal proceedings. Although OSHA knows this risk, it has happened from time-to-time that an aggressive compliance officer interviews a management witness without the company's representative present. If that happens, once the company learns of the situation, the company's legal counsel should be immediately involved, if he or she is not already. It is critical to elevate the issue and correct the situation before the investigation proceeds in any way. Because management employees are the company's agents, anything they say can and will be used against the company as if the company made an official statement. This is a critical right that all employers must exercise to protect the integrity of the investigation.

3.3 Right Not to Perform Work or Process Demonstrations

In a somewhat related situation, employers also have the right not to perform demonstrations of work or work processes for OSHA. OSHA is entitled to observe work as it is being performed, but cannot insist that it be shown how equipment operates, or how particular operations are performed.

Sometimes, however, it is to an employer's advantage to stage such demonstrations, as when it is necessary to clarify misunderstandings or simply to impress the compliance officer. Be mindful, however, that "Murphy's Law" is operative, and that even the best-planned demonstrations sometimes go astray at just the wrong time.

3.4 Right to Continue Operations in a Safe Manner

Some compliance officers will tell an employer to stop performing a certain operation because it is dangerous, or a violation. It is important to understand that a compliance officer has no authority to direct the cessation of work. Under the OSH Act, only a federal district court judge has the power to enjoin work, and then only where OSHA has shown that an imminent danger exists. Of course, if a compliance officer points out an obvious danger or hazard, it should be corrected immediately.

3.5 Right to Protect Trade Secret Information

The OSH Act's requirement that employers keep certain records and produce them to the government does not mean the employer must disclose its trade secrets or other confidential business information. During an investigation, if OSHA requests such information, the employer has the right to protection of its information and typically would designate such information as confidential. If the information is highly sensitive, the employer should have a thorough discussion with OSHA to determine if there are alternative means of disclosure, such as redacting portions of documents. If OSHA is not cooperative, the employer can also refuse to produce the information without a warrant or subpoena. If that happens, the employer can use the legal process to seek additional protection. Note, however, that an employer cannot claim trade secret protection on documents and information it is required to keep and produce under the OSH Act.

In our scenario, if Steve requests the company produce its contract and any subcontracts for the project and job site where the accident occurred, you would have a discussion with him about what information he is seeking. If he is looking only for provisions related to safety, you could likely redact the other business information in the contracts without issue.

3.6 Right to Establish Unpreventable Employee Misconduct as a Defense

In many investigations, there is an issue of whether an employee's misconduct caused the accident. To establish the affirmative defense of unpreventable employee misconduct, the employer must show (1) the employer has established work rules designed to prevent the violation; (2) it has adequately communicated those rules to its employees; (3) it has taken steps to discover violations; and (4) it has effectively enforced the rules when violations have occurred. However, these requirements are very strict in practice and employers should carefully and critically consider whether than can meet all of these requirements. Although an employer may have rules in place that it communicated to employees (the first two elements), showing the discovery (third) and enforcement (fourth) elements, as a practical matter, is much more difficult. This defense is also seen by OSHA as the "blame the victim" defense, so it is given a heightened level of scrutiny and requires strict proof. In practice, it is best to not utter the words "employee misconduct," but to demonstrate clearly the evidence supporting all the elements. Let OSHA reach its own conclusion based on the strength of the evidence. It certainly is not a good idea to raise this defense before the investigation has proceeded along far enough for OSHA to have seen sufficient evidence to reach its own conclusion.

In our scenario, you would likely not succeed with the employee misconduct defense because of the final two elements. Although John spoke about wearing seat belts, there is not sufficient evidence to show the company did anything to enforce the requirement.

Notes

1. 29 U.S.C. § 657(f)(1); 29 C.F.R. § 1903.11(a).

2. Section 11(c)(1), 29 U.S.C. § 660(c)(1).

3. 29 C.F.R. § 1977.12(b)(2).

4. 29 U.S.C. § 662(a).

5. 29 C.F.R. § 1903.2(a).

6. 29 C.F.R. § 1904.1.

7. 29 C.F.R. § 1910.1020; known as the "Access Rule."

8. 29 C.F.R. § 1910.1200 (general industry); 29 C.F.R. § 1926.59 (construction industry).

9.   29 U.S.C. § 657(a)(2), (e) and (f)(2).

10. 29 U.S.C. § 654.

11. 29 U.S.C. § 657(a).

Chapter 7

Refusal to Work and Whistleblower Protection

Kenneth B. Siepman, Esq.
Ogletree, Deakins, Nash, Smoak & Stewart, P.C.
Indianapolis, IN

1.0 Overview

This chapter discusses the protection afforded by the law to employees who either refuse to perform an assigned task on the grounds that it presents a danger to safety or health,[1] or who register a complaint, i.e., "blow the whistle" regarding a hazardous condition.

Employer-employee relationships in the United States historically have been governed by the doctrine of employment at will. Under this doctrine, in the absence of a contract of employment for a fixed term, an employee may leave his employment at any time, and the employer may discharge an employee at any time for any reason, good or bad, or indeed for no reason at all.

The at-will doctrine has undergone continuing erosion. Not only do state and federal statutes create exceptions, but courts have also found or created exceptions under various common law theories. Further, collective bargaining agreements usually provide that employees can be terminated only for cause.

2.0 Refusal to Work

2.1 Federal Statutes

A number of federal statutes protect employees, under certain circumstances, from being discharged or disciplined for refusing to perform an assigned task.[2] Our discussion is limited to the protection afforded under the Occupational Safety and Health Act ("OSHA" or "Act")[3] and the National Labor Relations Act ("NLRA"), as amended.[4]

2.1.1 Occupational Safety and Health Act

While OSHA does not expressly confer upon employees a right to refuse to work, Section 11(c) of the Act prohibits discrimination against an employee because he has filed a complaint, taken part in any legal proceeding brought under OSHA, or exercised "on behalf of himself or others...any right afforded" by the Act.[5]

The Secretary of Labor ("Secretary") interpreted this clause to entail a right to refuse to work and, accordingly, promulgated a regulation found at 29 C.F.R. § 1977.12. Under this regulation, an employee cannot be disciplined for refusing to perform work if (1) such refusal was made in good faith and not for some ulterior purpose; (2) the refusal is based upon the existence of a dangerous condition such that a reasonable person faced with the same situation would conclude that there is a real danger of death or serious injury; (3) there was insufficient time to deal with the hazard through the use of the ordinary enforcement mechanisms provided by the Act; and (4) "where possible," the employee has first tried to get his employer to eliminate or correct the perceived dangerous conditions.

The leading case upholding an employee's right to refuse to perform dangerous work under OSHA is *Whirlpool Corp. v. Marshall*, 445 U.S. 1 (1980). In *Whirlpool*, the Supreme Court upheld the Secretary's regulation and, through that prism, examined the claims of two workers who were reprimanded after refusing to walk out onto a wire mesh screen suspended 20 feet above the plant floor. The Court, applying the regulation, held that the two men had been improperly disciplined (just two weeks earlier a man had fallen through the screen to his death, and the employees had several times requested that the mesh screen be strengthened). But, the Court stressed that Whirlpool could have stayed within the law simply by reassigning the employees to other, non-hazardous work and could have disciplined the employees if they had refused to perform it.

The *Whirlpool* holding remains good law. Recognizing the potential for abuse, however, subsequent courts have required "objective evidence" demonstrating the existence of the hazardous condition to show that an employee's apprehension was reasonable.[6]

OSHA interprets broadly the "right[s] afforded" employees under Section 11(c) of the Act. For instance, according to OSHA, when an employee reports a work-related injury, the employee exercises a "right afforded" by the Act.[7] Consequently, an employer who disciplines or discharges an employee following the report of a work-related injury needs to ensure that it does so for a legitimate reason, *e.g.*, the employee's violation of a safety rule.[8]

2.1.1.1 Enforcing Rights under OSHA

OSHA does not give employees a private right of action.[9] Only the Secretary can bring an action to enforce an employee's Section 11(c) right to refuse to perform hazardous work.

An employee who believes he has been disciplined in violation of Section 11(c), must file a complaint with the Secretary within 30 days of the alleged violation.[10] This period may, however, be extended or "tolled" "on recognized equitable principles or because of strongly extenuating circumstances."[11] Upon receiving the complaint, the Secretary makes a determination whether Section 11(c) has been violated, and, upon that determination, may bring suit against the employer. Although the Secretary is supposed to notify the complaining employee of the determination within 90 days,[12] the Secretary often takes longer to decide whether to file an action. Moreover, the 90-day limit is intended for the benefit of the complainant, not the employer, and the Secretary can bring an action outside its limits with impunity.[13]

Workers' attempts to force the Secretary to sue on their behalf have proven unsuccessful.[14]

2.1.1.2 Secretary's Burden in Litigation

The Secretary has the burden to prove that an employer violated Section 11(c). To succeed, the Secretary must produce evidence that 1) the employee had reasonable fear of serious injury or death, 2) there was insufficient time to eliminate the danger through regular statutory enforcement mechanisms and 3) circumstances permitting, the employee tried unsuccessfully to have the employer correct the dangerous condition.[15]

2.1.1.3 Burden Shifting Analysis

Courts apply a "burden shifting" analysis, adopted from Title VII discrimination cases, to determine whether an employer violated Section 11(c).[16] Under this analysis, the Secretary must first demonstrate that the employee engaged in protected activity, and that as a result of doing so, the employee suffered "an adverse action."[17] In other words, the Secretary must prove that "but for" the employee's protected activity, the employee would not have suffered the adverse action.[18] If the Secretary makes this showing, the burden shifts to the employer to articulate a non-discriminatory reason for the adverse employment action, such as, for example, poor performance or insubordination. If the employer articulates a non-discriminatory reason, the burden shifts back to the Secretary to prove that this reason is a mere "pretext" for discrimination, i.e., a false reason designed to cover an illegal motive.[19]

2.1.1.4 Remedies

OSHA provides that where an employee has been discriminated against in violation of Section 11(c), the court may restrain all such violations and "order all appropriate relief including rehiring or reinstatement…with back pay."[20] "All appropriate relief" can include compensatory and punitive damages.[21]

2.1.2 National Labor Relations Act[22]

Although the principle purpose of the NLRA is to encourage the resolution of workplace disputes through collective bargaining, two sections of the NLRA protect employees who refuse to perform unsafe work.

The first is Section 7 of the Act, which provides that employees have the right to "engage in…concerted activities for the purpose of … other mutual aid or protection…"[23]

The second is Section 502 of the Labor Management Relations Act of 1947 (LMRA), which amended the NLRA.[24] Section 502 provides that the "quitting of labor by an employee or employees in good faith because of abnormally dangerous conditions for work at the place of employment [shall not] be deemed a strike under this Act." Section 502 principally protects unionized employees by creating a statutory override of collective bargaining agreements that contain no-strike clauses under which employees who cause a work interruption can be disciplined.

2.1.2.1 Protection under Section 7

Section 7 prohibits an employer from disciplining employees for concerted activity protesting unsafe working conditions. The seminal case is *NLRB v. Washington Aluminum*, 370 U.S. 9 (1962), in which employees were discharged for walking off the job in protest of extremely cold conditions in the work area after a furnace failed. The National Labor Relations Board (NLRB) and the Supreme Court found that firing the employees was an "unfair labor practice" because they were engaged in "concerted activity" for "mutual aid or protection". The Court in *Washington Aluminum* held that "reasonableness" does not play any role in the identification of protected concerted conduct—that "it has long been settled that the reasonableness of workers' decisions to engage in concerted activity is irrelevant to the determination of whether a labor dispute exists or not)."[25] The Court followed up by stating that when concerted activity in a labor dispute is covered by the NLRA, only steps that are "unlawful, violent or in breach of contract ... [or] characterized as 'indefensible' because they ... show a disloyalty to the workers' employer" are subject to penalty.[26]

Accordingly, the protection afforded by Section 7 depends not on the reasonableness of the employees' refusal to work, but on whether the refusal is "concerted" activity. But what actually counts as "concerted"?

The NLRB considered this precise question in its *Meyers* decisions.[27] In the first of these cases, the Board held that an activity is concerted if it is "engaged in with or on the authority of other employees, and not solely by and on behalf of the employee himself."[28] In the second case, the Board explained that its standard "encompasses those circumstances where individual employees seek to initiate or to induce or to prepare for group action, as well as individual employees bringing truly group complaints to the attention of management."[29] In *Meyers,* an employee was discharged in part because he refused to drive a truck that he claimed to be unsafe. The Board concluded that "concerted activity" is "collective" in nature and that the driver's refusal to drive the truck was for his own individual interest and was not collective in nature, thus not "concerted activity."

A union-represented employee whose refusal to work due to unsafe work conditions can be construed as an attempt to enforce a labor agreement will generally satisfy this "collective" requirement[30]; a single, non-union employee refusing to do work will not unless there is evidence the employee was acting "with or on the authority of other employees."

Note that the lack of *specific* authorization by the other employees does not necessarily cast a single employee's protest outside the protection of Section 7. An employee can satisfy the "with or on the authority of other employees" requirement by voicing shared, group concerns.[31]

2.1.2.2. Comparison of Section 7 and Section 502

Section 502 is significantly narrower than Section 7. Under Section 7, a good-faith belief that danger exists is sufficient to protect action, so long as it is "concerted." Under Section 502, in addition to a good-faith belief, there must be "ascertainable objective evidence" that an "*abnormally* dangerous" condition for work exists.[32] (Emphasis added). Many jobs are, for lack of a better term, "normally dangerous" to one degree or another, and Section 502 does not provide a general right for employees in dangerous professions to refuse to work. Finally, the employee's refusal must be based on an "identifiable, presently existing threat" not some vague general safety concern.[33]

2.1.2.3 Cooperation between OSHA and the NLRB

To avoid unnecessary litigation and conflicting enforcement efforts, the NLRB and OSHA have agreed that when a charge of retaliation that would be covered by Section 11(c) is filed with the NLRB and a corresponding complaint is filed with OSHA, the NLRB will either dismiss or defer the charge. Where such a charge is filed only with the NLRB, the Board will advise the charging party of his right to file a Section 11(c) complaint with OSHA. When it is determined the charge falls within the NLRB's exclusive jurisdiction but relates to discrimination based on safety and health issues, the Board and OSHA will confer.[34]

2.1.3 Arbitration and Collective Bargaining Agreements

2.1.3.1 Collective Bargaining Agreements

Most labor agreements require the employer to provide a safe workplace. Some agreements expressly provide that an employee may refuse to perform work that he believes to be unsafe. Usually such refusal can be made only after certain steps have been followed, e.g., requesting the union steward and supervisor to review the situation, and accepting alternative work until the dispute is resolved.

In addition, virtually all labor agreements provide that an employee can be disciplined only for cause and where the union disagrees with the employer's issuance of discipline, the matter will ultimately be resolved through arbitration.

Typically, an employee cannot unilaterally refuse to perform work. For example, an employee who believes that his seniority entitles him to a different job than the one he is being asked to perform is entitled to file a grievance under the agreement, but must perform the assigned task in the meantime. "Work now – grieve later."

In such a case, the employee who refused to work would be subject to discipline, including discharge "for cause," i.e., insubordination. But arbitrators have carved out an exception when the refusal is based on safety concerns. To qualify for this exception, "the employee must show that a safety or health hazard was the real reason for the refusal, and that the alleged hazard existed at the time of the employee's refusal." [35]

It has frequently been said that the arbitrator is a "creature of the contract," and as such cannot exceed the authority that parties have mutually agreed to give him. The Supreme Court has held that an arbitrator "does not have 'general authority to invoke public laws that conflict with the bargain between the parties.'"[36] Consequently, the language of the agreement and principles of contract interpretation govern an arbitration decision.

Nevertheless, where statutory rights may be in question, arbitrators will sometimes find that they have the authority, granted either explicitly or by implication, to consider statutory requirements in rendering their decision. Whether an arbitrator has such authority is a fact-specific question.

2.1.3.2 Arbitration not under a Collective Bargaining Agreement

In recent years, many employers, in an effort to limit litigation, have entered agreements with employees that require all complaints be resolved through arbitration. Other employers have provided for voluntary arbitration as a part of an internal employee complaint procedure, often set forth in an employee handbook or policy manual.

There are two issues: (1) can the employee who signed the agreement to arbitrate be foreclosed from initiating litigation to enforce a statutory right, and (2) can an agency like OSHA be precluded from exercising its enforcement powers by an employee's agreement to arbitrate. It is now well settled that, with limited exceptions, employees can be required to arbitrate claims arising under federal and state statutes. In a landmark decision in 1991, the Supreme Court upheld enforcement of an agreement requiring an employer to arbitrate discrimination claims arising under the Age Discrimination in Employment Act (ADEA).[37] Ten years later, in *Circuit City Stores, Inc. v. Adams*, 532 U.S. 105 (2001), the Supreme Court held that employment contracts, other than those for transportation "workers actually engaged in the movement of goods in interstate commerce," were covered by the Federal Arbitration Act (FAA); therefore, contractual agreements to arbitrate statutory employment discrimination claims were enforceable under the FAA.

On the other hand, the general rule is that agreements to arbitrate are only enforceable against the parties to the agreement, meaning that the Secretary is not bound by the terms of such agreements.[38] This conclusion was driven home by the Supreme Court in *EEOC v. Waffle House, Inc.*, 534 U.S. 279 (2002). In that case, the Supreme Court allowed the EEOC to bring an action under the Americans With Disabilities Act (ADA) in federal court for both injunctive relief and for back pay, reinstatement, and compensatory damages for an employee who was bound by an agreement to arbitrate his ADA claims. The Court noted that the EEOC was not a party to the arbitration agreement, it did not agree to arbitrate its claims, and its suit was not a derivative action—it had an independent right to prosecute a claim without the victim's consent and without direction from the victim in the EEOC's pursuit of the public interest.[39]

2.1.3.3 Deferral

A related issue is the extent of deference that an agency such as OSHA or the NLRB is likely to give to an arbitration award. In December 2014, the NLRB modified its deferral standard. The NLRB will defer to arbitration when (1) the underlying dispute arises from a collective bargaining relationship, (2) the employee does not allege that the employer is opposed to the employee's "exercise of protected rights," (3) the parties are willing to arbitrate, (4) the arbitration agreement covers the dispute, (5) the unfair labor practice centers on a collective bargaining agreement, and (6) the arbitrator is explicitly authorized to decide the statutory issue (either in the collective bargaining agreement or by agreement of the parties in a particular case).[40] The Board will also defer to an arbitrator's decision if certain conditions are met.[41] Of course, this doctrine will not be applied if the result of the arbitration award requires an unlawful act.[42] There is also a broad exception to deferral when its application would require a party to file the same grievance twice to resolve an ultimate issue.[43] In all cases the Board reserves the right to assume jurisdiction whenever it believes the purposes of the NLRA require it.[44]

The Secretary of Labor has promulgated a regulation authorizing OSHA to postpone its own determination and defer to the decision of an arbitrator or other agency where it concludes that the rights conferred by Section 11(c) have been adequately protected.[45] But, there is some judicial authority rejecting OSHA's authority to suspend action on a complaint pending the NLRB's decision on a corresponding charge.[46]

Under very limited circumstances, a court may review an arbitration award even if OSHA or the NLRB has deferred to the arbitration. The FAA allows courts to vacate arbitration awards when (1) the award was obtained through corruption, fraud, or other undue means, (2) an arbitrator was evidently partial or corrupt, (3) the arbitrators were guilty of some sort of misconduct of the proceedings, or (4) the arbitrators exceeded their powers or exercised them so imperfectly that no final award was made.[47] Further, some federal courts have held that arbitration awards can be vacated for "manifest disregard of the law."[48] Any party injured by the award can petition the court for review.[49]

The law in this area is intricate, and the nuances are beyond the scope of this chapter. The existence of an agreement to arbitrate should, at least, signal the need for informed legal guidance.

2.2 State Statutes

An employee's right not to be disciplined for refusing to perform allegedly hazardous work is protected in many state statutes or constitutions. Like OSHA, most state statutes provide that only the authorized state official may initiate litigation, but some state statutes create a private right of action.[50]

2.3 Common Law

In addition to statutory protections, many state courts recognize a tort action for wrongful discharge based on a "public policy" exception to the doctrine of employment at will. Under this exception, an employee discharged for refusing to perform illegally hazardous work may bring a civil action to get his job back and to recover compensatory or even punitive damages.[51] Note that even in states that recognize a common-law action for wrongful discharge, the action may be pre-empted by the existence of a separate, administrative remedy.[52]

3.0 Whistleblowing[53]

In many instances the protection for refusal to work or whistleblowing emanates from the same statutory language or common law principle, e.g., Section 11(c) of OSHA, 29 USC § 660(c), which prohibits an employer from discriminating in any way against an employee, either for filing a complaint or refusing to work.

In some instances, a statute prohibits discrimination based on either refusal to work or whistleblowing but not both. For example, § 502 of the Labor Management Relations Act, 29 USC § 143, protects only the right to refuse to work.

3.1 Federal Statutes

There are a number of federal statutes that expressly prohibit retaliation against employees who complain of hazardous conditions or illegal actions attributable to ether employers.[54] Only OSHA protections concern us.

3.1.1. Occupational Safety and Health Act

The anti-retaliation provisions of the OSH Act are found in Section 11(c), discussed above in connection with refusal to work.

Although on its face, Section 11(c) appears to be limited to protecting those who file complaints with OSHA or testify in OSHA proceedings, it has been given a broad interpretation. Hence, even employees who complain only to their employers are protected from retaliation under Section 11(c), so long as such complaints are made in good faith.[55]

Employees are also protected under Section 11(c) from retaliation based on their having informed journalists of safety and health concerns at the workplace.[56]

3.1.1.2 Preemption

In some cases, courts have ruled that a given state statute or legal action is preempted by one or more federal statutes.[57] An extended discussion of the complex question of preemption is beyond the scope of this chapter, but anyone involved in a potential state-whistleblowing case should be aware of the issue.

3.2 State Statutes

Most states have statutes that protect employees who complain of hazards that threaten employees or the public in general. Some of these protect both private and public employees who blow the whistle on unsafe conditions,[58] while others protect only public employees.[59]

3.3 Common Law

In addition to state statutes that protect the right of employees to complain about hazardous conditions, the courts of most states have created an exception to employment at will in the form of an action in tort for violation of public policy.

The essentials necessary to establish a successful case under the public policy exception vary from state to state but always involve "a causal-motivational link between a violation of public policy and the discharge."[60]

In New Mexico, an employee makes a prima facie case by showing 1) discharge for performing an act that public policy has authorized or encouraged, 2) the employer's knowledge or suspicion that the employee's act was protected, 3) causal connection between protected action and the discharge, and 4) that the employee suffered damage thereby.[61]

In some states, a separate tort action based on the public policy exception is available in addition to the statutory remedy.[62] Other states do not recognize the public policy exception and only an express statutory provision will suffice to support an action for retaliatory discharge for reporting hazardous conditions.[63]

4.0 Conclusion

Federal and state legislation and court decisions establish strong protection for workers faced with dangerous conditions. Employers should avoid disciplining or otherwise discriminating against employees who, in good faith, refuse to work under conditions they believe to be dangerous or who point out the existence of such conditions, unless there are unrelated, legitimate reasons to do so.

Notes

1. Hereinafter, "safety" includes "health."

2. See, e.g., Surface Transportation Assistance Act, 49 U.S.C. § 31105(a) (commercial driver may refuse to operate a vehicle when the operation would violate a regulation or safety standard or when he or she reasonably apprehends that the conditions of the vehicle makes injury to the driver or the public a risk); Federal Railroad Safety Act, 49 U.S.C. § 20109(b) (a railroad carrier may not discriminate against or discharge an employee who, in good faith, refuses to work when confronted with a condition that a reasonable employee would believe either presents an imminent risk of serious injury or death or that cannot be corrected through regular statutory means due to the urgency of the situation if the employee has notified his or her employer of the condition in a timely manner).

3. 29 U.S.C. § 651 et seq.

4. 29 U.S.C. § 141 et seq.

5. 29 U.S.C. § 660(c).

6. See, e.g., Marshall v. National Indus. Constructors, 8 OSHC 1117 (D.Neb. 1980)(finding no objective evidence to justify the employee's refusal to work).

7. See 29 C.F.R. § 1904.36.

8. Disciplining an employee following a work-related injury may also result in a worker's compensation retaliation claim.

9. Ellis v. Chase Communications, Inc. 63 F.3d 473, 477 (6th Cir. 1995) ("OSHA does not create a private right of action."); Donovan v. Square [?] Co., 709 F.2d 335, 338-39 (5th Cir. 1983) ("The government alone possesses the right to bring suit under section 11(c). A private cause of action does not exist.")

10. 29 U.S.C. § 660(c)(2).

11. 29 C.F.R. § 1977.15(d)(3). Essentially, the 30-day limit operates as a statute of limitations. If the period expires because the employer has "lulled" the employee into not exercising statutory rights, equitable rules may allow an extension. For example, in Donovan v. Hahner, Foreman & Harness, Inc., 11 OSHC (BNA) 1081 (D. KAN. 1982), aff'd 736 F.2d 1421 (10th Cir. 1984), the Court held that the 30-day period was tolled because the employer led the employee to believe he was laid off, when in fact he had been fired. On the other hand, the regulation expressly states that the pendency of a grievance under a collective bargaining agreement or a complaint before another agency does not justify tolling the 30-day period.

12. 29 U.S.C. § 660(c)(3).

13. See, e.g., Donovan v. Freeway Construction Co., 551 F.Supp. 869, 878 (D. R.I. 1982) (in rejecting an employer's argument that the Secretary's failure to institute an action within 90 days precluded the suit, the court noted "[t]he defendant may not pervert a statutory provision protecting employees against employer harassment in order to insulate itself against an action brought in response to a [Section 11(c)] grievance.").

14. Wood v. Department of Labor, 275 F.3d 107 (D.C. Cir. 2001).

15. See Chao v. Karamourtopoulos, 2006 DNH 40, 21 OSHC (BNA) 1474 (D. N.H. 2006).

16. See Gombash v. Vesuvius USA, Inc., 380 F.Supp. 2d 977 (N.D. Ill. 2005).

17. In a retaliation case under Title VII, the Supreme Court held that "a plaintiff must show that a reasonable employee would have found the challenged action materially adverse, which in this context means it well might have dissuaded a reasonable worker from making or supporting a charge of discrimination." Burlington N. & S. Railway Co. v. White, 126 S.Ct. 2405, 2415 (2006). See also Hendrix v. American Airlines, Inc., 2004-AIR-10, 2004 SOX-23, slip op. at 11-14 (ALJ Dec. 9, 2004)(analyzing divergent administrative decisions on the meaning of "adverse action" under whistleblower protection statutes).

18. 29 C.F.R. § 1977.6(b)

19. See, e.g., Reich v. Hoy Shoe Co., 32 F.3d 361, 363 (8th Cir. 1994).

20. 29 U.S.C. § 660(c)(2).

21. Reich v. Cambridgeport Air Systems, Inc., 26 F.3d 1187, 1191-1192 (1st Cir. 1994) (upholding district court's award of damages equal to twice the employees' lost back pay); see also Franklin v. Gwinnett County Pub. Sch., 503 U.S. 60 (1992) (holding that absent a clear indication from Congress, all remedies are presumed available in actions brought pursuant to a federal statute).

22. 29 U.S.C. § 151 et seq.

23. 29 U.S.C. § 157.

24. 29 U.S.C. § 143.

25. Washington Aluminum, 370 U.S. at 16.

26. Id at 17.

27. Meyers Indus., Inc., ("Meyers I") 268 N.L.R.B. 493 (1984), on remand, Meyers Indus., Inc., ("Meyers II") 281 N.L.R.B. 882 (1986), aff'd. 835 F.2d 1481 (D.C. Cir. 1987), cert. denied 487 U.S. 1205 (1988).

28. Meyers I at 497 (footnote omitted).

29. Meyers II at 887.

30. See NLRB v. City Disposal Sys., 465 U.S. 822 (1984)(employee that refused to drive a vehicle he believed was unsafe engaged in protected concerted activity where he relied on an express provision of the collective bargaining agreement to support his refusal).

31. See NLRB v. Talsol Corp., 155 F.3d 785 (6th Cir. 1998)(employee's expressions of safety concerns protected where made in presence of other employees, "concerned the safety of the plant and were not purely personal gripes").

32. Gateway Coal Co. v. U.M.W., 414 U.S. 368, 386-87 (1974).

33. Id. at 386.

34. Memorandum of Understanding between OSHA and NLRB. 40 Fed. Reg. 26,083-084.

35. Elkouri and Elkouri, How Arbitration Works, Seventh Ed. (BNA 2012) p. 16 31 ("Elkouri").

36. Gilmer v. Interstate/Johnson Lane, 500 U.S. 20, 34, (1991) (quoting Alexander v. Gardner-Denver Co., 415 U.S. 36, 53, (1974)).

37. Gilmer, 500 U.S. 20 (1991).

38. See, e.g., Reich v. Sysco Corp., 870 F.Supp 777 (S.D. Ohio 1994)(holding that an arbitration award concerning an employee's discharge did not preclude an action under Section 11(c) of OSHA by the Secretary).

39. The Court left open the question of whether a settlement or arbitration judgment in which the employee participated would impact the validity of the EEOC's claim. It did, however, note that it could and "should preclude double recovery by an individual." Id. at 297.

40. Babock & Wilcox Construction Co., 361 NLRB No. 132, slip op. at 12-13 (Dec. 15, 2014); Tri-Pak Mach., 325 NLRB 671, 672 (1998).

41. The NLRB will defer to an arbitrator's decision where the procedures are fair and regular, the parties agree to be bound, and the party urging deferral (typically the employer) demonstrates that: (1) the arbitrator was explicitly authorized to decide the unfair labor practice issue, (2) the arbitrator was presented with and considered the statutory issue, and (3) the National Labor Relations Act "reasonably permits" the arbitrator's award. Babcock & Wilcox, 361 NLRB No. 132, slip op. at 10.

42. LID Elec., Inc. v. IBEW, Local 134, 362 F.3d 940, 943 (7th Cir. 2004).

43. Elkouri at p. 10-47. This occurs, for example, where a union alleges an employer refused to provide information related to a grievance. See Earthgrains Baking Cos., Inc., 327 NLRB 605, 611 (1999) (the Board stated it "has generally refused to defer issues that would result in a two-tiered system requiring a union to file a grievance to obtain information potentially relevant to its processing of a second underlying grievance.").

44. 29 U.S.C. § 160(a).

45. 29 C.F.R. § 1977.23.

46. See Newport News Shipbuilding v. Marshall, 1980 U.S. Dist. LEXIS 15906, 90 Lab. Cas. (CCH) P12,427 (E.D. Va. 1980).

47. 9 U.S.C. §10(a)

48. Red Apple Supermarkets/Supermarkets Acquisitions v. Local 338, RWDSU, 1999 U.S. Dist. LEXIS 12199, 162 L.R.R.M. 2365 (S.D. N.Y. 1999) (citations omitted).

49. 9 U.S.C. 10(c).

50. Compare, e.g., Brevik v. Kite Painting, Inc., 416 N.W. 2d 714 (Minn. 1987) (Minnesota State OSH Act creates individual cause of action for discriminatory discharge) with Silkworth v. Ryder Truck Rental, 70 Md. App. 264 (1987) (Maryland state act tracks federal law and thus creates no private right of action.)

51. See, e.g. Cabesuela v. Browning-Ferris Industries of Calif. Inc., 68 Cal. App. 4th 101, 80 Cal. Rptr. 2d 60 (6th Dist. 1998), (holding that an employer who fires an employee for protesting unsafe work conditions violates public policy).

52. See Kornischuk v. Con-Way Cent. Express, 2003 U.S. Dist. LEXIS 14459 (S.D. Iowa 2003) (a common-law claim for retaliatory discharge in violation of public policy existed but was inapplicable because a statute set out an administrative remedy for truck drivers discharged in violation of public policy).

53. For a comprehensive discussion of court decisions on whistleblowing, see 75 ALR 4th, 13, Gregory G. Sarno, J.D., Annotation: Liability for Retaliation Against At-Will Employees for Public Complaints or Efforts Relating to Health or Safety. See also Monique C. Lillard, Exploring Paths to Recovery for OSHA Whistleblowers: Section 11(c) of the OSH Act and the Public Policy Tort , 6 Empl. Rts. & Pol'y J. 329 ('Lillard''); William Dorsey, An Overview of Whistleblower Protection Claims at the United States Department of Labor, 26 NAALJ 43 (2006).

54. Surface Transportation Assistance Act, 49 U.S.C. § 31105(b) (commencing, participating or testifying in proceedings relating to violations of commercial motor vehicle safety provisions); Federal Railroad Safety Act, 49 U.S.C. § 20101 (refusal to work or reporting safety violations); Energy Reorganization Act, 32 U.S.C. § 5851(a) (commencing, participating, or testifying in proceedings under the ERA or Atomic Energy Act).

55. See, e.g., Marshall v. Springville Poultry, 445 F.Supp. 2, 5 OSHC 1761 (M.D. Pa. 1977).

56. See, e.g., Donovan v. R.D. Andersen Constr. Co., 552 F.Supp. 259, 10 OSHC 2025 (D. Kan. 1982).

57. In the aftermath of Taylor v. Brighton Corp., 616 F2.d 256, which held that Section 11(c) creates no private right of action, some state courts were reluctant to allow a de facto private right of action under Section 11(c) through the mechanism of the common-law. See generally Lillard for an expanded discussion of this issue.

58. See, e.g., Ct. Gen. Stat. § 31-51m, Ohio Rev. Code § 4113.52.

59. See, e.g., Alaska Stat. 39.90.100, Del. Code Ann. Tit. 29 § 5115.

60. Lillard at 351.

61. Id., citing Weilder v. Big J. Enters., Inc., 953 P.2d 1089, 1096-97 (N.M. Ct. App. 1997).

62. Shawcross v. Pyro Prods., Inc., 916 S.W.2d 342 (Mo. Ct. App. 1995).

63. See Walsh v. Consolidated Freightways, Inc., 563 P.2d 1205 (Or. 1977), Grzyb v. Evans, 700 S.W.2d 399 (Ky. 1985).

Chapter 8

Hazard Communication: Implementation of the Globally Harmonized System in the 21st Century

John B. Flood and Zachary S. Stinson
Ogletree, Deakins, Nash, Smoak & Stewart, P.C.
Washington, D.C.

1.0 Overview

That producers and importers of hazardous chemicals and the employers who use such chemicals are obligated to evaluate and communicate their hazards is certainly well known, as these obligations are firmly rooted within the business practices of industry.[1]   Since 1983, when the Occupational Safety and Health Administration (OSHA) first implemented its standard regarding hazard communication, producers, importers, employers, and workers have grown accustomed to complying with these requirements and have relied upon the resulting increased flow of information to deal properly with hazardous chemicals in the workplace.[2] As OSHA described it, "[t]here is a whole generation of employers and employees now who have never worked in a situation where information about the chemicals in their workplace is not available."[3]

But the previous regulatory scheme of OSHA's Hazard Communication Standard (HCS) is now being displaced because of ongoing efforts to harmonize America's standards with those used internationally. As the world becomes increasingly flat, and globalization and international standardization become the norm instead of the exception,[4] significant changes for American industry are now underway. These changes were not unanticipated given the indications by OSHA made over the years, and the length of time of the rulemaking process. Indeed, the earliest indications of a potential move toward global harmonization in this arena were made by OSHA as early as 1983, when the agency first signaled that it was committed to global unification of hazard communication.[5] Concerns over regulatory change and increased compliance costs prompted OSHA to proceed slowly and judiciously down the path toward global harmony while recognizing the concern that there may be a great economic burden that the changes to the HCS would bring.[6] Nevertheless, that change has now arrived.

After issuing its Advanced Notice of Proposed Rulemaking in 2006,[7] OSHA promulgated its Revised Hazard Communication Standard (RHCS) six years later on March 26, 2012,[8] to conform to the United Nations' Globally Harmonized System of Classification and Labelling of Chemicals (GHS),[9] and the transition period for full compliance continues through June 1, 2016.[10] Although the basic scope and regulatory framework have remained largely the same, the definitions of what constitutes a hazard have changed, along with the requirements for labels and safety data sheets (formerly, "material safety data sheets"), among other changes.[11]

In this chapter, we provide an overview of the original HCS and the regulatory move toward the globally harmonized RHCS. We next review the key purpose and scope of the RHCS, its key requirements, and all of the significant changes as compared with the prior approaches from the HCS. We also consider the regulatory timeframe for implementation and compliance, because, at the time of publication, this transition is still underway, and finally, we consider areas of ongoing concern.

2.0 The Hazard Communication Standard and Push Toward Global Harmonization

2.1 The Hazard Communication Standard

In 1982, OSHA established its Hazard Communication Standard (HCS) based on a purpose best described in a single word—*awareness*. HCS sought to promote workplace safety by ensuring that those employees who were exposed or reasonably like to be exposed to chemicals in the workplace were to be made aware of the hazards, and, accordingly, know how to protect themselves effectively.[12] To that end, the standard required chemical manufacturers and importers to evaluate those hazards and ensure that this information was communicated to the wholesale consumers and end users of hazardous chemicals—the employers and employees.[13]

Although when HCS was first promulgated it applied only to manufacturing companies, OSHA expanded coverage to all employers whose workers could be potentially exposed to hazardous chemicals—a significant extension.[14] OSHA's estimates over time have also revealed the broad reach of hazardous materials into the modern workplace. In 1988, OSHA estimated that about 32 million workers were potentially exposed to chemical hazards in the workplace, and that there were approximately 575,000 existing chemical products.[15] The agency's reports from the last decade showed an increase in these numbers, estimating that:

- As many as 650,000 hazardous chemical products were present;

- In more than three million workplaces; and

- The HCS protected more than 30 million workers potentially exposed to hazardous chemicals.[16]

Accordingly, the rule was that most (if not all) manufacturers and importers of chemicals, as well as employers whose workplaces feature such chemicals, were to be covered by the HCS.

2.2 Problems with HCS

Although many viewed the HCS as a success in terms of the increased awareness and protections it afforded workers since its inception, it was also viewed as deficient in several key ways. First and foremost, the HCS was not fully consistent with the requirements for hazard communication used by other countries—which ranged from countries with advanced hazard communication requirements to many others with only minimal or no requirements at all. Second, the requirements for hazard communication reflected in the HCS were not fully consistent with requirements from other governmental agencies in the United States that regulate the communications requirements for hazards that fall within their purview. Finally, the accuracy and reliability of information contained on Material Safety Data Sheets (MSDSs) and labels for hazardous chemicals had been called into question.

A key reason for the move toward harmonization within the United States and elsewhere was that the HCS was not fully consistent with the requirements for hazard communication used by other countries. Many countries have laws with different requirements concerning hazard communication for chemicals, and naturally, such laws vary greatly depending on their country of origin.[17] Thus, varying requirements existed on an international scale concerning which substances were covered, what was and was not a hazard, what information was required on labels and MSDSs, and what symbols and pictograms were required to convey appropriate warnings.[18] One example is the acute oral toxicity level (LD_{50}) for liquids and substances under the HCS, which differed from the levels established by other countries and international bodies (*e.g.*, Australia, Mexico, Japan, South Korea, and the European Union).[19] Another example includes the flammability of liquids. Different international standards regarding flammability resulted in certain liquids being deemed hazardous under the hazard communication systems of some countries and accordingly labeled as such, while not so deemed and labeled under the systems of others.[20]

Another source of concern that served as a basis for the move toward global harmonization related to the quality of the information that was being disseminated regarding hazardous chemicals. For example, questions persisted about the accuracy of information on MSDSs that were being distributed by employers and manufacturers. Although compliance with the HCS is a high priority for OSHA in terms of enforcement, there is no regulatory body that reviews each MSDS to ensure its accuracy.[21] To be sure, the sheer volume of MSDSs in the stream of commerce warranted the conclusion that some of the information on MSDSs was out of date or incomplete, despite the standard's requirement to update MSDSs within three months of significant changes.

A study commissioned by OSHA in 1991 regarding the accuracy of information contained on 150 MSDSs found that:

- 37% accurately identified health effects data,

- 76% gave accurate information regarding first aid procedures,

- 47% accurately identified appropriate personal protective equipment to be utilized, and

- 47% accurately identified all relevant occupational exposure limits.[22]

While this study indicated the potential need for change, it was limited in terms of its scope. OSHA has noted, perhaps paradoxically to those concerned with the costs associated with moving to a globally harmonized system, that a comprehensive study of the potential problem had never been undertaken because such a study or investigation would have to be "far-reaching, costly, and time-consuming."[23] Because no investigation was undertaken, it is nearly impossible to determine the extent of the problem of inaccurate information in MSDSs.

In addition to concerns about accuracy of information, concerns also existed about whether or not end users of chemicals could readily understand the information contained in MSDSs.[24] Many MSDSs contain complicated technical explanations that may be lost on much of the intended audience.[25] Although the primary audience for MSDSs is safety and health professionals who presumably are trained to understand highly technical explanations and information, MSDSs, like labels, need to convey their information clearly and comprehensively so that it can be understood by experts and non-experts alike.

The MSDS requirements also imposed substantial paperwork burdens on the companies creating them, although the burden lessened as many companies moved toward electronic communications.[26] Despite such increased efficiencies, these burdens continue to exist particularly for multinational companies, as well as medium and smaller companies that engage in international trade of chemicals, which must comply with multiple rules and regulations within the United States and abroad. Certainly, compliance with such requirements requires significant investments of time and money, and served as a basis for consideration of the move toward a globally harmonized system.[27]

Similar concerns also existed about the labeling requirements of the HCS.[28] Employers had a choice as to how they will convey information on a label, as the HCS standard allowed the information on the labels to be "performance oriented."[29] One problem resulting from this is that workers saw many different labels that attempted to convey the same information, thus leading to confusion. There was no specific standard as to the format of labeling, so one symbol may convey different meanings depending on the manufacturer or distributor.[30]

Concerns also existed about the qualifications of those who prepared MSDSs. The HCS did not specify any requirements regarding the qualifications of persons who prepare MSDSs.[31] Naturally, this may have been a less significant problem for larger chemical producers and distributors, which have greater resources available to employ persons with advanced technical degrees and training regarding chemical hazards and the requirements of the HCS or at least to pay outside providers of such services.[32] By contrast, smaller companies may have far fewer resources for these endeavors. In OSHA's view, this was a significant factor in the differing quality of information provided on MSDSs.[33] If such was true within the United States, the same would certainly be true, to a far greater degree, on a global scale.

These problems made the concept of a globally harmonized system for hazard communication very appealing, if such a system would create uniformity in labeling and MSDS requirements.

2.3 Move toward GHS and RHCS

When the HCS was first enacted in 1983, its preamble articulated OSHA's commitment to move toward global harmonization. Despite OSHA's early statement of support for a globally harmonized system, the Globally Harmonized System of Classification and Labeling of Chemicals (GHS) did not formally progress until June of 1992, when a mandate from the United Nations Conference on Environment and Development called for the development of a globally harmonized system concerning chemical hazards.[34] This mandate led to the creation of the Coordinating Group on the Harmonization of Chemical Classification Systems under the umbrella of the Interorganization Programme for the Sound Management of Chemicals. OSHA chaired the coordinating group, and took the lead for the United States on classification of mixtures and hazard communications.[35]

Through this combined effort, the Globally Harmonized System of Classification and Labeling of Chemicals was created and adopted by the United Nations Committee of Experts on the Transport of Dangerous Goods and the Globally Harmonized System of Classification and Labeling of Chemicals in December of 2002.[36] The Globally Harmonized System of Classification and Labeling of Chemicals is a system for standardizing and harmonizing the classification and labeling of chemicals on a global/international basis.[37] The intended benefits of the GHS include the provision of an internationally comprehensible system for hazard communication and providing a "recognized framework for those countries without an existing system."[38] These benefits further include the reduction of the need for testing and evaluation of chemicals, and the promotion of international trade in chemicals "whose hazards have been properly assessed and identified on an international basis."[39]

Importantly, the GHS is not a regulation or a standard. Rather, it provides a "framework" from which competent authorities "may select the appropriate harmonized classification and communication elements. Competent authorities of participating nations will decide how to apply the various elements of the GHS within their systems based on their needs and the target audience."[40] As explained herein, the flexibility inherent within the GHS may ultimately serve to limit its ability to produce the desired benefits of a globally harmonized system.

3.0 Revised Hazard Communication Standard

3.1 Key Purpose and Scope

Like manufacturers and importers of hazardous chemicals, employers have significant obligations under the RHCS, the first of which is to develop and maintain a written hazard communication program in the workplace.[41] As discussed below, such programs encompass the employer's obligations to make sure that their employees are aware of chemical hazards in the workplace including the potential effects of exposure to those chemicals, through the proper use and maintenance of labels and Safety Data Sheets and through the provision of information and training.[42]

3.2 Key Requirements

The standard's initial responsibilities lie with the chemical manufacturers and importers of chemicals, who must classify chemical hazards before sending these products into the stream of commerce within the United States. Chemical manufacturers are employers "with a workplace where chemical(s) are produced for use or distribution," and an importer is "the first business with employees within the Customs Territory of the United States which receives hazardous chemicals produced in other countries for the purpose of supplying them to distributors or employers within the

United States."[43] Importantly, distributors of hazardous chemicals are also covered by the HCS and also have a duty to transmit information concerning hazardous chemicals.[44]

The hazard-classification approach has departed from the HCS substantially in the RHCS. Manufacturers and importers must first determine the hazard classes, and the category of each class that apply to the specific chemical.[45] Whether a given chemical is hazardous is based upon the "available scientific evidence concerning such hazards."[46] Beyond this, manufacturers and employers must identify and consider the "full range of available scientific literature and other evidence concerning the potential hazards."[47] Manufacturers and employers are under no requirements to test the chemical in order to make a determination on how to classify its hazards.[48] Moreover, the RHCS has added the definition of "classification," which means to identify and review relevant data regarding a chemical's hazards, and then to decide whether the chemical will be classified as such.[49] The new definition also includes making a determination on the degree of the hazard, where one exists.[50]

Beyond this initial evaluative process, the RHCS strives to promote awareness of chemical hazards in the workplace at various stages of their progression in the stream of commerce through three key components: the use of labels, Safety Data Sheets, and the provision of information and training to employees against the backdrop of a larger hazard communication program.

3.2.1 Labels

The RHCS imposes additional obligations upon manufacturers, importers, and distributors once a chemical is deemed hazardous through labeling requirements. Labels are a critical component of the RHCS in terms of ensuring that employers and employees are aware of the presence of hazardous chemicals in the workplace, because labels are usually the first tool to alert users or handlers to the presence of hazardous chemicals. OSHA has aptly described the role of a label as that of a "snapshot" intended to serve as the initial reminder to workers that materials may be hazardous.[51]

Whereas the HCS had a "simple and performance-based approach" regarding the labeling system, the RHCS has overhauled it with detailed and specific provisions for labeling. Specifically, chemical manufacturers and importers must provide detailed labels that include product and supplier identifiers, as well as signal words, pictograms, and hazard statements for each class, in addition to "precautionary statements," as set forth below.[52]

Before a container with hazardous chemicals leaves the manufacturing plant or the place of export, it must be "labeled, tagged, or marked" to identify the hazardous chemical contained therein.[53] Hazards not otherwise classified do not need to be included on the label.[54] Generally speaking, the labeling requirement applies to "each container of hazardous chemicals."[55] Manufacturers, importers, and distributors must also ensure that labels comply with the requirements of any other OSHA health standards which specifically apply to a given hazardous chemical.[56]

The HCS did not mandate that any specific format for labels be used, although the standard did require certain minimal information such as the identity of the hazardous chemical, appropriate hazard warnings, and the name and address of the chemical manufacturer, importer, or other responsible party.[57] A commonly used format for labels was found in the national consensus standard adopted by the American National Standards Institute (ANSI) for the preparation of labels.[58] Under the RHCS, product identifiers, signal words, hazard statements, pictograms, precautionary statements, and the name, address and telephone of the manufacturer, importer, or other responsible party must now be included on the label.[59]

Employers also have obligations for labels for hazardous chemicals once they arrive at the workplace. Employers must ensure that each container of hazardous chemicals is "labeled, tagged, or marked" to convey the information required of the manufacturer or importer, or must further provide the general information regarding the chemical hazards.[60] To satisfy this second obligation, employers can use "words, pictures, symbols, or a combination thereof," so long as they provide at least "general information regarding the hazards of the chemicals" and "provide employees with the specific information regarding the physical and health hazards of the hazardous chemical" when combined with other information given to employees as part of the employer's hazard communication program.[61] However, there is a new mandate that specific information for each hazard class and chemical be used.[62] The RHCS also provides employers with flexibility in terms of the labeling requirements concerning the use of individual stationary process containers, as well as for portable containers into which employees place hazardous chemicals for their own immediate use during a single shift.[63] Employers cannot remove or deface existing labels on containers of hazardous chemicals which they receive, unless they immediately mark the container with another appropriate label. The labels or other types of warnings must be legible and written in English.[64]

Manufacturers, importers, distributors, and employers all have a continuing obligation to provide accurate information on labels for hazardous chemicals. The RHCS also requires revisions to labels for chemicals within three months of becoming aware of any new, significant information regarding the hazards of chemicals, and updates to labels accordingly on subsequent shipments of the affected chemicals.[65]

3.2.2 Safety Data Sheets

If labels are intended to serve as the initial reminder to employers and employees of the presence of hazardous chemicals, then Safety Data Sheets (SDSs) are designed to serve as the RHCS's more substantive means of ensuring awareness of hazardous chemicals in the workplace. To that end, the RHCS requires chemical manufacturers, importers, and employers to develop and maintain SDSs for hazardous chemicals.[66] This is one of the more noticeable and significant areas of change from the HCS's old Material Safety Data Sheets. In an effort to bring standardization to the SDSs, OSHA has adopted the approach that is nearly identical to the ANSI standards for the sheets.[67]

The RHCS sets forth the key components that must be contained within every SDS, which include:

- Identification, including identification of hazards;

- Composition/information on ingredients;

- Measures to take, including first aid, fire-fighting procedures, and accidental release protocols;

- Handling and storage, including exposure controls and personal protection;

- Physical and chemical properties of the substance, as well as its stability and reactivity;

- Toxicological, and ecological information, and considerations regarding its disposal;

- Transport information;

- Regulatory information; and

- Other information, including date of preparation or last revision.[68]

This information is required on every SDS, to the extent the information is known, and the RHCS requires that the specific format set forth in the Standard be used when presenting this information. OSHA has explained that the previous HCS was designed to be a "performance-oriented" standard, and that it did not mandate the use of a specific format for the MSDSs in order to accommodate companies in various industries that already used MSDSs of varying formats.[69] While OSHA had recommended formats for MSDSs, under the new standard, the focus on uniformity for the SDSs now requires adherence to the Standard and specific additional requirements from the appendices.[70] Indeed, much of the previous HCS requirements for the MSDS categories of information are now found, in their revised and more detailed form, in Appendix D. OSHA's stated rationale for the new standard's SDS requirements is to improve the effectiveness of its hazard communications through homogeneity, and allow for easier compliance from employers. OSHA's previous recommendation for format compliance, ANSI, now forms the basis for the new requirements for SDSs.

Chemical manufacturers or importers must provide distributors and employers with an appropriate SDS with the initial shipment of a hazardous chemical and with the first shipment of such after an SDS is updated.[71] Similarly, distributors must provide SDSs to other distributors and employers with their first shipment of a hazardous chemical and with their first shipment of such after updated information is available.[72] Employers must maintain SDSs in the workplace for each hazardous chemical and must ensure that they are "readily accessible" during each shift to employees in the areas where they work.[73]

As with labels, each SDS must be written in English, although the use of other languages in addition to English is permissible.[74] Manufacturers, importers, distributors, and employers all have a continuing obligation to provide accurate information on SDSs for hazardous chemicals. The RHCS requires that employers and distributors revise SDSs for chemicals within three months of becoming aware of any new, significant information regarding the hazards of chemicals.[75]

3.2.3 Training and Information for Employees

The third key component of the RHCS to ensure awareness of hazardous chemicals in the workplace is the requirement that employers provide information directly to employees and train them about hazardous chemicals in the workplace. Employers must provide employees with "effective information and training" regarding hazardous chemicals in the workplace at the time of their first assignment to a work area and whenever a chemical hazard is introduced into the area for which the employee has not been trained.[76] Information that must be provided to employees includes:

- The requirements of the RHCS,

- Any operations in their work area where hazardous chemicals are present, and

- The location and availability of the employer's written hazard communication program (including the required list of hazardous chemicals and SDSs).[77]

Employers must train employees on ways to detect the presence or release of hazardous chemicals in their work areas and must provide employees with the knowledge of the physical, health, simple asphyxiation, combustible dust, and pyrophoric gas hazards, as well as hazards not otherwise classified, of the chemicals in the work area.[78] Employers must also train employees on measures that employees can take to protect themselves against these hazards and inform employees about procedures the employer has adopted to protect employees against exposure. Employers must also provide a hazard communication program that includes an explanation of the labels received on shipped containers and the workplace labeling system used by their employer; the safety data sheet, including the order of information and how employees can obtain and use the appropriate hazard information..[79]

3.2.4 Hazard Communication Program

The RHCS requires employers to develop and maintain a hazard communication program at each work place to provide information to their employees about the chemical hazards present in the workplace.[80] The hazard communication program must explain how the employer will satisfy its obligations concerning labels, SDSs, and the provision of information and training under the RHCS.[81] The program must also include a list of the hazardous chemicals that are known to be present in the workplace. Chemicals should be identified with a product identifier that is referenced on the appropriate SDS, and the employer will inform employees of the "hazards of non-routine tasks . . . and the hazards associated with chemicals contained in unlabeled pipes in their work areas."[82]

The RHCS also contains special provisions concerning the hazard communication program of employers on multi-employer worksites. It requires an employer on a multi-employer worksite (e.g., "Employer A") to ensure that its hazard communication program addresses how it will provide other employers at the same worksite with information and access to its SDSs for hazardous chemicals to which these other employees may be exposed. The RHCS also requires Employer A to provide information on how it will advise other employers of any necessary precautionary measures to protect all employees on the worksite and information on how Employer A will advise other employers of the labeling system used at the site.[83]

3.2.5 Other Changes under RHCS

Other substantial changes between the HCS and RHCS include the evaluation of mixtures. Chemical manufacturers, importers, or employers evaluating chemicals are now required to follow the procedures described in Appendices A and B to §1910.1200 to classify the hazards of the chemicals, including determinations regarding when mixtures of the classified chemicals are covered by this section.[84]

Another key concern with implementation of the RHCS is the costs associated with the changes.

3.3 Implementation of RCHS

The United States generally agreed to work toward implementation consistent with the goals of these international groups.[85]

Key deadlines for implementation of RHCS are as follows:[86]

- December 1, 2013 – Employers must train employees on the new label elements and SDS format.

- June 1, 2015 – Chemical manufacturers, importers, distributors and employers must comply with all modified provisions of the final rule for RHCS, except:

- December 1, 2015 – Distributors shall not ship containers labeled by the chemical manufacturer or importer unless it is a GHS label.

- June 1, 2016 – Employers must update alternative workplace labeling and hazard communications programs as necessary, and provide additional employee training for newly identified physical or health hazards.

During the transition period for implementation, and prior to the aforementioned implementation dates, manufacturers, importers, distributors, and employers may comply with the provisions of the HCS or the RHCS (or both).[87] When a violation exists under either the HCS or the RHCS, OSHA will issue citations under both standards.[88]

On July 9, 2015, OSHA's newly issued enforcement memorandum for the RCHS went into effect.[89] The directive instructs that manufacturers and importers are to make good-faith efforts to obtain and integrate updated information to comply with the RCHS, and that inspectors may request documentation of the efforts made to obtain information from upstream suppliers, alternative sources, and classify data themselves.[90] Furthermore, inspectors have been instructed to examine employers' written programs to ensure compliance and implementation.[91] Moreover, the directive instructs inspectors to consider citations for violations of the RHCS under OSHA's general-duty clause.[92] These trends from OSHA could increase further the regulatory concerns from industry with respect to the implementation and enforcement of the RHCS.

OSHA's recent enforcement activities reveal that it has been, and will continue to take enforcement action for violations of the core provisions of the RHCS, largely consistent with historical standards. For Fiscal Year 2014, for example, violations of the core provisions of the RHCS were the second-most cited violations by OSHA.[93] However, as these statistics predate the final implementation of the standard and the recently issued enforcement directive, the ultimate impact remains to be determined.

3.4 Areas of Continuing Concern

In response to its notice of proposed rulemaking in 2006, the initial step to promulgating the RHCS, OSHA received more than 100 comments from a variety of sources, including large multinational corporations, smaller corporations and businesses from a variety of industries, and various groups and associations within the field of safety and health.[94] These comments serve to highlight a number of key concerns with the implementation of the RHCS.

One key concern is the impact that the flexibility that the RHCS (and the international GHS) will bring to the table. While flexibility can be a positive attribute of a regulation or regulatory system, the inherent flexibility within the GHS framework for these standards may prove detrimental to the achievement of its fundamental goal of global harmony concerning chemical hazards if nations choose to implement GHS in different ways.[95]

Under the GHS, *harmonization* is defined as "establishing a common and coherent basis for chemical hazard classification and communication, from which the appropriate elements relevant to means of transport, consumer, worker, and environment protection can be selected."[96] The goal of harmonization clearly should not be mistaken for a fully uniform system of hazard communication, however, as the flexibility inherent in the GHS framework means that not all countries and competent authorities that adopt the GHS will do so in a uniform manner. The lack of full harmony that the GHS would bring because of its inherent flexibility may ultimately prove to be a barrier to achieving the goals of harmonization, despite the perceived benefits that the move to the RCHS will bring.

A second significant concern relates to the costs that the implementation of the RHCS will bring, which would be most significant during the process of implementing its requirements, which is ongoing.[97] The costs of moving to RHCS would vary significantly depending upon the role of companies in relation to hazardous chemicals, with the highest costs falling to producers of such chemicals. Long-term costs may also be impacted to the extent that the GHS is not truly harmonized on an international scale.

4.0 Conclusion

Since 1983, producers and importers of hazardous chemicals, as well as employers utilizing such chemicals in the workplace, have increasingly relied upon the requirements of the HCS as a source of stability for protecting workers exposed to hazardous chemicals. This stability did not, however, negate several problems that existed with the HCS and its requirements, many of which grew increasingly problematic as the flow of hazardous chemicals has continued to increase within the stream of international commerce. While changes associated with the RHCS may ultimately prove beneficial, the process of implanting its requirements and the enforcement by OSHA will almost certainly prove costly and confusing, nd the flexibility of the RHCS (and the international GHS) could hinder the primary goal of globally harmonizing hazard communication if the system is not adopted among different states and competent authorities in a truly harmonized manner.

Notes

1. OSHA, Hazard Communication in the 21st Century Workplace, March 2004, Executive Summary, available at https://www.osha.gov/dsg/hazcom/finalmsdsreport.html.

2. The first HAZCOM Standard went into effect on November 25, 1983. 48 Fed. Reg. 53, 280 (1983) (codified as amended at 29 C.F.R. § 1910.1200). In 1988 and 1989, OSHA expanded coverage of the HCS from only the manufacturing industry to all employers and estimated that the standard would cover an additional 18 million workers in more than 3.5 million worksites, at estimated costs of about $687 million for the first year alone. U.S. Department of Labor, Fact Sheet No. OSHA 89-26, June 26, 1989, on file with the author).

3. OSHA, Hazard Communication in the 21st Century Workplace, March 2004, pp. 2, 3, available at https://www.osha.gov/dsg/hazcom/finalmsdsreport.html.

4. See generally Friedman, Thomas L., The World is Flat: A Brief History of the Twenty-First Century (Farrar, Straus and Giroux, 2005).

5. See OSHA, Hazard Communication in the 21st Century Workplace, March 2004, p. 10, available at https://www.osha.gov/dsg/hazcom/finalmsdsreport.html (noting that the preamble to the HCS in 1983 "included a commitment by OSHA to pursue international harmonization of hazard communication requirements").

6. See OSHA, Hazard Communication in the 21st Century Workplace, March 2004, p. 3, available at https://www.osha.gov/dsg/hazcom/finalmsdsreport.html. In 1988 and 1989, OSHA expanded coverage of the HCS from only the manufacturing industry to all employers and estimated that the standard would cover an additional 18 million workers in more than 3.5 million worksites, at estimated costs of about $687 million for the first year alone. U.S. Department of Labor, Fact Sheet No. OSHA 89-26, June 26, 1989, on file with the author).

    See also Peter A. Susser, The OSHA Standard and State "Right-to-Know" Laws: The Preemption Battle Continues, Employee Relations Law Journal (Spring 1985), p. 615, noting that OSHA estimated the rulemaking process for the HCS as the costliest rulemaking in OSHA's history.

7. 71 Fed. Reg. 176, 53618 (Sept. 12, 2006).

8. 77 Fed. Reg. 17,574 (2012) (codified at 29 C.F.R. pts. 1910, 1915 & 1926).

9. 77 Fed. Reg. 17,574 (2012) (codified at 29 C.F.R. pts. 1910, 1915 & 1926).

10. 77 Fed. Reg. 17,582; see also OSHA, Effective Dates https://www.osha.gov/dsg/hazcom/effectivedates.html.

11. See OSHA, Modification of the Hazard Communication Standard, https://www.osha.gov/dsg/hazcom/ hazcom-faq.html#2.

12. 29 C.F.R. § 1910.1200 (a)(1) provides that the "purpose of this section is to ensure that the hazards of all chemicals produced or imported are evaluated, and that information concerning their hazards is transmitted to employers and employees." (Emphasis added.) See also, OSHA Publication 3104, Hazard Communication, A Compliance Kit, 1988, p. A-1.

13. 29 C.F.R. § 1910.1200 (a)(1).

14. OSHA, Hazard Communications in the 21st Century Workplace, March 2004, Executive Summary, p. 4, available at http://www.osha.gov/dsg/hazcom/finalmsdsreport.html.

15. OSHA Publication 3104, Hazard Communication: A Compliance Kit, p. A-1.

16. OSHA, Hazard Communications in the 21st Century Workplace, March 2004, Executive Summary, p. 4, available at http://www.osha.gov/dsg/hazcom/finalmsdsreport.html.

17. OSHA Publication, Hazard Communication in the 21st Century Workplace, March 2004, p. 10, available at http://www.osha.gov/dsg/hazcom/finalmsdsreport.html.

18. OSHA Publication, Hazard Communication in the 21st Century Workplace, March 2004, p. 10, available at http://www.osha.gov/dsg/hazcom/finalmsdsreport.html (cite to UN document, p. 3).

19. OSHA Publication, A Guide to The Globally Harmonized System of Classification and Labeling of Chemicals (GHS), Figure 1.2, available at www.osha.gov/dsg/hazcom/ghs.html.

20. OSHA Publication, A Guide to The Globally Harmonized System of Classification and Labeling of Chemicals (GHS), Figure 1.3, available at www.osha.gov/dsg/hazcom/ghs.html.

21. OSHA Publication, Hazard Communication in the 21st Century Workplace, March 2004, p. 13, available at http://www.osha.gov/dsg/hazcom/finalmsdsreport.html (noting that in fiscal year 2003, OSHA issued over 7,000 citations for violations of the HCS, making it the second most frequently cited OSHA standard and resulting in over $1.3 million in assessed penalties).

22. OSHA Publication, Hazard Communication in the 21st Century Workplace, March 2004, available at available at http://www.osha.gov/dsg/hazcom/finalmsdsreport.html (citations omitted).

23. OSHA Publication, Hazard Communication in the 21st Century Workplace, March 2004, available at available http://www.osha.gov/dsg/hazcom/msds-osha174/msdsform.html.

24. OSHA Publication, Hazard Communication in the 21st Century Workplace, March 2004, p. 6, available at available at http://www.osha.gov/dsg/hazcom/finalmsdsreport.html.

25. OSHA Publication, Hazard Communication in the 21st Century Workplace, March 2004, p. 6, available at available at http://www.osha.gov/dsg/hazcom/finalmsdsreport.html.

26. OSHA Publication, Hazard Communication in the 21st Century Workplace, March 2004, p. 6, available at http://www.osha.gov/dsg/hazcom/finalmsdsreport.html. OSHA noted that the complexity of information provided on MSDSs may also be due in part to products liability lawsuits within the United States based upon a theory of "failure to warn."

27. OSHA Publication, A Guide to the Globally Harmonized System of Classification and Labeling of Chemicals, pp. 2–3, available at http://www.osha.gov/dsg/hazcom/ghs.html.

28. OSHA Publication, Hazard Communication in the 21st Century Workplace, March 2004, p. 8, available at http://www.osha.gov/dsg/hazcom/finalmsdsreport.html.

29. OSHA Publication, Hazard Communication in the 21st Century Workplace, March 2004, available at available at http://www.osha.gov/dsg/hazcom/finalmsdsreport.html.

30. OSHA Publication, Hazard Communication in the 21st Century Workplace, March 2004, p. 8, available at available at http://www.osha.gov/dsg/hazcom/finalmsdsreport.html.

31. OSHA Publication, Hazard Communication in the 21st Century Workplace, March 2004, p. 7, available at available at http://www.osha.gov/dsg/hazcom/finalmsdsreport.html.

32. OSHA Publication, Hazard Communication in the 21st Century Workplace, March 2004, p. 7, available at available at http://www.osha.gov/dsg/hazcom/finalmsdsreport.html.

33. OSHA Publication, Hazard Communication in the 21st Century Workplace, March 2004, p. 7, available at available at http://www.osha.gov/dsg/hazcom/finalmsdsreport.html.

34. Advanced Notice of Proposed Rulemaking, Federal Register/Vol. 71, No. 176/Tuesday, September 12, 2006/Proposed Rules.

35. Advanced Notice of Proposed Rulemaking, Federal Register/Vol. 71, No. 176/Tuesday, September 12, 2006/Proposed Rules, pp. 53618–53619.

36. Advanced Notice of Proposed Rulemaking, Federal Register/Vol. 71, No. 176/Tuesday, September 12, 2006/Proposed Rules.

37. OSHA Publication, A Guide to The Globally Harmonized System of Classification and Labeling of Chemicals (GHS), available at www.osha.gov/dsg/hazcom/ghs.html. The full official text of the GHS (also referred to as "The Purple Book") is available at http://www.unece.org/trans/danger/publi/ghs/ghs_rev00/00files_e.html.

38. OSHA Publication, A Guide to The Globally Harmonized System of Classification and Labeling of Chemicals (GHS), p. 7, available at http://www.unece.org/trans/danger/publi/ghs/ghs_rev00/00files_e.html.

39. OSHA Publication, A Guide to The Globally Harmonized System of Classification and Labeling of Chemicals (GHS), p. 7, available at http://www.unece.org/trans/danger/publi/ghs/ghs_rev00/00files_e.html.

40. OSHA Publication, A Guide to The Globally Harmonized System of Classification and Labeling of Chemicals (GHS), p. 10, available at www.osha.gov/dsg/hazcom/ghs.html. Competent authority means "any national body(ies) or authority(ies) designated or otherwise recognized as such in connection with the Globally Harmonized System of Classification and Labelling of Chemicals (GHS)." Within the United States, for example, "competent authorities" would include OSHA, the Environmental Protection Agency, and the Department of Transportation. Similarly bodies in other countries would also be competent authorities.

41. 29 C.F.R. § 1910.1200 (b)(1).

42. 29 C.F.R. § 1910.1200 (b)(1).

43. 29 C.F.R. § 1910.1200 (c).

44. 29 C.F.R. § 1910.1200 (b)(1).

45. 29 C.F.R. § 1910.1200(d)(1).

46. 29 C.F.R. § 1910.1200 (b)(1); 29 C.F.R. § 1910.1200 (d)(2). Employers who choose not to rely upon the hazard analysis conducted by chemical manufacturers and importers must also comply with this provision.

47. Id. § (d)(2).

48. Id.

49. Id. § (c).

50. Id.

51. OSHA, Hazard Communication in the 21st Century Workplace, March 2004, p. 4, available at http://www.osha.gov/dsg/hazcom/finalmsdsreport.html.

52. Comparison of RHCS and HS, www.osha.gov/dsg/hazcom/side-by-side.html.

53. 29 C.F.R. § 1910.1200 (f)(1).

54. 29 C.F.R. § 1910.1200 (f)(1).

55. 29 C.F.R. § 1910.1200 (f). An exception to this requirement exists for the labeling of solid metal, solid wood, or plastic items that are not exempted from the labeling requirements because of their later use, or for whole grain shipments, for which a label only has to be provided to the customer at the time of the first shipment, so long as the information on the label does not change. 29 C.F.R. § 1910.1200 (f)(2).

56. For example, perchloroethylene used in dry cleaning.

57. 29 C.F.R. §1910.1200.

58. This standard is ANSI Z129.1—1994.

59. 29 C.F.R. §1910.1200(f)(1)(i)–(vi).

60. 29 C.F.R. § 1910.1200 (f)(5).

61. 29 C.F.R. § 1910.1200 (f)(5)(ii).

62. App'x C.

63. 29 C.F.R. § 1910.1200 (c),(f)(8).

64. 29 C.F.R. § 1910.1200 (f)(9), (10). Employers with employees who speak languages other than English may add information in those other languages, so long as it is still provided in English.

65. 29 C.F.R. § 1910.1200 (f)(11).

66. 29 C.F.R. § 1910.1200 (g).

67. See Hazardous Workplace Chemicals-Hazard Evaluation and Safety Data Sheets and Precautionary Labeling Preparation (ANSI Z400.1 & Z129.1 - 2010).

68. 29 C.F.R. § 1910.1200 (g)(2). Additional protections are afforded to help protect trade secrets when identifying hazardous chemicals on an SDS. See 29 C.F.R.§ 1910.1200 (i).

69. OSHA, Hazard Communication in the 21st Century Workplace, March 2004, p. 4, available at http://www.osha.gov/dsg/hazcom/finalmsdsreport.html.

70. This standard is ANSI Standard Z400.1.

71. 29 C.F.R. § 1910.1200 (g)(6)(i).

72. 29 C.F.R. § 1910.1200 (g)(7)(i).

73. 29 C.F.R. § 1910.1200 (g)(8).

74. 29 C.F.R. § 1910.1200 (g)(2).

75. 29 C.F.R. § 1910.1200 (g)(5).

76. 29 C.F.R. § 1910.1200 (h)(1).

77. 29 C.F.R. § 1910.1200 (h)(2).

78. 29 C.F.R. § 1910.1200 (h)(3)(i), (ii).

79. 29 C.F.R. § 1910.1200 (h)(3)(iii), (iv).

80. 29 C.F.R. § 1910.1200 (b)(1), (e)(1).

81. 29 C.F.R. § 1910.1200 (e)(1).

82. 29 C.F.R. § 1910.1200 (e)(1)(i), (ii).

83. 29 C.F.R. § 1910.1200 (e)(2)(i–iii).

84. 29 C.F.R. § 1910.1200 (d)(3)(i).

85. Advanced Notice of Proposed Rulemaking, Federal Register/Vol. 71, No. 176/Tuesday, September 12, 2006/Proposed Rules, p. 53619.

86. OSHA Publication, Effective Dates, available at
https://www.osha.gov/dsg/hazcom/effectivedates.html.

87. See Inspection Procedures for the Hazard Communication Standard (HCS 2012),
OSHA Enforcement Directive Number CPL 02-02-079 (July 9, 2015), available at
http://www.osha.gov/dep/index.html.

88. Id. at 7.

89. See generally id.

90. See id. at 43.

91. Id. at 30.

92. Id. at 86–88.

93. See Top 10 Most Frequently Cited Standards for Fiscal 2014 (Oct. 1, 2013, to Sept. 30,
2014), available at https://www.osha.gov/Top_Ten_Standards.html.

94. See Hazard Communication Docket H022K, available at
http://ecomments.osha.gov/search/browseEx-hibits.asp.

95. See Hazard Communication Docket H022K, Exhibit 2-7, p. 3, "However, if the GHS
is not truly harmonized and states/countries can continue to add requirements and
modify the GHS, then we have traded one confusing system for another."

96. OSHA Publication, GHS-OSHA HCS Comparison, Comparison of Hazard
Communication Requirements, available at
http://www.osha.gov/dsg/hazcom/GHSOSHAComparison.html.

97. See, for example, Hazard Communication Docket H022K, Exhibit 3-26, p. 2
(estimating the costs of training and adaptation to the GHS at over $8 million within
the newspaper and printing industries); Exhibit 3-9 (estimating transition costs of at
least $426,000 for Chevron Phillips Chemical Co. for software enhancements, updating
MSDSs and labels, and publishing new documents in multiple languages); and Exhibit
2-2 (noting that the costs required to update their new chemical database system used
to prepare shipping documents for chemicals and the costs of training on the new
requirements would be significant).

Chapter 9

Voluntary Safety and Health Self-Audits

Margaret S. Lopez
Gwendolyn K. Nightengale
Ogletree, Deakins, Nash, Smoak & Stewart, P.C.
Washington, D.C.

1.0 Overview

With the complex array of federal and state regulations that reach nearly every aspect of every company's operations, it is not surprising that many companies find self-auditing of their regulatory compliance to be an essential business practice.[1] This is certainly true with regard to workplace safety and health.

Quite clearly, safety and health audits have become an integral procedure for a majority of enterprises. The fact is, well-planned and executed periodic audits can provide company management with an excellent means of identifying safety and health hazards and correcting such conditions promptly. Audits may serve as the central monitoring device for a company's safety and health program. Auditing and diligent follow-up can enhance workplace safety and health and can serve to demonstrate a company's commitment to safety. Audits also allow companies, over time, to identify trends in their long-term efforts to improve safety and compliance with regulations.

With the increasing employer utilization of self-auditing procedures, there has been some concern about the ways in which audit information may be used against the employer in regulatory agency enforcement activities as well as in criminal and civil litigation. This chapter briefly discusses the broad significance of voluntary self-auditing as a component of a company's safety and health program. The chapter then considers the legitimate concerns about the potential for self-audit information to be used as adverse evidence against the company. The chapter describes the Occupational Safety and Health Administration's (OSHA's) voluntary self-audit policy, which the agency developed to respond to those concerns. Finally, other potential legal protections for audit information are discussed.

2.0 The Significance of Voluntary Safety and Health Auditing

Safety and health audits are a comprehensive approach to monitoring and assessing safety protections and ensuring compliance with government regulations. There are important reasons why almost any employer might choose to employ auditing in its safety program. Reasonable screening for violations and prompt elimination of problem conditions not only serve to protect safety and health, but also put the employer in a position of being able to demonstrate diligence in its regulatory compliance efforts. Such a showing of diligence could help counter allegations of negligence in legal proceedings against the employer. In this connection, it is noteworthy that some insurance providers strongly encourage the use of auditing by reducing premiums for companies with strong safety and health programs that include auditing.

As safety and health regulations have come to cover more and more aspects of the workplace, and as workplaces themselves, in many instances, have become more complex, auditing has become a significant tool for monitoring regulatory compliance and identifying potential safety and health hazards in the workplace. Auditing gives the company the ability literally to inspect itself and thereby find and resolve problems before they become legal liabilities.

2.1 Overview of Audits

2.1.1 The Audit Team

Safety and health audits can take many forms. Often they are conducted by teams of company personnel who specialize in safety and health. Many companies also use outside consultants to audit company operations. Outside consultants offer two advantages over exclusive use of in-house personnel. First, consultants can provide technical expertise needed to identify hazards company personnel may miss. Second, they can provide an independent review of the company's operation.

Legal counsel are also often involved in conducting safety and health audits. The company's lawyers can assist in evaluating whether a condition identified by technical members of the audit team actually would constitute a violation under legal precedent. They also can assist, from a legal perspective, in assessing the severity of the conditions observed and can provide guidance, again from a legal perspective, for evaluating potential remedies for the hazards identified. As is discussed in more detail later in this chapter, the involvement of legal counsel in the audit also may provide the basis for application of evidentiary privileges for protecting certain audit information from mandated disclosure to third parties.

2.1.2 Scope of the Audit

Audits vary considerably in their scope. They may extend from a limited audit (focusing on a part of an operation or compliance with a single regulation) to a comprehensive audit covering the entire operation and the full panoply of OSHA regulations. They may be conducted infrequently, or they may be conducted on a regular, scheduled basis. The frequency of audits may be a function of the particular company's size and ability to dedicate resources. Larger companies may be more dependent on audit procedures than smaller companies to ensure that safety requirements are being met. No matter how frequently an employer conducts audits, the important thing is to ensure that a procedure is put in place to promptly respond to any deficiencies, violations, or hazards that are discovered. Generally speaking, the scope of an audit should never be greater than the employer's capability for immediate response.

2.1.3 Audit Information

Depending upon their scope and the methods used to conduct them, audits may produce a considerable amount of information. A variety of tools may be used to gather this information. Some of the more common are interviews, first-hand observations, technical tests and measurements, and questionnaires. This information is usually preserved in the form of notes of the audit team members, test reports and other technical data, diagrams, photographs, and videos. Before the audit begins, there are often memoranda and perhaps an audit-planning document that detail the scope and outline the steps for conduct of the audit.

Once the information-gathering phase is over and the audit team has had a chance to evaluate the data, most audit teams will create a formal report. The report will detail the audit team's findings and recommendations for correction of hazards. The report, therefore, typically contains not only descriptions of what the team found, but also the team's analysis of those findings and its opinions about corrective action to be taken.

After the audit is finished, other documents may be created that reflect the company's response to the audit findings and recommendations. These documents may show management's consideration of alternative actions, either proposed in the audit itself or beyond those the audit team recommended. These documents may also describe the manner and timing of correction of the problems identified in the audit.

Certainly, much of the information generated in the course of an audit is important to the audit's serving its intended function of improving safety and health at the operation. In many instances, a company's audit information demonstrates that the company has a solid safety and health program in place. It shows that the company regularly conducts thorough audits and in a timely manner addresses the deficiencies found in those audits.

2.2 Auditing Tips

An exhaustive list of "dos" and "don'ts" for auditing are beyond the scope of this chapter. However, a few tips to avoid legal pitfalls are offered here for general guidance.

2.2.1 Take Steps to Protect Confidentiality of Audit Information

In some circumstances, it may be in the company's best interest to resist efforts of outside parties seeking to force disclosure of the company's audit information—not necessarily because the company has done anything wrong, but because the audit may be used in a way that may turn the company's good intentions and efforts against itself.[2] Thus, from a legal perspective, part of the audit planning should include consideration of what documents will be created and how their confidentiality will be maintained. The reasons and methods for this are discussed in further detail later in this chapter. We mention it here, however, because disclosure issues should always be in the forefront of counsel's mind when advising a company on conducting a safety and health audit.

2.2.2 Be Prepared to Promptly Respond to Every Hazard Identified in the Audit

It is essential that the company commit, in advance of the audit, to remedy every hazard that will be identified in the audit in a timely way. For obvious reasons, conducting a voluntary audit and then ignoring the problems found is worse than not auditing at all. Care must also be taken if upper management disagrees with an audit team's conclusion that something is unsafe. Prudence would dictate that management accept and address the conclusions of the audit team without reservation. Once an issue is raised in an audit, it cannot be ignored without peril. More than one company has found itself facing charges after a serious accident because audit warnings were ignored or rejected.

2.2.3 Document Every Significant Step Taken to Respond to Hazards

Whenever a company has documented problems found, it is important to also document promptly that the problems were resolved in a timely manner. If the resolution will take some time, the company should create a contemporaneous record of each step taken to resolve the problem. In this way, the company will be able to show its prudence and good faith in responding to hazards that have come to its attention.

2.2.4 Do Not Censor the Auditors

From the preliminary planning phase through preparation of the last document concerning the audit findings and recommendations or remediation, management must not censor audit documents. It is one thing to decide in advance that, for valid reasons, the audit will only address a limited list of compliance issues, but it is quite another thing to instruct an auditor to ignore a hazard that was found in the audit (or to edit a valid finding out of a draft report).

Again, once the company has committed to conducting an audit, the company must be willing to face up to the findings, whatever they are. Covering up findings will only escalate legal problems for the company should the censorship later come to light in an enforcement or litigation action.

It is also imperative that audit documents not be destroyed as a means of covering up evidence of a violative condition. To avoid even the appearance of possible destruction of evidence, audit documents should only be destroyed in the ordinary course of document destruction in accordance with the company's record retention and destruction program.

2.2.5 Attribute Appropriate Gravity to Audit Findings and Recommendations

The audit report and all other audit-related documents should neither exaggerate nor understate problems found in the audit. Audit documents should treat the findings and recommendations in a manner commensurate with their actual importance. This will provide the proper foundation for later measurement of the appropriateness of the company's response, should that be necessary.

3.0 OSHA's Voluntary Self-Audit Policy

3.1 Purpose

In recognition of the important role that voluntary audits play in maintaining effective workplace safety and health programs (and in response to considerable political pressure[3]), OSHA issued a formal policy on agency use of company audit information.[4] The policy statement, titled *Final Policy Concerning the Occupational Safety and Health Administration's Treatment of Voluntary Employer Safety and Health Self-Audits*, was published in the *Federal Register* in July 2000. The policy states that "the Agency will not routinely request self-audit reports at the initiation of an inspection, and the Agency will not use self-audit reports as a means of identifying hazards upon which to focus during an inspection."[5] It does not appear that the policy has greatly affected companies' use of self-audits in their safety programs, however. As noted above, many companies choose to involve their legal counsel in the self-audit process for evidentiary privilege purposes. However, even documents prepared or reviewed by an attorney will not be protected to the extent that the audit is required by law or regulation. Still, OSHA's policy is at least an official commitment (albeit with caveats) toward enforcement restraint in support of the agency's stated goal to encourage voluntary auditing.

3.2 Scope

OSHA's audit policy only applies to audits that are "systematic, documented, and objective reviews conducted by, or for, employers to review their operations and practices to ascertain compliance with the Act."

Further, the policy only applies to voluntary audits. It does not apply to audits that are required to be conducted by the Act, by agency regulations, or by a settlement agreement.[6]

The policy extends to information gathered in the course of a voluntary self-audit, as well as to "analyses, conclusions, and recommendations" resulting from the audit. Thus, the term *voluntary self-audit report*, as used in the policy, is given a broad definition, including not only the audit report itself but, presumably, other documents containing audit-related information.

3.3 Provisions

3.3.1 Use of Self-Audits in Agency Inspections

There are four major declarations in the policy. The first states that "OSHA will not routinely request voluntary self-audit reports at the initiation of an inspection." In other words, "OSHA will not use such reports as a means of identifying hazards upon which to focus inspection activity." This declaration is OSHA's attempt to respond to a major concern of employers that they were essentially creating a roadmap in company audit records of areas of the operation for OSHA inspectors to focus on in an enforcement inspection. Even so, the word *routinely* has still left some employers nervous that OSHA has left itself a large loophole. In fact, it appears that this policy has operated as a check-and-balance on the compliance officer in the field, prohibiting an inspector from using self-audit results as a roadmap during his inspection unless he first receives permission from the OSHA area director.

The policy goes on to state that "if the Agency has an independent basis to believe that a specific safety or health hazard warranting investigation exists, OSHA may exercise its authority to obtain the relevant portions of voluntary self-audit reports relating to the hazard." Here lies a larger loophole.[7]

3.3.2 No Citation for Corrected Conditions

In the second declaration, OSHA states that it "will not issue a citation for a violative condition that an employer has discovered as a result of a voluntary self-audit, if the employer has corrected the violative condition prior to the initiation of an inspection (and prior to a related accident, illness, or injury that triggers the inspection) and has taken appropriate steps to prevent a recurrence of the violative condition that was discovered during the voluntary self-audit." This statement illustrates the importance of documenting actions taken to respond to hazardous conditions identified in an audit, particularly actions that are aimed at preventing recurrence of the hazard.

3.3.3 Protection from Use of Self-Audits to Show Willfulness

The third declaration, the so-called safe harbor provision, provides that "if an employer is responding in good faith to a violative condition discovered through a voluntary self-audit and OSHA detects the condition during an inspection, OSHA will not use the voluntary self-audit report as evidence that the violation is willful." (A willful violation is one in which the employer intentionally violated a requirement of the Act and has either shown reckless disregard for the possibility of a violation or has shown plain indifference to safety and health of employees. Penalties for such violations may range from a minimum of $5,000 to $70,000.)[8]

OSHA explains that "this policy is intended to apply when, through a voluntary self-audit, the employer learns that a violative condition exists and promptly takes diligent steps to correct the violative condition and bring itself into compliance, while providing effective interim employee protection, as necessary." Again, note the importance of good documentation with respect to intermediate responsive steps.

3.3.4 Penalty Reduction for Good Faith

The final declaration provides that the agency will "treat a voluntary self-audit that results in prompt action to correct violations found [as described in the safe harbor provision], and appropriate steps to prevent similar violations, as strong evidence of an employer's good faith with respect to the matters covered by the voluntary self-audit." Accordingly, the agency will reduce by up to 25% the amount of the penalty that would otherwise be assessed for the violation.[9] The policy defines *good faith* as "an objectively reasonable, timely and diligent effort to comply with the requirements of the Act and OSHA standards."[10]

3.4 Limitations

In issuing the policy, the agency was careful to state in the policy itself that it is internal guidance only and is not legally binding: "The policy statement is not a final Agency action. It is intended only as a general, internal OSHA guidance, and is to be applied flexibly, in light of all appropriate circumstances. It does not create any legal rights, duties, obligations, or defenses, implied or otherwise, for any party, or bind the Agency." Although one court in the Western District of Wisconsin required the agency to comply with this policy on the grounds that it created a reasonable expectation of privacy that businesses rely on in conducting internal safety audits, an employer may find it difficult to successfully hold the agency to the provisions of this policy in any form of legal action.[11]

3.5 Critique

3.5.1 "Routine" Use

Since the policy has been issued, the regulated community has been concerned about OSHA's commitment to refrain only from "routinely" requesting audit information at the outset of an inspection.[12] The policy is completely silent as to what the agency would consider a "nonroutine" situation, in which OSHA would ask to see a company's audit report.[13] Indeed, the agency asserts in the supplementary information to the policy that compliance officers must be given discretion in this regard. This leaves employers without guidance as to when the agency will or will not request to see company audit information.

3.5.2 Use of Audit Information to Supplement Other Evidence Already Found

The policy states that OSHA may seek company audit information in circumstances in which the agency already has other evidence of a violation. Thus, there remain many situations in which the agency will still use company audits as further evidence to prove a violation or to establish the gravity of a violation—all the more reason companies must be diligent about timely correcting hazards identified in audits.

3.5.3 Penalty Reduction

The 25 percent penalty reduction offered for good faith demonstrated by self-auditing is no more than the agency already offered prior to implementation of the policy. The policy simply publicizes the fact that voluntary audits may be used to find good faith. It would have been more effective in terms of encouraging auditing to have offered a greater penalty reduction than that already in place.

4.0 Privileges & Protections from Disclosure of Audit Information

4.1 Introduction

There are certainly many instances in which it is in the company's interest to disclose safety and health audit information to OSHA and other third parties. As already discussed, audit information may be used to establish good faith and thereby avoid a finding of a willful violation. Under OSHA's self-audit policy, the fact the company has performed an audit and taken prompt and effective action to correct the hazards found may result in a 25% penalty reduction. Audit information may also be helpful to the company in the civil and criminal litigation context in demonstrating that the company has a strong commitment to safety and health in the workplace.

Unfortunately, there are also many situations in which it is not in the company's best interest to be forced to disclose its audit information. For this reason, company counsel must be knowledgeable about the protections available, how to secure and preserve them, and what their limitations are. This section discusses three of these protections: the self-audit privilege, the attorney/client privilege, and the attorney work product doctrine.

4.2 The Self-Audit Privilege

The self-audit privilege, at least in its common law version, is seldom found by courts to be available to protect safety and health audits. As is explained in greater detail later in this section, this is partly because courts tend to regard the privilege as unnecessary to encouraging employers to conduct audits and because courts also tend to find that the public good is better served by requiring disclosure of audit information. The rare statutory version of this privilege (where such a statute exists) would presumably be a far more reliable protector of audit information. Whether a company has available to it a statutory audit privilege or must rely on the less dependable common law version, it is important for legal counsel to know what this privilege is and to be thoroughly familiar with its limitations.

4.2.1 The Common Law Audit Privilege

4.2.1.1 General Description

By *common law* audit privilege, we mean the privilege that is created by the courts, as opposed to the privilege that is created by the legislature. This is the form of the privilege that we look exclusively to the courts to define.[14]

Courts are very reluctant to create new privileges. This is because generally they do not want to stand in the way of full discovery in litigation.

In the federal courts, privileges are recognized pursuant to Rule 26 of the Federal Rules of Civil Procedure and Rule 501 of the Federal Rules of Evidence.[15] Rule 26 of the Federal Rules of Civil Procedure provides that privileged information is protected from discovery.[16] In other words, a court cannot require a party to reveal to another party privileged information. Rule 501 of the Federal Rules of Evidence provides that evidentiary privileges shall be created and interpreted in the common law, subject to exceptions such as where Congress has created a privilege through statute.

The purpose of the common law audit privilege is to encourage companies to audit their operations concerning subject areas in which there is a strong public interest that accurate and candid self-audits be conducted. In situations where the privilege is applicable, some or all of a company's audit information will be protected from disclosure. In other words, the company will not be required to reveal privileged audit information to an opponent in discovery.

4.2.1.2 Factors Used in Determining Whether to Apply the Privilege

Unfortunately, the audit privilege has been inconsistently defined by the courts that do apply it, and many courts refuse to even recognize that such a privilege exists. Those courts that do recognize the privilege (often in contexts other than safety and health) will usually consider the following factors in determining whether the privilege will apply: (1) whether the information at issue was generated in the course of a self-audit conducted by the company; (2) whether the company intentionally preserved the confidentiality of the information; (3) whether there is a strong public interest in encouraging audits of this type to be conducted; and (4) whether there is a strong likelihood that not applying the privilege in this context will discourage companies from conducting these types of audits.[17]

This first factor, whether the information at issue was generated in the course of a self-audit conducted by the company, may appear to be an easy one to satisfy, but often it is not. Some courts have found that audits of an investigatory nature (as opposed to a pure compliance review audit) do not qualify for the privilege because the courts find that the company would have conducted these audits regardless of whether a privilege was available.[18] Some courts have refused to apply the privilege to protect self-critical information that was generated in the normal course of business of the company.[19] Again, the courts reason that since the company finds it to be a good business practice to regularly audit its operation, the company does not need a privilege to encourage auditing. Auditing is simply a necessity to running the business.

Confidentiality is a critical factor for securing the privilege. Like other evidentiary privileges, this privilege is considered waived if the information at issue has been disclosed to third parties.[20] Therefore, it is imperative that the company restrict distribution of audit information to those who need that information to assist the company in achieving its objectives in conducting the audit. This may include outside consultants, but even within the company, distribution must be limited. A simple aid for preserving confidentiality and (should the need arise) demonstrating the company's intent to keep the information confidential is to mark all audit documents "confidential."

The third and fourth factors are related to the purpose of the privilege, which is to protect the public interest in having companies conduct such audits. As discussed above, under these factors, courts will consider the nature of the audit and whether employers will refrain from conducting such audits in the absence of the privilege.[21]

4.2.1.3 Other Limitations in Application of the Audit Privilege

Even where this privilege is found to apply, it is generally limited to protection of opinions and subjective analysis, not facts.[22] Therefore, the common law privilege would most likely not protect information showing that a hazard existed or that management personnel knew the hazard existed, but it may protect opinions of auditors about the scope of the hazard and recommendations for corrective action.

This is a qualified privilege. The privilege can be overcome by the opposing party showing that it has a compelling need for the information—for example, by showing that the information is critical evidence and that there is no other reasonably available means of independently obtaining such information.

The audit privilege generally does not apply to protect against discovery by the government.[23] This is obviously a severe limitation in the safety and health context. This limitation is based on the principal purpose of the common law privilege, which is to further a public interest. Unlike other privileges, this privilege does not serve to protect the owner of the information. Courts, therefore, generally conclude that it is not in the public interest to preclude access to such information by a government agency.

4.2.2 Statutory Audit Privilege

There is no federal statute that provides an audit privilege in the safety and health context. However, there have been multiple bills introduced in Congress over the years that, had they been enacted, would have created such a statutory privilege.[24] At the state level, at the present time, at least two states have enacted legislation creating an audit privilege for safety and health audits.[25]

4.3 The Attorney/Client Privilege

Because the self-audit privilege is generally unavailable to protect company safety audits, companies must look to other protections from disclosure. The only protections likely to be available require attorney involvement in the audit for the protections to apply. If the company wants to rely on these other protections, it is necessary before the audit commences to secure the involvement of company counsel in the audit.

The purpose of the attorney/client privilege is to facilitate candid communication between attorneys and their clients, unfettered by concerns that such communications may be discoverable. This goal is accomplished by the availability of the attorney/client evidentiary privilege, which protects from required disclosure confidential communications to provide legal services between client and attorney.[26]

This privilege will apply to protect all such communications between the client and attorney, whether those communications consist of fact or opinion. It is important to note, however, that merely sending a document to counsel does not cloak that document with the privilege. The document must have been created for the purpose of assisting the attorney in providing legal advice. Thus, merely copying company counsel on the final audit report, without further involvement from counsel in the planning and conduct of the audit and in making recommendations based on that report, will not be sufficient to protect the report from disclosure.[27]

Counsel must be actively involved in the audit process for audit information to be protected by the attorney/client privilege.[28] It is best if company counsel actually leads the audit, at least by directing that the audit be performed, reviewing the audit findings and recommendations, and adding legal conclusions and advice to the audit report. Since the privilege does not apply to lawyers providing business advice, it is critical that the lawyer's role in the audit process be to provide legal advice to the company.[29]

This does not mean that counsel must participate in all aspects of the audit. The privilege will extend to communications with and between members of the audit team (even if outside consultants are used) as long as the ultimate purpose of those communications is to enable company counsel to provide legal advice to the company (presumably on compliance and legal liability issues).[30] In this regard, it is best if company counsel is actively involved in the selection and direction of audit team members.[31]

As with the self-audit privilege, the attorney/client privilege will be considered to have been waived if audit information is not kept strictly confidential. The same methods to preserve the confidentiality of audit information for purposes of the audit privilege apply to preserve the attorney/client privilege. Limiting distribution within the company may be as important as preventing disclosure outside the company.

Unlike the audit privilege, the attorney/client privilege is absolute. It cannot be overcome by a claim that the party seeking the information has a compelling need for that information.

4.4 Attorney Work Product Doctrine

The work product doctrine is more limited than the attorney/client privilege. It will protect confidential opinions and analyses of the attorney that appear in audit information only if such material was created in anticipation of litigation.[32] This is a substantial limitation for safety and health audits because they often are not conducted in anticipation of litigation. The fact that they may be conducted with an eye to avoiding the remote possibility of litigation in the future has been found by some courts to not be enough to justify application of the work product doctrine.[33]

The work product doctrine generally does not protect factual information, only opinions and analysis of counsel.[34] Therefore, the work product doctrine will not protect all of a company's audit information.[35]

Like the audit privilege, the protection from disclosure that the work product doctrine otherwise may provide can be overcome by a showing of substantial need on the part of the party seeking access to the information.[36] If that party can show that it can obtain the information only through extraordinary means or that it cannot obtain the information from any other source and that the information is critical to its case, a court may find that the information must be handed over.

5.0 Conclusion

Given their reliance on workplace audits to improve safety and health and to monitor regulatory compliance, employers need to consider what protections may be available in the event they have a need to protect audit information from being used as adverse evidence in an agency enforcement action or in litigation. In most circumstances, safety and health audits will not be protected by a self-audit privilege. Other protections are also limited and usually necessitate substantial involvement of counsel. These include the attorney/client privilege and the attorney work product doctrine.

In view of the foregoing, it is in the interest of every employer to treat all audit activities and reports with a view to their possibly becoming public or at least available to adverse parties in litigation. Therefore, it is critical that companies conducting audits follow through appropriately and ensure prompt correction of every deficiency found. In this way, companies will fully benefit from the auditing process, and the safety and health of their workplaces will be enhanced.

Notes

1. Of course, companies also conduct self-audits for reasons not directly related to monitoring regulatory compliance (e.g., to monitor compliance with company policies or to look for potential problems not related to regulatory concerns). Since this book concerns OSHA law, this chapter concentrates on audits in the OSHA regulatory compliance context.

2. For example, the company may not have had sufficient time to correct an identified condition before it would be required to disclose its audit documents in a government inspection or investigation or in a legal proceeding.

3. Numerous bills have been introduced in Congress in the past that would have created an audit privilege in one form or another and would have required OSHA to employ other incentives to encourage auditing. *See, for example,* Safety and Health Audit Promotion Act of 1999, H.R. 1438, 106th Cong. (1999); Safety and Health Audit Promotion and Whistleblower Improvement Act of 1999, H.R. 1439, 106th Cong. (1999); Safety and Health Improvement and Regulatory Reform Act of 1995, H.R. 1834, 104th Cong. (1995).

4. OSHA, Final Policy Concerning the Occupational Safety and Health Administration's Treatment of Voluntary Employer Safety and Health Self-Audits, 65 *Fed. Reg.* 46,498-46,503, July 28, 2000.

5. *Id.*

6. Many standards require companies to self-inspect to monitor compliance and retain and make available to the agency related records of monitoring. *See, for example,* 29 C.F.R. §§ 1910.119, 1910.120, 1910.1025(d), (e), 1926.20(b).

7. The authority that OSHA is purporting to rely on must be carefully examined in each case. If there is no OSHA regulation requiring a document, such as an audit report, OSHA compliance officers may not be able to compel its production without a search warrant or subpoena. At the same time, it must be remembered that even if the agency does not seek to obtain audit information at the time of inspection, the audit report may be demanded later in discovery during litigation concerning any citations that may be issued based on other evidence.

8. 29 U.S.C. § 666(a).

9. The Act provides that an employer's good faith is to be considered by the Occupational Safety and Health Review Commission in assessing civil penalties. 29 U.S.C. § 666(j). OSHA's Field Inspection Reference Manual authorizes the agency to reduce a civil penalty by up to 25% when the agency finds the employer acted in good faith.

10. The Act provides that an employer's good faith is to be considered by the Occupational Safety and Health Review Commission in assessing civil penalties. 29 U.S.C § 666(j). OSHA's Field Operations Manual authorizes the agency to reduce a civil penalty by up to 25 percent when the agency finds the employer acted in good faith.

11. *See Grede Wisconsin Subsidiaries,* 24 OSHC 1061, 1063 (W.D. Wis. 2013).

12. See "OSHA Employer Audit Policy Applauded, but Some Questions Remain, ASSE Says," *Occupational Safety and Health* (BNA) Vol. 30, No. 33 Aug. 17, 2000 at 757.

13. In the draft of the self-audit policy statement that OSHA published for comment in October, 1999, the agency provided examples of situations in which it might seek audit information. Proposed Policy Statement Concerning the Occupational Safety and Health Administration's Use of Voluntary Employer Safety and Health Self-Audits, 64 *Fed. Reg.* 54,358-54,361 (October 6, 1999). The examples provided were where the agency was investigating a fatal or catastrophic accident or where the agency has reason to believe a hazardous violation exists and is investigating the extent of the hazard.

14. Since many of the significant cases concerning evidentiary privileges have arisen outside the safety and health context but are nevertheless applicable to safety and health audit privilege issues, this section cites cases that do not necessarily concern safety and health audits per se.

15. The Occupational Safety and Health Review Commission's Rules of Procedure provide that the Commission will apply the Federal Rules of Civil Procedure, unless the Commission Rules have a specific provision covering a subject, and they provide that the Federal Rules of Evidence apply to Commission proceedings. 29 C.F.R. § 2200.2(b), 2200.71.

16. The analogous Review Commission Rule provides, "The information or response sought through discovery may concern any matter that is not privileged." 29 C.F.R. § 2200.52(b).

17. *See* Peter A. Gish, "The Self-Critical Analysis Privilege and Environmental Audit Reports," 25 *Envtl. L.* 73, 80-82 (Winter 1995); Donald P. Vandegrift Jr., "Legal Development: The Privilege of Self-Critical Analysis: A Survey of the Law," 60 *Alb. L. Ref.* 171, 187 (1996).

18. *See Davidson v. Light*, 79 F.R.D. 137, 139-40 (D. Colo. 1978). *But see Dowling v. American Hawaii Cruises, Inc.*, 971 F.2d 423, 427 (9th Cir. 1992).

19. *See Dowling*, 971 F.2d at 426. *But see Hickman v. Whirlpool Corp.*, 186 F.R.D. 362 (N.D. Ohio 1999).

20. *See Peterson v. Chesapeake & Ohio Ry. Co.*, 112 F.R.D. 360, 363 (W.D. Mich. 1986).

21. *See ASARCO v. NLRB*, 805 F.2d 194 (6th Cir. 1986).

22. *See Price v. County of San Diego*, 165 F.R.D. 614, 619 (S.D. Cal. 1996).

23. *See Reich v. Hercules, Inc.*, 857 F. Supp. 367, 371 (D.N.J. 1994).

24. See, e.g., Safety and Health Audit Promotion Act of 1999, H.R. 1438, 106th Cong. (1999); Safety and Health Audit Promotion and Whistleblower Improvement Act of 1999, H.R. 1439, 106th Cong. (1999); Safety and Health Improvement and Regulatory Reform Act of 1995, H.R. 1834, 104th Cong. (1995).

25. Ariz. Rev. Stat. Ann. § 12-2323 (2013); Environmental, Health, and Safety Audit Privilege Act, Tex. Rev. Civ. Stat. Art. 4447cc(2005).

26. *See Upjohn v. United States*, 449 U.S. 383, 389-90 (1981).

27. *See Hercules, Inc.*, 857 F. Supp. 367, 372 (D.N.J. 1994).

28. *Id.*

29. *Milk Specialties Co.*, 854 F. Supp. 2d 629, 632 (E.D. Wis. 2012).

30. *See In re: Grand Jury Matter*, 147 F.R.D. 82, 84–86 (E.D. Pa. 1992).

31. *See In re: Grand Jury Matter*, 147 F.R.D. 82, 84-86 (E.D. Pa. 1992); *Delek Refining*, 23 OSHC 1567, 1568-1569 (July 2011).

32. *See Hickman v. Taylor*, 329 U.S. 495, 510–511 (1947); Fed. R. Civ. P. 26(b)(3)

33. *See Martin v. Bally's Park Place Hotel & Casino*, 983 F.2d 1252, 1260–1261 (3rd Cir. 1993).

34. *See Hickman*, 329 U.S. at 511 (1947); *Hercules*, 857 F. Supp. at 373. *But see Bally's Park*, 983 F.2d at 1261–1262.

35. *See Hickman v. Taylor*, 329 U.S. 495, 511 (1947); *Hercules*, 857 F. Supp. at 373.

36. *See Bally's Park*, 983 F.2d at 1255.

Chapter 10

Inspections and Investigations

R. Lance Witcher
Ogletree, Deakins, Nash, Smoak & Stewart, P.C.
St. Louis, MO

1.0 Introduction

The Occupational Safety and Health Administration (OSHA) utilizes workplace inspections and investigations as its primary means of enforcing safety standards. Such inspections are necessary because they are the only effective way in which OSHA can monitor compliance under the Occupational Safety and Health Act. Although OSHA has relatively broad authority with respect to conducting workplace inspections and investigations, that authority is not without limits. Both employers and employees have constitutional and statutory rights that must be observed during any inspection or investigation.

Although the motivations behind each OSHA inspection are different, the basic process of any OSHA visit is largely the same. Upon arriving at the worksite, the OSHA compliance safety and health officer (compliance officer) (1) holds an opening conference, describing the purpose and scope of the inspection; (2) performs a physical walkthrough of the facility, specifically the areas of the facility within the scope of the inspection; and (3) conducts a closing meeting in which s/he describes his/her findings and reviews any potential citations.

This chapter (1) discusses the types of inspections and investigations conducted by OSHA; (2) describes the process with which the compliance officer will conduct the inspection, including the rights of both OSHA and the employer during each of these steps, and (3) details what an employer can expect following the inspection, along with the employers ability to contest any findings.

2.0 Types of Inspections and Investigations

Generally, OSHA investigations fall into one of five general categories: (1) Programmed Inspections; (2) Complaint Investigations; (3) Referral Investigations; (4) Imminent Danger Investigations; and (5) Fatality/Catastrophe (FAT/CAT) Investigations. Although an OSHA visit may take on different designations, it is best to recognize that most OSHA events are actually investigations, regardless of their official designation.

2.1 Programmed Inspections

OSHA generally conducts these inspections of facilities on a random basis within industry sectors. The decision regarding which sectors OSHA will focus its resources is usually driven by accident rates, with those sectors with high accident rates getting the most attention. Additionally, if an employer or industry poses a special risk or hazard, OSHA may conduct more inspections with respect to those particular hazards. Programmed inspections may either be part of a special or local emphasis program, or conducted to follow up on a priority enforcement case under Federal OSHA's Enhanced Enforcement Policy.

2.2 Complaint Investigations

OSHA is required, under Section 8(f)(1) of the Act, to follow up on complaints filed by employees or other concerned parties. Complaints can be submitted to OSHA in person or by phone, fax, mail, or through the agency's internet site. When a complaint complies with the formality requirements of section 8(f) (i.e., it is reduced to writing and set forth with reasonable particularity the grounds for the notice, and is signed by the employee) it is considered a formal complaint.[1] Complaints that do not comply with these formality requirements are considered informal complaints. Formal complaints will generally trigger an investigation while informal complaints generally result in OSHA sending a letter to the employer called a Notice of Alleged Safety or Health Hazard. The employer is obligated to investigate and respond to the complaint quickly, typically within five days. In a response letter, the employer must describe the results of its investigation and note corrective action taken or planned. A failure to meet these obligations will likely trigger an OSHA inspection.

2.3 Referral Inspections

OSHA sometimes conducts a "referral" inspection following receipt of information from another federal or state agency, a union, any non-employee, or based upon media coverage of an accident or other event.

2.4 Imminent Danger Inspections

OSHA conducts an on-site investigation whenever it receives a report of imminent danger (an immediate threat that an employee will suffer death or serious physical harm). Similar to complaint investigations, these are usually triggered by an employee complaint under section 8(f)(1) of the Act. However, because the complaint has alleged that a workplace condition poses an imminent danger it is given the highest priority by OSHA. Imminent danger is defined in Section 13(a) of the Act to mean a danger "which could reasonably be expected to cause death or serious physical harm immediately or before the imminence of such danger can be eliminated through the enforcement procedures otherwise provided by the Act."[2]

2.5 Fatality/Catastrophe (FAT/CAT) Investigations

OSHA conducts an inspection following a workplace accident resulting in the death of an employee or the in-patient hospitalization of three or more employees.[3] The Act requires employers to report such incidents to OSHA within eight hours of their occurrence.[4] Once the employer reports such an incident, it can likely expect both an investigation and, subsequently, a citation. A failure to report such an incident is itself a violation.[5]

3.0 OSHA at the Doorstep

The unannounced arrival of an OSHA compliance officer is often a surprise to the employer and presents a couple of options. The employer can (1) consent to the inspection and allow the compliance officer onto the premises, (2) refuse entrance onto the premises and require the compliance officer to obtain a warrant prior to entry (assuming none of the exceptions discussed below are applicable), or (3) challenge the warrant in federal court.

3.1 Consenting to the Inspection

Due to the relative ease with which a compliance officer can obtain a warrant and the multiple exceptions to the warrant requirement, employers generally consent to allow the compliance officer on site to perform the inspection.[6] However, consenting to an inspection does not mean that an employer has abrogated all rights and the compliance officer can perform whatever inspections s/he chooses. For example, in the context of a complaint inspection, most courts allow the employer to limit the scope of the inspection to the violations alleged in the complaint.[7] If the compliance officer exceeds the scope of the complaint, or the scope of employer's consent, the employer may terminate the inspection and/or suppress evidence that went beyond such scope.[8]

Employers also retain the ability to challenge the conduct of the inspection, regardless of whether it is consensual or performed pursuant to a warrant. As discussed more fully below, employers have a right to accompany the compliance officer on his inspection. If that right is denied, the employer can challenge the inspection on the grounds that the officer violated Section 8(e) of the Act and the denial impaired the employer's ability to defend itself at the hearing.[9] Finally, employers can challenge the inspection on the grounds that it was unreasonable and disrupted the employer's business.[10]

3.2 Requiring a Warrant

Section 8(a) of the Occupational Safety and Health Act allows OSHA to "enter without delay and at reasonable times any ... workplace ... and ... to inspect and investigate during regular working hours and at other reasonable times, and within such reasonable limits and in a reasonable manner, any such place of employment and all pertinent conditions."[11] However, this right has been abridged by the U.S. Supreme Court, which held that OSHA's right of entry without a warrant was an unconstitutional infringement on the Fourth Amendment's protection from warrantless searches.[12]

Employers should be cautioned, however, in requiring a compliance officer to obtain a warrant prior to entry because the probable cause standard for OSHA is much lower than that required of a traditional police warrant.[13] Additionally, there are multiple exceptions to this warrant requirement that allow OSHA to enter and inspect the premises without a warrant. To meet this relaxed probable cause standard, OSHA need only show that either (1) there is a reasonable suspicion of a violation[14] or (2) the inspection is being requested under a neutral and reasonable administrative plan.[15]

In the context of an OSHA investigation, reasonable suspicion, or "administrative probable cause," also presents a relatively low bar that the agency must meet. While a mere complaint or allegation from an employee is not sufficient to obtain a warrant, OSHA must only establish "reasonable grounds to believe that a violation or danger exists."[16] In fact, the complaint need not allege a violation of any specific standard, and a violation of the general duty clause is sufficient for OSHA to establish a reasonable suspicion.[17] In light of these low standards for obtaining a warrant, employers generally consent to OSHA inspections.

3.3 Exceptions to the Warrant Requirement

There are three exceptions that allow OSHA to perform an inspection without a warrant: (1) consent, (2) imminent dangers, and (3) plain view. Consent, in the context of an OSHA investigation, is much broader than it otherwise would be. It need not be given voluntarily[18] or affirmatively,[19] and can even be given by a third party on behalf of the employer.[20]

The second exception to OSHA's need for a warrant is an imminent danger. "Imminent dangers" are as defined by the act as "conditions or practices in any place of employment which are such that a danger exists which could reasonably be expected to cause death or serious physical harm immediately or before the imminence of such danger can be eliminated through the enforcement procedures otherwise provided by this Act."[21] Said otherwise, the threat must be such that it poses an immediate risk that an employee will suffer death or serious physical harm.

The third exception is known as the "plain view" or "open fields" exception. Simply stated, a compliance officer does not need a warrant to issue a citation for any violations that s/he observes from a place where s/he has the right to be. Typically this refers to areas that are accessible to the public,[22] but can also include private land if the public is not excluded from it.[23] Under this doctrine, the compliance officer can also use devices such as binoculars or telephoto lenses to make observations, without obtaining a warrant.[24] Any pictures or videos taken while in plain view can be used as evidence on which to base a citation regardless of whether the compliance officer has a warrant or the employer consents to inspect the premises.[25] Also falling within the plain view exception is anything that the compliance officer may observe while performing a consensual inspection.[26] That is, if a compliance officer is touring a specific area of the facility, in response to a complaint, and observes a violation occurring in another area outside of the scope of the inspection, he can issue a citation without a needing to obtain a warrant. This is true even if the area in which he observes the violation is controlled by a separate employer.[27]

3.4 Challenging the Warrant

There are several ways in which an employer may challenge an inspection warrant. First, the employer can go to federal court and get a warrant quashed before it is executed. For example, if the employer turns away a compliance officer who does not have a warrant, it may "beat OSHA to court" and preemptively move to quash the issuance of a warrant. Second, the employer can refuse entry to the compliance officer with a warrant, and then challenge the validity of the warrant in its defense of OSHA's motion for contempt. Third, the employer can allow the inspection to proceed, under objection, and challenge the validity of the warrant in front of the commission.[28] Once the inspection has occurred, the employer, in most jurisdictions, must first exhaust its administrative remedies before the Commission as opposed to initially seeking judicial recourse.[29]

The riskiest avenue for an employer to pursue is refusing to allow a compliance inspector, who already has a warrant in hand, onto the premises. Such a refusal can almost guarantee that OSHA will file a motion for contempt and the employer risks being found in contempt as well as potential liability for the associated attorney fees and costs incurred by OSHA in bringing that motion.[30]

However, allowing the inspection to proceed under a potentially invalid warrant also poses a risk to the employer. First, regardless of the validity of the warrant or the finding of citations, wall to wall inspections can be time consuming for the employer and may tie up management for extended periods of time while the compliance officer is on site inspecting the facility. Second, as discussed above, in the majority of jurisdictions, employers must first exhaust all administrative remedies before the Commission prior to pursuing any action in court.[31] Furthermore, if evidence of violations is found, some courts have found that the exclusionary rule under an invalid warrant applies only to penalties issued against the employer, and not to the issuance of any abatement requirements.[32] Third, the evidence obtained under an invalid warrant might not be excluded if such warrant was obtained in good faith.[33]

4.0 The Inspection Tour

As an initial formality, a compliance officer is required to present his credentials to "the owner, operator, or agent in charge" upon arriving on the premises.[34] There are four basic components to any OSHA visit, regardless of the type: (1) opening conference, (2) tour of the facility or relevant area, (3) reviewing of records, and (4) closing conference.

4.1 Opening Conference

Following the presentation of credentials, the compliance officer is required to "explain the nature and purpose of the inspection; and indicate generally the scope of the inspection and the records...they wish to review."[35] This information is usually presented during an opening conference in the presence of the employer along with a representative of the employees, or union official.[36] With respect to the scope of the inspection the compliance officer will identify the documents and physical location he would like to inspect as well as any other tasks he will be performing during the inspection (such as taking photographs, videos, or interviewing employees). The compliance officer will also present a copy of the complaint, if applicable.[37]

4.2 Physical Inspection

After the opening conference the compliance officer will begin the physical inspection of the areas of the facility s/he designated during the opening conference. Both the employer and employees are permitted to have a representative accompany the compliance officer during the inspection, the latter of which generally comes into play only for unionized facilities.[38] The physical inspection is the stage in which the compliance investigator will obtain the majority of evidence upon which s/he will base citations. Therefore, it is crucial that employers effectively negotiate the scope of this inspection during the opening conference. Moreover, during the walkaround, the compliance officer should be shown only those areas which he/she previously identified during the opening conference. The management representative should not volunteer to expand the scope of the inspection. Moreover, the management representative should answer the compliance officer's questions but should not volunteer extra information. In keeping with the old adage that no good deed goes unpunished, providing the compliance officer with copies of the employer's safety and health program, safety rules or other extraneous information as evidence of the employer's good faith may simply provide the compliance officer with additional fodder for a citation.

Although generally confined to the scope of the inspection identified in the opening conference, the compliance officer has broad authority with respect to types of information s/he can collect and how s/he goes about collecting that information.[39] During the inspection, the compliance officer is authorized to record any and all pertinent information. This can include taking photographs, video, and even attaching monitoring equipment to employees in order to obtain exposure sampling.[40] To the extent that trade secrets are captured during any such recording, protection from public disclosure may be ordered.[41] The employer must be sure to take the same pictures, record the same video, and observe the same readings as the compliance officer. The evidence gathered during this tour is what will be used for any future citations. If the employer relies on the officer's observations, and those observations are inaccurate, it will have no basis to later dispute citations. If possible, and if the search is consensual, the employer should attempt to negotiate that its representative takes the pictures and provide OSHA with copies.

As part of the walk around inspection the compliance officer is permitted to talk privately with non-management employees.[42] Although the employer cannot prohibit its employees from speaking with the compliance officer, it can advise its employees of their rights not to speak with the compliance officer. Employees who do elect to speak with the compliance officer are protected from discrimination and reprisal. Employee interviews can be conducted at any reasonable time or in any reasonable location, so long as they do not "create a risk of injury or unduly disrupt production."[43] Although OSHA has the right to conduct the interviews, employees also have the right to have legal representation present for such an interview[44] or refuse the interview all together. By consenting to an interview, it then becomes voluntary and the employee has a significant amount of control over how it is conducted. The employee can refuse to answer certain questions, refuse to be recorded, or even request that s/he bring someone with him to the interview (including members of the management team or legal counsel).[45] However, if an employee completely refuses to be interviewed, Section 8(b) of the Act grants OSHA the ability to issue subpoenas to compel that testimony. In order to obtain a subpoena, OSHA need only show: "(1) the subpoena must be within OSHA's authority; (2) it must satisfy due process; (3) the information sought must be relevant and material to an OSHA investigation; and (4) the subpoena must not be unduly burdensome."[46]

Management has the right to be present and/or to have its legal counsel present during OSHA interviews of management and supervisory employees, whether on or off company time. Again, however, a management or supervisory employee who provides a statement to the compliance officer is protected from discrimination and reprisal with one exception: if the employee admits to a safety violation or working unsafely, the employer may counsel or discipline the employee accordingly. In general, a management employee is one who has any type of supervisory responsibilities and is paid on a salaried basis. Management's right to be present should not be given up. Most Compliance Officers recognize this right, but even if resistance is encountered, it is not advisable to give up this right. The reason is that statements made by management employees are nearly always binding on the company. They are considered legally as admissions against the employer's interest. As such, it is worth investing the time to prepare management employees for their interviews. They must, of course, be instructed to answer questions truthfully, but carefully. Management employees should be urged to answer only the question asked without volunteering information and to avoid admitting that a certain condition or practice violates an OSHA standard. There is no requirement that an employer allow a manager sign a statement or permit recording of the interview.

The management representative who participates in the walkaround should take detailed notes of what transpires during the walkaround, including comments made and measurements taken by the compliance officer and statements made by any witnesses. The management representative should answer the compliance officer's questions and explain the work that is being performed. The management representative should not argue with the compliance officer but should firmly and politely explain his/her rationale as to why the targeted conduct/condition is not a violation. Additionally, the management representative should not volunteer extra information. In keeping with the old adage that no good deed goes unpunished, providing the compliance officer with copies of the employer's safety and health program, safety rules or other extraneous information as evidence of the employer's good faith may simply provide the compliance officer with additional fodder for a citation. At no time should the compliance officer be left unattended. This is an invitation to the compliance officer to wander around the plant or jobsite looking for other potential violations.

4.3 Document Review

Section 8(c)(1) of the Act requires that employers "make available" those documents that it is obligated to maintain under the Act, such as OSHA 300 logs, the supplemental record, and the annual summary. If the inspection is being performed pursuant to a warrant, the employer should limit the production of documents to those described in the warrant. If the inspection is consensual the employer should utilize the opening conference to negotiate which documents the compliance officer needs and attempt to limit production to those documents that closely relate to the complaint being investigated. However, Section 8(b) of the Act authorizes OSHA to subpoena documents and physical evidence if the employer is unwilling to cooperate with specific requests.[47] While OSHA does not have "an entirely free hand" to issue "fishing trip subpoenas," it must only show "that the documents it requests are relevant to the purpose of an authorized investigation."[48] Whatever the circumstance under which a document is requested, it is best that the employer require the compliance officer to present his/her request in writing. This will help to both organize the requests and ensure there are no mistakes regarding which documents are being sought.

Examples of documents that OSHA will typically request include, but are not limited to, the following:

- The facility's OSHA 300 log for the current and past three years;

- Completed first report of injury forms for each entry on the 300 logs;

- The annual summary of work-related injuries and illnesses;

- Employee safety training records; and

- The company's written safety and health program, including written programs required by OSHA standards, such as a hazard communication program, lockout/tagout program, emergency action plan, and other safety programs that are relevant to the scope of the inspection.

A compliance officer may also make a request for hours worked by employees at a facility, which is a legitimate request for information that OSHA will use to compute injury and illness incidence rates for the facility. Nevertheless, absent a search warrant, the only records employers must produce are the OSHA injury and illness logs and annual summaries.

Accordingly, requests for other documents should be limited to issues identified in the complaint and negotiated with the compliance officer.

4.4 The Closing Meeting

Following the inspection tour and document review, the compliance officer will almost certainly hold a closing meeting with the employer and employee representatives.[49] Unlike failure to hold an opening conference, the compliance officer's failure to conduct a closing conference or to inform the employer of the alleged violations is not considered a denial of due process or grounds for voiding the citations. The closing meeting offers the employer the opportunity to confer with the compliance officer prior to the issuing of any citations. At this meeting the compliance officer will inform the employer of any apparent violations and possible abatement measures the employer could take to correct the violations. The compliance officer will also discuss other matters related to enforcement such as abatement dates, possible penalties or citations, and how to file a notice of contest.[50] The closing meeting also offers the employer to clear up any mistakes or misunderstandings and discuss with the compliance officer the steps it has already taken towards abatement. Because citations may not be received immediately following an inspection, the closing meeting allows the employer to have a good understanding of the possible or probable violations the citation(s) will contain and enable it to begin preparing evidence and/or addressing the areas of concern.

5.0 After the Closing Meeting

Section 9(a) of the Act requires that a citation: (1) describe the violations with particularity; (2) make reference to the standards violated; (3) affix reasonable times for abatement; and (4) impose monetary penalties where appropriate.[51] Upon receiving the citation, the employer should pay particular attention to the time limits as it must contest the citation within 15 working days, if it chooses to do so. The employer should consider contesting everything as doing so allows the employer to bargain, or attempt to settle, with OSHA while all time limits are suspended. Additionally, the employer should strongly consider entering into a voluntary settlement conference with OSHA as the Regional or Assistant Regional Director is authorized to settle contested citations.[52]

Notes

1.   29 C.F.R. § 1903.11.

2.   29 C.F.R. § 1903.13

3.   29 C.F.R. § 1904.39(a).

4.   *Id.*

5.   *Id.*

6.   *See* discussion supra notes 11-27 and accompanying text.

7.   *Marshall v. North American Car Co.*, 626 F.2d 320, 8 OSHC (BNA) 1722 (3rd Cir. 1980); But see Donovan v. Burlington Northern, Inc., 11 OSHC (BNA) 1055 (9th Cir. 1982), cert. denied, 963 U.S. 1207 (1983).

8.   *In re: Inspection of the Workplace Located at 526 Catalan St.*, 741 F.2d 172, 177 (8th Cir. 1984) (finding that an employee complaint did not justify a wall to wall inspection).

9.   *Frank Lill & Son, Inc. v. Sec'y of Labor*, 362 F.3d 840, 846 (D.C. Cir. 2004) (holding that, where the employer was not prejudiced by such a denial, vacatur of the citation is not appropriate).

10. *The National Coal Museum*, 19 OSHC (BNA) 1748, 1752–1753 (Rev. Comm'n 2001) (finding that in order to establish an affirmative defense, the employer must have evidence of unreasonable conduct by OSHA during the Inspection).

11. *See* 29 C.F.R. § 1903.3.

12. *Marshall v. Barlow's Inc.*, 436 U.S. 307 (1978) ("The businessman, like the occupant of a residence, has a constitutional right to go about his business free from unreasonable official entries upon his private commercial property").

13. *Marshall v. Barlow's, Inc.*, 436 U.S. 307, 316 (1978).

14. *See Matter of Midwest Instruments Co.*, 900 F.2d 1150, 1153 (7th Cir. 1990).

15. *See Irving v. United States*, 162 F.3d 154, 168 (1st Cir. 1998) ("OSHA may legitimately devote its limited enforcement resources to monitoring workplaces and working conditions that pose the most serious threats to worker health and safety. OSHA has done so in part, for example, by adopting inspection priorities and an administrative plan to govern programmed general inspections").

16. *Matter of Establishment Inspection of Kelly-Springfield Tire Co.*, 13 F.3d 1160, 1166 (7th Cir. 1994).

17. *Id.*

18. *See Lakeland Enterprises of Rhinelander, Inc. v. Secretary of Labor*, 402 F.3d 739, 745, 21 BNA OSHC 1001 (7th Cir. 2005) ("The ALJ correctly concluded that any Fourth Amendment objection was waived because [the employer] did not object to [the compliance officer's] inspection and request a warrant at the scene").

19. *See Parsons Co.*, 18 BNA OSHC 1462 (Rev. Comm'n 1997) (citing Secretary of Labor v. Sanders Lead Company, 15 BNA OSHC 1640 [1992]) (holding that there is no requirement that a compliance officer inform the employer of its rights to object and demand a warrant).

20. *See LaForge & Budd Constr.*, 16 BNA OSHC 2002 (1994) (consent deemed effective where the city manager, who was the manager of the municipal authority, which owned the site, consented to the inspection over the objection of the contractor who was working on the site).

21. 29 U.S.C.A. § 662.

22. *See Lakeland*, 402 F.3d at 745 (holding that where the site is open and readily observable to the public, there is no expectation of privacy and thus no warrant required).

23. *See Tri-State Steel Constr. Inc.*, 15 BNA OSHC 1093, 1910 (1992).

24. *See L.R. Wilson & Sons, Inc. v. OSHRC*, 134 F.3d 1235 (4th Cir. 1998).

25. *See L.R. Wilson & Sons, Inc.*, 17 BNA OSHC 2059 (No. 94-1546, 1997), aff 'd, 134 F.3d 1235, 18 OSHC (BNA) 1129 (4th Cir. 1998) (no credential-presentation violation). *See also Regional Scaffolding & Hoisting Co.*, 17 BNA OSHC 2067 (No. 93-577, 1997) (no walkaround-accompaniment violation).

26. *See Noble Steel, Inc. v. OSHRC*, 17 OSHC 1465, 1467 (10th Cir. 1995).

27. *Id.*

28. Davis Metal Stamping, Inc., 1982 WL 22726 (OSHRC 1982) (holding that in challenging executed warrants employers must first exhaust their administrative remedies); See also Sturm, Ruger & Co., Inc. v. Occupational Safety & Health Admin., 186 F.3d 63, 64 (1st Cir. 1999) (holding that once an administrative inspection has been completed, courts have generally insisted that administrative remedies be exhausted).

29. The majority position holds that a combination of factors favors the exhaustion doctrine such as the protection of administrative autonomy, deferral to administrative expertise, development of a factual record facilitating subsequent judicial review, and easing the burden on the courts by weeding out cases that become moot. *See e.g. Baldwin Metals Co., Inc. v. Donovan*, 642 F.2d 768, 771 (5th Cir. 1981). However, the minority position, such as that in the Seventh Circuit, does not foreclose the initial review of warrant challenges to inspections that have already been completed. *See e.g. Chromalloy Am. Corp. v. Donovan*, 684 F.2d 504 (7th Cir. 1982) (fourth amendment challenge to inspection based upon warrant executed two years after its issuance).

30. *See e.g. West Point-Pepperell, Inc. v. Donovan*, 689 F.2d 950, 10 O.S.H. Cas. (BNA) 2057, 1982 O.S.H. Dec. (CCH) ¶26275 (11th Cir. 1982).

31. *Smith Steel Casting Co. v. Brock*, 800 F.2d 1329, 1334 (5th Cir. 1986).

32. *Id.*

33. *See In re: Sturm Ruger & Co.*, 20 BNA OSHC 1720, 1726 (holding that evidence gathered pursuant to a warrant obtained in good faith will not be suppressed, even if the warrant is later invalidated); Brock v. Brooks Woolen Co., 782 F.2d 1066 (1st Cir. 1986) (finding that the Secretary acted in objectively reasonable good faith in securing and executing the warrant).

34. *See* 29 U.S.C.A. § 657(a).

35. 29 C.F.R. § 1903.7.

36. 29 C.F.R. § 1903.8(a).

37. *OSHA Field Information Reference Manual*, CPL 2.103 (FIRM) Chapter II (A)(3).

38. 29 U.S.C.A. § 657(e).

39. 29 U.S.C.A. § 657(e).

40. 29 C.F.R. § 1903.7(b).

41. *In re: Establishment Inspection of Kelly-Springfield Tire Co.*; 13 F.3d 110, 16 BNA OSHC 1561 (7th Cir. 1994).

42. *See U.S.C.A.* § 657(a)(2); 29 C.F.R. § 1903.7(b).; 29 C.F.R. § 1903.3(a).

43. *See Urick Foundry v. Donovan*, 542 F. Supp. 82 (W.D. Pa. 1982).

44. *See Secretary of Labor v. Muth*, 34 F.3d 240, 16 BNA OSHC 1984, 1985 (4th Cir. 1994).

45. *Id.*

46. *Id.*

47. See U.S.C.A. § 657(b).

48. *Reich v. Montana Sulphur & Chem. Co.*, 32 F.3d 440, 447 (9th Cir. 1994).

49. 29 C.F.R. § 1903.7(e).

50. FIRM Chapter III (D)(9)(b).

51. 29 C.F.R. § 1903.19.

52. *Id.*

Chapter 11

Understanding and Contesting OSHA Citations

John F. Martin
Ogletree, Deakins, Nash, Smoak & Stewart, P.C.
Washington, D.C.

1.0 Overview

This chapter summarizes the elements of an OSHA citation and explains the process for contesting a citation.

2.0 Understanding the OSHA Citation

At the conclusion of an inspection or investigation, OSHA may, in its discretion, issue one or more citations to an employer for any violations of specific OSHA standards or the General Duty Clause.[1] The agency will send a written citation to the employer, usually by certified mail or, in certain situations, personal delivery. The citation will come from the local OSHA area office that conducted the inspection, and contain an inspection number, date(s) of inspection, the name of the employer, a description of the worksite inspected, and a description of the employer's obligations and rights of the violation(s), and a description of the violation.

U.S. Department of Labor
Occupational Safety and Health Administration
Suite A
5807 Breckenridge Parkway
Tampa, FL 33610
Phone: (813)626-1177 FAX: (813)626-7015

Citation and Notification of Penalty

To:
Sea World of Florida LLC
and its successors
6600 Sea Harbor Drive
Orlando, FL 32821

Inspection Site:
7007 Sea Harbor Drive
Orlando, FL 32821

Inspection Number: 314336850
Inspection Date(s): 02/24/2010 - 08/23/2010
Issuance Date: 08/23/2010

The violation(s) described in this Citation and Notification of Penalty is (are) alleged to have occurred on or about the day(s) the inspection was made unless otherwise indicated within the description given below.

Figure 1: Example of an OSHA citation.

The OSH Act requires OSHA to issue citations with "reasonable promptness." In no event may OSHA issue a citation after the expiration of six months following the occurrence of a violation.

2.1 The Violation

The OSH Act requires the agency to "describe with particularity" the violation(s), which consists of a reference to the standard(s) believed to be violated, and usually a one-to-two paragraph description of the non-compliant condition or conduct found.

Citation 1 Item 1 Type of Violation: **Serious**

29 CFR 1910.23(d)(1)(iii): Flight(s) of stairs with 4 or more risers, less than 44 inches wide and having both sides open were not equipped with one standard stair railing on each side:

Instance a) Employees were exposed to a 10'3" fall hazard in that, a stairway railing system was not installed on the front side left bridge of the Believe stage in Shamu Stadium.

Instance b) Employees were exposed to a 10'3" fall hazard in that, a stairway railing system was not installed on the front side right bridge of the Believe stage in Shamu Stadium

ABATEMENT DOCUMENTATION REQUIRED

Date By Which Violation Must be Abated: 09/02/2010
Proposed Penalty: $ 5000.00

Figure 2: Description of an OSHA citation.

2.2 The Classification

OSHA classifies citations into four categories: (1) Willful; (2) Repeat; (3) Serious; and (4) Other.

2.2.1 Willful Violations

Willful violations are the most severe classification of violation. A violation is "willful" if it is "an act done voluntarily with either an intentional disregard of, or plain indifference to, the Act's requirements."[2]

2.2.2 Repeat Violations

While the OSH Act does not define the term "repeat violation," courts typically require proof that the respondent violated the same standard on an earlier occasion in a substantially similar fashion.[3] The OSH Act contains no proscription on how far back OSHA may look for a Repeat violation, but the agency currently confines itself to a five-year look-back of an employer's citation history nationwide.[4]

2.2.3 Serious Violations

Serious violations are the most commonly-cited violation. OSHA classifies a violation as Serious when the hazard created by the violated standard has a substantial probability of causing death or serious physical harm to an employee.[5]

2.2.4 Other Violations

"Other" violations, sometimes also referred to as "Other Than Serious," are cited when an employee violates a standard, but the hazard is *not* capable of causing death or serious physical harm. OSHA typically reserves Other violations for paperwork violations, such as the failure to keep proper OSHA 300 logs.

2.2.5 De Minimis Violations

The "de minimis" violations borrows from the Latin phrase *de minimis non curat lex*, which literally translates into "the law does take account of trifles."[6] For OSHA, a "de minimis" violation has "no direct or immediate relationship to safety or health."[7] OSHA generally does not issue citations for "de minimis" violations. Instead the agency may send a notice advising the employer of the issue and urging them to correct it, although abatement is not legally required, and no penalty is assessed.

2.3 The Penalty

A citation will also contain a proposed penalty. The amount can vary, depending on the gravity and severity of the violation, including the number of employees exposed to hazard. The OSH Act, however, caps penalty amounts at $70,000 per violation for Willful and Repeat violations, and $7,000 for Serious and Other Than Serious violations.

2.3.1 Factors Considered by OSHA in Calculating Proposed Penalties

OSHA bases their penalty amounts on four factors: (1) gravity of the violation, (2) size of the business, (3) the employer's good faith, and (4) the employer's history of violations. OSHA's proposed penalties in the citation, however, are not binding on the Review Commission.[8] The Review Commission may, in their discretion, defer to OSHA's calculation, but they are free to reject it and devise their own.

2.3.1.1 Gravity

Gravity is the primary factor in OSHA's assessment of a penalty.[9] OSHA decides gravity by analyzing two factors:

- the severity of the injury or illness that could result from a violation; and

- the probability that such an injury or illness will occur.[10]

OSHA classifies severity as either:

- High (Death; permanent disability; chronic, irreversible illnesses);

- Medium (Injuries or temporary, reversible illnesses resulting in hospitalization or a variable but limited period of disability);

- Low (Injuries or temporary, reversible illnesses not resulting in hospitalization and requiring only minor supportive treatment); or

- Minimal (not low, medium or high severity and would not cause death or serious physical harm);[11]

OSHA classifies probability as:

- Greater (likelihood of injury or illness is relatively high); or

- Lesser (likelihood of injury or illness is relatively low).[12]

In assessing probability, the agency reviews a variety of factors, including number of employees exposed, frequency of exposure or duration of employee over-exposure to contaminants, employee proximity to the hazardous conditions, use of appropriate personal protective equipment, and youth and inexperience of employees, especially those under 18 years old.[13]

2.3.1.2 Employer Size

For small businesses, OSHA may reduce the proposed penalty up to 60 percent in most cases, or up to 80 percent in willful cases.[14] The size of the business is determined by the number of employees controlled by the employer in all of its workplaces over a 12-month period.[15] OSHA will not reduce a proposed penalty for employers with more than 250 employees.[16]

2.3.1.3 Employer Good Faith

To reward and incentivize good-faith efforts to implement an effective safety and health management system in the workplace, the agency may reduce the penalty. In considering good faith, OSHA reviews the employer's overall safety and health program. If the agency feels the employer has an effective written safety and health program, covering all relevant OSHA standards, OSHA may reduce the proposed penalty by up to 25percent.[17] OSHA will not reduce penalties for willful and repeat violations.[18]

2.3.1.4 Employer Violation History

If the employer has a "clean" record (no serious, willful, or repeated violations in the last three years), OSHA may reduce a proposed penalty by ten percent.[19]

2.3.1.5 "Egregious" Violations

OSHA ordinarily proposes one penalty for each standard violated, even though multiple employees may be exposed to the violation and even though an employer may have multiple instances of the same violation. For example, an employer's failure to install safety guards on ten identical machines is ordinarily cited and penalized as one violation, not ten. Furthermore, OSHA normally would propose one penalty for a machine guarding violation, regardless of the number of employees using the machine. However, under a policy OSHA developed for "egregious" violations, the agency may treat each instance of a violation as a separate violation, thus multiplying the potential penalty amount. At least one federal court, however, has rejected the idea of citing employers on a per-employee basis.[20]

2.4 Abatement Requirements

The OSH Act requires employers to abate, or correct, violative conditions.[21] The citation will provide a deadline for the employer to abate the citation. It will not include any directive or recommendation for abatement, unless the citation is for a violation of the General Duty Clause, also known as Section 5(a)(1) of the OSH Act. Employers must provide proof of abatement to OSHA by the deadline, although extensions can be obtained, if the employer can persuade OSHA or the Review Commission to grant one. OSHA commonly accepts photographs of corrective action, but sometimes a statement from the employer detailing the abatement methods can suffice. When in doubt, a telephone call to the Compliance Officer can often clarify the ambiguity and provide guidance on required actions needed to satisfy abatement.

3.0 Addressing an OSHA Citation

Any employer receiving a citation must first post the citation in a prominent place, at or near the worksite referenced in the citation. Typically, posting on the employee bulletin board containing the mandatory U.S. Department of Labor postings relating to minimum wage, etc. will suffice. OSHA permits an employer to cover up or redact the penalty amounts.

4.0 What are my Options?

Companies not familiar with OSHA are often unsure of their rights to contest a citation. An invoice is attached to the citation; many assume they cannot appeal the matter, and simply pay the penalty. While accepting the citation is one possible resolution, the employer has the right to contest any citation it receives.

4.1 Accepting the Citation

If the employer decides the citation is valid, or at least not worthy of litigating, the employer need do nothing but abate the hazard within the time specified on the citation and remit the penalty amount to OSHA. Most OSHA citations are uncontested.

4.2 Informal Conference and Settlement

If the employer requests one, OSHA will conduct an informal conference. These conferences are typically scheduled during the fifteen-working-day contest period, and allow the employer and the agency to discuss the citations in an informal setting, usually with an area director or assistant area director. The agency will usually make a settlement offer, generally consisting of lower penalty amounts; sometimes the agency withdraws one or more violations.

Importantly, the informal conference does not suspend or delay the fifteen-working-day deadline for an employer to contest a citation.

4.3 Formal Settlement

If the employer sends OSHA a notice of contest, it may still settle the case at any time, very similar in the manner done at informal settlement. The only prerequisite required is that OSHA and the employer reach an agreement on all material terms and reduce their understanding to a signed, written document. The settlement agreement is submitted to the Administrative Law Judge (and ultimately the Review Commission) for approval. OSHRC approval is usually a formality; the Review Commission rarely rejects settlement agreements.

4.3 Contesting OSHA Citations

If the employer decides to contest the citation, it must service a Notice of Contest on the OSHA office that issued the citation, and within fifteen working days of receipt of the citation. See sections 6.0 and 6.1 *infra* for details.

5.0 Why Should an Employer Contest a Citation?

5.1 Abatement Can Be Significant and Long Term

While the penalty amount may be minor, abatement can become cost-prohibitive. OSHA officials are typically not experts in the employer's industry, so the abatement methods they may request can oftentimes be broad and burdensome. Abatement may require, for example, substantial changes to manufacturing equipment. Or the purchase of new, expensive equipment, which must be well-maintained. Or change processes affecting the employer's other facilities. These costs may be quickly spiral an employer into competitive disadvantage.

5.2 Citations Can Result in Repeat Violations Later

Any citation on an employer's record may be used by OSHA as the basis for a repeat violation. Repeat violations will typically subject an employer to a multiple of five or ten times the previous citation.

5.3 Each Citation on the Employer's Record Increases the Chances of a Willful Violation

If an employer has multiple citations and violations in a brief amount of time, the odds increase that, for subsequent inspections and citations, an OSHA inspector believes the company is willfully or intentionally violating the OSH Act. Who is, after all, more likely to receive a Willful violation: an employer with one violation in the past five years, or an employer with twenty-five violations in the past five years?

5.4 Jail Time

Willful violations of the OSH Act that lead to death can result in criminal indictments. The OSH Act authorizes the United States Government to prosecute employers for such acts, with up to six months in prison for the first offense and up to twelve months for subsequent convictions.[22] The Department of Justice can indict on-site supervisors, but have also previously indicted company officers or directors with no direct oversight of the work that led to the fatality.[23]

5.5 Citations Can Sometimes Be Used Against an Employer in Litigation

Depending on state law, OSHA citations may be used in a variety of ways in civil lawsuits, such as a wrongful death or personal injury actions. For example, in some states, violations of safety standards can be introduced to prove that the employer was negligent per se.[24] In other states, violations may be used as evidence of the employer's negligence.[25]

Workers' compensation laws generally protect an employer from tort claims arising out of employee injuries, providing the exclusive means for an injured employee to recover from his employer. But contractor employees are not restrained by the exclusivity of the workers' compensation system in the claims they may assert. And several states contain exceptions for intentional acts or gross negligence.[26] The harsh language in a willful citation can often used effectively by plaintiffs' attorneys to argue that his or her tort case falls within this exception.

5.6 Citations Can Hinder Business Opportunities and Damage Reputations

When soliciting business and new contracts, prospective customers are more frequently scrutinizing vendors' safety record, including a review of OSHA citations issued to the employer. Citations cannot be concealed; each one is published on OSHA's website, dating all the way back to 1971. Vendors with certain violations or several violations may be disqualified from soliciting business.

In recent years OSHA has implemented a Severe Violators Enforcement Program, known as SVEP, placing employers with significant citations into a "bad actors" program and increasing inspections against those employers. Once in SVEP, it is very difficult for employers to be removed from the program.

In addition, each citation on an employer's record increases the likelihood of damage to the employer's goodwill and business reputation. The more violations on an employer's record, the more likely it is for the employer to be perceived as an unsafe company, scaring away business, lowering morale, and increasing additional scrutiny from OSHA.

6.0 The Process of an OSHA Contest

6.1 Notice of Contest

The Notice of Contest must be in writing. Federal OSHA has no form for a notice of contest. While there are no formalities or no magic words to intone, an employer must adequately identify all aspects of the citation that it wishes to contest -- the alleged violation, the characterization of the violation, the penalty, the abatement, the abatement date, or all of the above. . The notice must be adequate to put OSHA on notice that the employer is contesting either all or at least some part of the citation.

A notice of contest challenging only the penalty essentially waives the ability to challenge the violation, the characterization, and the abatement.[27] A notice of contest challenging only the abatement date is treated as a petition for modification of the abatement period.[28] It is common for employers to contest all items listed in the citation, and all elements of each item – the violation itself, the classification, the proposed penalty, the abatement, and the abatement date.

6.1.1 Fifteen Working Day Deadline to Contest

The Notice of Contest must be served on OSHA within fifteen working days of receipt of the citation. With very few exceptions, a citation not timely contested becomes a final order of the Review Commission, which may not be reviewed by any court or agency.[29]

OSHA starts counting the fifteen-day clock on the day when the citation is received by any agent of the employer. The agency typically sends the citation via certified mail to the closest local office where the alleged violation occurred, but sometimes OSHA will serve citations in person. In large companies, this can create confusion as to when a citation was received, as the citation moves from local offices to the legal and HSE departments. Rather than waste time guessing when the citation was received by the company, the safest practice is to assume OSHA hand-served the citation on the company on the date of issuance listed on the citation, and count fifteen working days from then.

Engaging in settlement discussions with OSHA does not stop the clock on the contest period. In many employers' minds, the contest period create a way-too-short deadline to negotiate a settlement with OSHA. But keep in mind settlement talks can always continue after an employer submits its notice of contest.

On occasion the Review Commission has excused a late of a notice under Federal Rule of Civil Procedure 60(b).[49] Rule 60(b) allows a party to obtain relief from a final order on the following grounds: (1) a "mistake, inadvertence, surprise, excusable neglect, fraud, misrepresentation, or other misconduct of an adverse party has occurred"; (2) it is no longer equitable that the order should have prospective application; or (3) there is not any other reason justifying relief. But such cases are rare. Ignorance of the law, being too busy, or sloppy mail intake do not constitute a "mistake" or excusable neglect warranting relief from a final order under this rule. However, an employer's lateness in filing its notice of contest may be excused if the delay was due to some misconduct by OSHA.[30]

6.2 Review by OSHRC Administrative Law Judge

Once OSHA receives an employer's notice of contest, the agency must immediately forward the citation and contest to the Occupational Safety and Health Review Commission in Washington, D.C.[31] The agency, known as the Review Commission or OSHRC, is frequently mistaken as being part of OSHA. It is an independent federal agency tasked by Congress to resolve contested OSHA citations. Upon receipt of the contest materials, an OSHRC clerk will docket the matter and pass the case on to the Chief Administrative Law Judge. The Chief ALJ will then assign the case to one of the Review Commission's ALJs in Washington, Atlanta, or Denver, and decide whether the case qualifies for Simplified Proceedings.

6.2.1 Docketing

OSHRC dockets a case by assigning it a case number (e.g., the first case docketed on January 2, 2015 will receive the number 15-0001). The Review Commission will send the employer a two-part docketing card with the number. The employer must detach and post one-half of the card, which contains a notice to employees, informing them that the citation is under contest and of their rights to participate in the proceedings. The employer must then date and sign the other half of the card and mail it back to the Review Commission. This second half of the card notifies the Review Commission of the posting. If the employer fails to return the card, the Review Commission will send a reminder. If OSHRC receives no card back from the employer, it reserves the right to dismiss the employer's contest.

6.2.2 Rules of Procedure and Evidence

The Review Commission has their own rules of procedure, commonly known as the Commission Rules.[32] The Commission Rules are a custom-tweaked version of the Federal Rules of Civil Procedure. The Review Commission largely adopts many of the Federal Rules of Civil Procedure, with some exceptions. Discovery, for example, is streamlined in OSHRC proceedings.[33] Depositions may only be taken with leave of court or by consent of all parties.[34]

OSHRC has no separate rules of evidence. Instead, the Review Commission uses and incorporates the Federal Rules of Evidence.[35]

6.2.3 Prehearing

Once the case is docketed by OSHRC, OSHA, through the Office of the Solicitor of the U.S. Department of Labor, files a Complaint.[36] Very similar to a federal court complaint, the pleading details the allegations against the employer, now known as the Respondent. Typically, the Complaint alleges that the Respondent violated the OSH Act, and attaches a copy of the OSHA citations and incorporates them by reference.

The Respondent (employer) must file an Answer to the Complaint, admitting or denying the allegations (or professing a lack of knowledge about allegations, which operates as a denial).[37] The Respondent must also assert any affirmative defenses he wishes to argue at the hearing. Additional affirmative defenses may be amended to the Answer, with leave. The failure to file an answer will typically result in a quick motion for default judgment by OSHA.

Employers do not need to retain an attorney to represent themselves before OSHRC. Any company official can represent the company, which is known as *pro se* representation.[38] But it is strongly advisable to retain counsel.[39] OSHRC proceedings largely mirror federal court procedure. OSHA is always represented by counsel, who generally obtains great success against *pro se* parties.

Once an ALJ is assigned to the matter, he or she will schedule a status telephone conference with the parties to discuss discovery parameters and timing, and to set pretrial deadlines. The ALJ will then issue a prehearing order, containing pretrial deadlines such as a deadline to complete all discovery, a deadline to amend the complaint or answer,

6.2.4 The Hearing

The ALJ will schedule a hearing to hear witnesses and receive evidence from all parties. OSHA will proceed first, and typically call the compliance officer who conducted the inspection as its first witness. The employer/respondent will have the opportunity to cross-examine any of OSHA's witnesses, just as OSHA will have an opportunity to cross-examine the respondent's witnesses. The hearing can continue for days or weeks, until both sides have presented their full cases to the ALJ.

6.2.5 Post-Hearing

At the conclusion of the hearing, the parties have the right to submit post-hearing briefs, summarizing their position with citation to the testimony and evidence admitted at the hearing.[40] Once the briefs are submitted, the ALJ will analyze the entire case and prepare a written ruling. There is no deadline for the ALJ to issue a ruling. Rulings are commonly issued within three to six months of the hearing, but the more complex the case, the longer the time it can take for the ALJ to issue a ruling.

6.2.6 Simplified Proceedings

The Review Commission has streamlined procedures for fairly uncomplicated issues, relatively few citation items, and penalty amounts no higher than $30,000.[41] The procedures curtail and automate pleadings and discovery, with expedited hearing dates. Hearings under Simplified Proceedings are informal; the Federal Rules of Evidence do not apply.

Cases involving fatalities are not eligible for simplified proceedings. Any party may request in writing to use the simplified proceedings, and the opposing party has only a limited right to object to the use of the simplified proceedings rules. Under the simplified proceedings, the parties are required to meet and attempt to resolve their differences. If the matter cannot be resolved among the parties, a hearing will be held before an administrative law judge.

6.3 OSHRC Discretionary Review

6.3.1 Interlocutory Review

A party may seek interlocutory (intermediate) review of an ALJ's ruling by filing a petition with the Review Commission within five days of the ruling.[42] The Review Commission has discretion to grant or deny interlocutory review, and grants such review only for important questions of law or policy and when necessary to prevent disclosure of privileged information.[43] Denial of interlocutory review does not preclude a party from raising the issue again at a later time.

6.3.2 Normal Review

A party adversely affected by an ALJ's decision has no automatic appellate review by the Commission. Instead, the Commission has complete discretion whether to order review of an ALJ decision.[44] Any commissioner may order review of an ALJ decision on his own motion or upon request of a party.[45]

A petition for discretionary review (PDR) must be filed within twenty days following entry of the ALJ's decision.[46] A PDR has no particular form, should concisely explain what portions of the ALJ decision are being challenged and why the Commission should grant review.[47] Common reasons the Commission will grant review are where: the Judge's decision raises an important question of law, policy or discretion; review will resolve a question about which the Commission's Judges have rendered differing opinions; the Judge's decision is contrary to law or Commission precedent; a finding of material fact is not supported by a preponderance of the evidence; a prejudicial error of procedure or an abuse of discretion was committed.[48]

A party generally loses the right to seek judicial review of the Commission's final order if it does not file a PDR.[49] If the Commission directs review, it typically requests briefing. It may also hold oral arguments, upon motion of the parties or its own order, but rarely does so.

Although the Commission has authority to review an entire case, the Commission will ordinarily decline to review issues not specifically raised in the PDR or upon which the ALJ did not have the opportunity to rule.[50] Review by the Commission is *de novo*, meaning that the Commission is not bound by the ALJ's conclusions of fact or law.[51] The Commission does, however, usually defer to the ALJ's determinations about the credibility of witnesses.[52]

6.4 Judicial Review

Within 60 days following the issuance of a final order of the Commission—whether reached as a result of the Commission's refusal to review an ALJ decision or as a result of the Commission's review—an aggrieved party may petition for judicial review in a United States Court of Appeals.[53] Private parties who are adversely affected by a Commission's order may obtain judicial review in the United States Court of Appeals for the circuit either in which the violation is alleged to have occurred, where the employer has its principal office, or in the United States Court of Appeals for the District of Columbia Circuit.[54] The Secretary may appeal only to the United States Court of Appeals for the circuit in which the violation is alleged to have occurred or where the employer has its principal office, not to the D.C. Circuit.[55] As with all cases decided by the U.S. Courts of Appeals, decisions of the appellate court may be reviewed only by the Supreme Court of the United States on petition for writ of certiorari.

The Court of Appeals must affirm the Commission's findings of fact if supported by "substantial evidence on the record considered as a whole."[56] This means the court of appeals usually defers to the Commission's determinations on whether the burden of proof has been met, and the Commission's assessment of witness credibility.

In contrast, the Commission's penalty assessment is reviewed under an "arbitrary and capricious" or "abuse of discretion" standard.[57] This form of review presumes that the Commission's decision is correct.[58] Rarely will an appellate court overturn the Commission's penalty assessments.

6.5 Recover of Costs and Attorney Fees

Under the Equal Access to Justice Act, a private party who prevails in an OSHRC proceeding may be awarded attorney fees, expert witness fees, and other costs unless the Commission or court finds that the agency's prosecution of the action was substantially justified.[59] A party seeking an award of fees must apply to the Commission or court within thirty days of the final disposition of the case.[60] Fees may be recovered only for the portion of time during which OSHA's action was not substantially justified.[61]

7.0 Employee Participation

Affected employees and unions possess various rights to participate in OSHA and Review Commission proceedings, including ALJ hearings. These rights include:

- Compelling employers to post citations for review by employees;[62]

- Objecting to the abatement period of a citation on the ground that it is too long;[63]

- Participation in OSHRC litigation resulting from their employer's contest of a citation or filing of a petition for modification of abatement;[64]

- Conducting discovery, presenting and cross-examining witnesses, and submitting briefs on all issues raised in a case;[65]

- Appealing a Review Commission decision holding that no violation occurred (if OSHA does not object);[66]

- Compelling the Secretary to apprise employees or unions (if they have elected party status) of any settlement negotiations between the employer and the Secretary as to any issue, and to have an opportunity to offer input concerning the proposed settlement before it is presented to the judge for approval;[67] and

- Receiving a copy of any settlement agreement and having ten days to object to the reasonableness of the abatement dates provided in it.[68]

Unions and employees, however, lack the right to object to OSHA's withdrawal of a citation or to object to a settlement of a citation on any grounds other than the reasonableness of the abatement date.[69]

Notes

1. 29 U.S.C. § 658(a).

2. *Dayton Tire v. Sec'y of Labor*, 671 F.3d 1249, 1254 (D.C. Cir. 2012).

3. *P. Gioioso & Sons v. OSHRC*, 115 F.3d 100, 104 (1st Cir. 1997).

4. In previous years OSHA limited its look-back period for Repeat violations to three years, and limited geographic scope to two-to-eight state regions.

5. 29 U.S.C. §666(k).

6. *Sandifer v. U.S. Steel Corp.*, 134 S. Ct. 870, 873 (2014).

7. 29 U.S.C. §658(a).

8. 29 U.S.C. §666(j); *see also Roberts Pipeline Constr., Inc.*, 16 BNA OSHC 2029, 2030, 1993-95 CCH OSHD P30,576, pp. 42,331-32 (No. 91-2051, 1994), aff'd, 85 F.3d 632 (7th Cir. 1996).

9. *See OSHA's Field Operations Manual*, CPL 02-00-148, at 6-3.

10. *Id.* at 6-4.

11. *Id.*

12. *Id.*

13. *Id.* at 6-5.

14. *Id.* at 6-9.

15. *Id.*

16. *Id.* at 6-10.

17. *Id.* at 6-10 to 6-12.

18. *Id.* at 6-10.

19. *Id.* at 6-13.

20. *Reich v. Arcadian Corp.*, 110 F.3d 1192 (5th Cir. 1997).

21. 29 U.S.C. §659(a).

22. 29 U.S.C. §666(e).

23. *United States v. Doig*, 950 F.2d 411, 415 (7th Cir. 1991).

24. *See, e.g., Vickers v. Hanover Constr. Co.*, 125 Idaho 832, 835 (1994).

25. *See, e.g., Toll Bros. v. Considine*, 706 A.2d 493, 498 (Del. 1998).

26. *See, e.g., Tex. Labor Code* §408.001(b).

27. *See Florida East Coast Properties, Inc.*, 1 BNA OSHC 1532 (OSHRC 1974).

28. *Maxwell Wirebound Box Co.*, 8 BNA OSHC 1995, 1997 (OSHRC 1980).

29. 29 U.S.C. §659(a).

30. *Atlantic Marine Co. v. OSHRC*, 524 F.2d 476, 478 (5th Cir. 1975).

31. 29 U.S.C. §659(c).

32. *See* 29 C.F.R. Part 2200. A copy of the Commission Rules are available at the Review Commission's web site, at www.oshrc.gov.

33. *See* 29 C.F.R. Part 2200, Subpart D.

34. *See* 29 C.F.R. §2200.56 (Commission Rule 56).

35. 29 C.F.R. §2200.71 (Commission Rule 71).

36. 29 C.F.R. §2200.34(a) (Commission Rule 34(a)).

37. 29 C.F.R. §2200.34(b) (Commission Rule 34(b)).

38. 29 C.F.R. §2200.22(a) (Commission Rule 22(a)). This is another area where OSHRC litigation differs from federal court litigation. In federal court, a corporation must retain counsel, and a non-attorney company official cannot represent the company.

39. The old adage "the man who hires himself as his attorney has a fool for a client" is especially adept in OSHRC proceedings.

40. 29 C.F.R. §2200.74 (Commission Rule 74).

41. *See* 29 C.F.R. Part 2200, Subpart M.

42. 29 C.F.R. §2200.73 (Commission Rule 73).

43. *Id.*

44. 29 C.F.R. §2200.91 (Commission Rule 91).

45. *Id.*

46. 29 C.F.R. §2200.91(c) (Commission Rule 91(c)).

47. 29 C.F.R. §2200.91(d) (Commission Rule 91(d)).

48. 29 C.F.R. §2200.91(d) (Commission Rule 91(d)).

49. 29 C.F.R. §2200.91(f) (Commission Rule 91(f)); Keystone Roofing Co. v. Dunlop, 539 F.2d 960 (3d Cir. 1976).

50. 29 C.F.R. §2200.92(c) (Commission Rule 92(c)).

51. *See Accu Namics, Inc. v. OSHRC*, 515 F.2d 828, 834 (5th Cir. 1975) ("The statutory scheme contemplates that the Commission is the finder of fact.").

52. *See Wiley Organics, Inc. d/b/a Organic Techs.*, 17 BNA OSHC 1586 (OSHRC 1996); Okland Constr. Co., 3 BNA OSHC 2023, 2024 (OSHRC 1976).

53. 29 U.S.C. §660(b).

54. 29 U.S.C. §660(a).

55. 29 U.S.C. §660(b).

56. 29 U.S.C. §660(a).

57. *Brennan v. OSHRC*, 487 F.2d 438, 442 (8th Cir. 1973).

58. *See id.*

59. 5 U.S.C. §504; 29 C.F.R. §2204.106.

60. 29 C.F.R. §2204.302(a).

61. *Contour Erection & Siding Sys., Inc.*, 18 BNA OSHC 1714 (OSHRC 1999).

62. 29 U.S.C. §§ 658(b), 666(i).

63. 29 U.S.C. §659(b).

64. 29 U.S.C. §659(c).

65. 29 C.F.R. §2200.20 (Commission Rule 20).

66. *See Oil, Chemical & Atomic Workers v. OSHRC*, 671 F.2d 643 (D.C. Cir. 1982).

67. *See General Elec. Co.*, 14 BNA OSHC 1763 (OSHRC 1990).

68. 29 C.F.R. §2200.100(c) (Commission Rule 100(c)).

69. *See Cuyahoga Valley Ry. V. United Transportation Union*, 474 U.S. 3, 6 (1985) ("The Secretary has unreviewable discretion to withdraw a citation charging an employer with violating the Occupational Health and Safety Act."); *Pan American World Airways, Inc.*, 11 BNA OSHC 2003, 2004 (OSHRC 1984) (holding that "a union lacks the right to object to the adequacy of the abatement methods specified in a settlement agreement between the Secretary and an employer, and that a union may object only to the reasonableness of the abatement period specified by the agreement.").

Chapter 12

Criminal Enforcement of Violations

Marshall Lee Miller, Esq.[1]
Baise & Miller, P.C.
Washington, D.C.

1.0 Overview

It is well known that employers who are charged with violating Occupational Safety and Health Administration (OSHA) laws and regulations can face substantial civil penalties. Less well known is that certain violations may also be subject to criminal sanctions.

For other environmental laws, criminal penalties are threatened for a wide variety of offenses. For federal OSHA, however, the applicability of criminal charges is restricted to certain narrow circumstances. Two of these are fairly technical and administrative: First, if an individual gives advanced warning of an inspection,[2] and second, if an individual or employer knowingly makes any false statement, representation, or certification under the OSH Act.[3]

The third category is central to the penalty process itself. For OSHA, criminal prosecution is possible under the law only if an employer's willful violation of an OSHA standard results in the death of an employee,[4] Notice two things about this simple sentence. First, the violation must be "willful" as OSHA defines it, not "knowing" or one of the other almost indistinguishable legal terms that courts are forced to adjudicate Second, only a death triggers this violation, not disabling injury or impairment or other misfortunes that can apply to OSHA's serious level of penalty. That means that if a willful violation "only" causes paralysis, brain damage, amputation of several limbs, or a comatose state without leading to death,[5] the OSH Act criminal sanction do not apply and instead only civil penalties are authorized.

Moreover, unlike civil complaints, which OSHA pursues with its Labor Department staff, criminal charges may only be prosecuted by the United States Department of Justice (DOJ), following a formal referral by OSHA. That requires high-level sign off and is not treated as a routine matter. Then DOJ officials have to concur that a particular case is worthy of prosecutions, merits using scarce departmental resources, and is winnable. Not surprisingly, only a few cases get through this screening process, which is why you don't hear about OSHA criminal sanctions that often.

As a result, in an ironic reversal of roles, unions and other activist groups have pressed for more actions by state prosecutors. While these attempts were originally successful in state courts, some argue that the OSH Act preempts any state efforts to regulate the workplace through criminal enforcement actions. The Supreme Court apparently agrees, finding in *Gade v. National Solid Waste Management Association*[6] that a state's ability to regulate occupational safety and health standards is preempted by the OSH Act, unless approved by OSHA.[7]

These legal constraints, and the requirement of death prior to criminal prosecution, have led to increased pressure on OSHA's sister agency, the United States Environmental Protection Agency (EPA) to prosecute for workplace violations under the various environmental statutes, including the Clean Water Act, Clean Air Act, and the Resource Conservation and Recovery Act. This has also led to repeated attempts to persuade Congress to strengthen criminal enforcement provisions under the OSH Act.

2.0 Willful Violations Causing Death to Employee

The most serious penalties follow from the willful violation of an OSHA standard or rule when that violation results in death to an employee. While for first-time offenders the penalty is identical to that for making false statements—$10,000 and six months imprisonment, maximum penalties— for a repeat offender these penalties double to $20,000 and one year in prison.[8]

A "willful violation" is a knowing violation.[9] An employer must know, prior to the fatal accident, the essential facts and legal requirements.[10] However, to be prosecuted for a willful violation, the employer needs not to have exhibited "intentional disregard," but just "plain indifference," towards the safety requirements promulgated under the OSH Act.

The 1998 conviction of Roy G. Stoops is an example. Mr. Stoops, the owner of C&S Erectors of Nolesville, Indiana, allegedly failed to correct safety hazards and provide fall protection at a job site, even after receiving complaints from contractors. Following the fatal fall of an employee, an Indiana court found that Mr. Stoops' failure to act and listen to concerns amounted to a willful act. Taken together with C&S's substantial history of OSHA violations, the court decided to sentence Mr. Stoops to four months in prison.[11]

Knowing has also been interpreted as "know or *should* have known." In other words, a person cannot insulate himself from liability by either willful blindness or negligent conduct. As a strict liability statute, the normal criminal elements of malice or specific intent are removed.[12] While an omission or failure to act may be willful if voluntary or intentional,[18] the exact nature and extent of the employers' knowledge and understanding is an issue to be factored and examined by a jury considering criminal penalties.[19]

3.0 Federal Prosecution

OSHA does not handle criminal enforcement under the OSH Act. As noted above, the agency instead must refer potential cases to the DOJ for review. The actual process of referral can be quite drawn out and complicated.[8]

3.1 Referral and Review

In most situations, the process starts when the OSHA area director responsible for the inspection sends a recommended criminal action to his boss, the appropriate regional administrator. He will then hand the case to an Department of Labor solicitor, either regional or national,[9] who then has to decide whether the case meets the factual and legal grounds, and on sensitive matter the OSHA head office decides whether the case warrants referring to the DOJ for possible prosecution.[13]

There are a lot of steps. In fact OSHA spent 20 years devising an internal system to effectively keep track of referrals.[14] Actually, there is not that much to keep track of. Despite years of declaring that OSHA will place major emphasis on referrals, it just never seems to happen. Back in the

1980s, for example, OSHA announced that it intended to increase the number of referrals to Justice. Nothing much happened. Over three decades later an analyst fearlessly predicted that finally, "OSHA criminal charges on the rise" As a matter of policy, he opined that "OSHA now makes a criminal referral in every case involving an employee fatality and a willful violation."[15]

But remember, the referral is only the first half of the equation. The Justice Department has to act on the referral by bringing actual cases to court. Little attention has been paid to the snag here. For whatever reason – workload, the nebulous nature of the offenses, the difficulty of showing intent, or other factors – local prosecutors have not been eager to bring these cases.

Let the numbers speak for themselves. In the going-on five decades of OSHA's existence, there have been approximately 400,000 fatalities. Many, probably most of these were due to just plain "accidents," with no particular employer culpability, but take that figure as a base. During that same period, there have been less than a hundred OSHA actual criminal prosecutions. (Even with this, there is some definition problems about what should be counted that we will address below.) In the five year period 2009-2013, there have been consistently 800-900 fatality investigations per year by OSHA. From these, there have been an average of less than ten criminal referrals to Justice each year.[16] How many of these referrals have, or will, result in DoJ prosecutions is still uncertain.

3.2 Criminal Prosecutions Do Happen

But don't assume that federal criminal prosecutions never occur. Though rare, they do.

Adams Thermal Systems, a radiator company in South Dakota had directed its employees to bypass the machine guarding protection to save time. As a result, a worker was crushed to death. In OSHA and the Department of Justice charged criminal violations and secured $1.33 million in settlement, along with other penalties and controls, and $450,000 for the worker's spouse.[17]

You can't, however, always predict. In 2013 San Francisco trainees and unqualified workers were allowed to operate unsupervised mass transit trains at high speed while electrical workers were performing track repairs. Two of the electricians died . The California state equivalent, Cal/OSHA, charged the company's "simple approval" safety procedures were not only seriously inadequate but weren't being followed anyway. That flawed system had led to earlier deaths in 2001 and 2008. You would think this would be an ideal case for criminal prosecution, especially given the previous violations. But there was only a proposed fine of $210,000. There was surprisingly no prosecution for criminal negligence. The reason may have been that the BART transportation system was a quasi-governmental regional organization of the counties in the area and not a private company, but that too could show how the scales of justice are weighted.[18]

4.0 False Statements and Advance Notice

In addition to those cases involving a fatality, OSHA can also seek criminal sanctions for some other administrative or technical violations. Since these are not substantially different from those involving federal agencies generally, there is little need to dwell on them here.

One violation, somewhat particular to OSHA, should be pointed out, because not everyone may realize that it is indeed an offense. That is, the providing or obtaining of advance notice of an OSHA inspection. The OSH Act also makes the unauthorized disclosure of forthcoming inspections punishable by a fine of $1,000 and imprisonment for up to six months.[21]

Perjury, or lying under oath, as well as false statements to federal officers are illegal under federal criminal law.[20] Because of the increased OSHA emphasis on paperwork and various types of required record keeping, this one could pop up in what otherwise might seem routine clerical activities. In keeping records or making official filings required under the OSH Act, the signatory is affirming, under oath, the truth of any statements and documents. Knowing violation of this oath is a crime punishable by up to six months in prison and/or a $10,000 fine. In other words, since perjury is already a crime, with more substantial punishment under U.S. criminal statutes, there is nothing exceptional in specifying that this law also applies to workplace situations.

5.0 Who is an "Employer"?

If an employee gets killed on a work site, who might be held responsible. In the case of small businesses or traditional big companies, the answer is relatively straight-forward. But in this age of temporary workers, independent contractors, and subcontracting, debate over who is an employee under the OSH Act has grown. Traditionally, an employee is someone directly under control of the employer. Under the common law of agency, independent contractors are not necessarily agents and therefore not necessarily considered employees. Neither are the employees of a subcontractor or temporary agency, especially in relation to many mandated employee benefits.

However, under the Multi-Employer Doctrine (MED), any employer who creates a safety hazard and willfully violates OSHA standards on a multi-employer work site may be criminally liable for the death of a worker, whether their own or that of another employer.[11] In *Pitt-Des Moines*, an accident resulting from a deviation from industry standards and inadequate training caused a structure to collapse, killing both a Pitt-Des Moines employee and a subcontractor's employee. Pitt-Des Moines was found guilty of a willful violation due to their failure to follow standard industry procedures in securing steel beams and fined $1 million. On appeal, the 7th Circuit Court of Appeals upheld application of the MED principle.

The doctrine, however, is not without controversy. In addition to complaints concerning this drift away from the usual law of agency, in the case of *IPB, Inc. v. Herman*, the Court of Appeals for the D.C. Circuit has criticized its broad expansion of liability outside the express language of the OSH Act.[19]

6.0 State Enforcement

Just because federal OSHA has not been able to bring many federal prosecutions, legal activists have increasingly sought to use common law or state laws to bring their own actions against employers whose conduct resulted in death or serious injury to employees.

For the most part, these efforts were initially successful. Starting with *People v. Chicago Magnet Wire Corp.*,[20] several state supreme courts have found that OSHA did *not* preempt criminal prosecution of corporate officials. In *Chicago Wire*, an Illinois manufacturer was indicted for reckless endangerment by failing to provide adequate safety precautions to prevent exposure to the chemicals used in manufacturing wire. Likewise, in the leading case *People v. Pymm Thermometer*, the New York Court of Appeals affirmed a decision that criminal prosecution for reckless endangerment associated with employee exposure to mercury was not preempted by the federal OSH Act.[21]

The state courts, looking at the OSH Act's definition of standards, have generally found that criminal laws are not standards. Instead, they find that criminal laws serve a general purpose separate from regulation of the workplace. In *Pymm*, the court held that unlike the OSH Act, which established standards to prevent death or injury from occurring, New York's criminal laws were reactive, designed for punishment of acts already committed, and this serves a purpose separate from regulation of the workplace.

In *Gade*, however, the U.S. Supreme Court looked at whether OSHA preempted an Illinois licensing standard applicable to hazardous waste workers. The Court held that where a federal standard is in effect, the unauthorized state regulation of occupational, safety, and health issues is in conflict with the purposes and objectives of the OSH Act and is therefore impliedly preempted.[22] Justice O'Conner, writing for the Court, found the statement in § 18(b), that a state "shall" submit a plan if it wishes to "assume responsibility" for development of standards, indicating Congress' intent to preempt state law. As a result, a state may not enforce its own standards without federal approval.[23]

The Court found that those dual-impact regulations, which embody several purposes in addition to regulation of occupational, safety, and health, cannot avoid preemption.[24] However, the Court acknowledged that general applicability statutes, such as fire and traffic safety laws, that do not conflict with OSHA regulations and regulate the conduct of workers and non-workers alike, are not generally preempted.[25] To the extent that state prosecution is carried out under already existing general criminal and safety provisions, they should be allowed.

7.0 Prosecution under Environmental Statutes

Because of the limitations on criminal prosecution under OSHA and the concomitant problems under state and common law doctrines, some prosecutors have sought to deal with workplace situations by finding violations under the more flexible environmental laws.

At the federal level, the EPA is not constrained by the fatality requirement of the OSH Act. It has therefore been able to apply criminal laws in a wide range of circumstances, including some that touch on occupational situations. The major environmental statutes provide a host of criminal violations and penalties, and almost every environmental law provides some criminal liability. In most cases, violations can exist for both "knowing" and "negligent" conduct. It is not uncommon to find penalties of over $25,000 a day per violation and prison terms in excess of 15 years.

In light of this, EPA and OSHA have entered into memorandums of understanding (MOUs) whereby the two agencies work together to enforce both environmental and health and safety regulations in the workplace. These MOUs provide for joint inspections by EPA and OSHA investigators, a system for referrals of violations between the agencies, exchanges of data and other evidence uncovered during investigations, and cross-training.

For example, the Resource Conservation and Recovery Act (RCRA) provides "cradle to grave" controls and requirements on generators and transporters, and treatment, storage, and disposal facilities of hazardous wastes. RCRA imposes criminal penalties of up to $50,000 and two years imprisonment for knowing violations of the act. It also has put in place a system of heightened criminal liability for knowingly placing persons in danger of imminent death or serious bodily injury. Fines can run as high as $250,000 for an individual and $1,000,000 for a corporation, in addition to prison terms of up to 15 years.

The application of this approach is shown most clearly in the case of the McWane Company, a heavy industry company with foundries and factories under various names nation-wide. Over a number of years, the company was featured unfavorably on TV programs such as "Sixty Minutes" and in features in the *New York Times* for the many deaths and serious injuries in its operation. The stories were also all critical of OSHA, which, despite the deaths, seemed unable to take decisive action under its criminal powers. However, criminal prosecutions and convictions were finally obtained under the environmental laws, most notably the Clean Air Act but also the Clean Water Act and RCRA.[26]

8.0 Legislative Proposals and Prospects

In light of the inability of the OSH Act to provide criminal penalties for violations causing serious bodily harm, in addition to a belief by some that the criminal penalties provided are inadequate to serve a deterrence effect, calls have been made to toughen the OSH Act by expanding the range of activities that result in criminal liability as well as increase the penalties provided under the Act.

So far, these demands have been on the political periphery. Consumer activist Ralph Nader made criminal enforcement of workers deaths a major point in his little-noticed third party campaign for the presidency of the United States back in 1992. In 1999 the late liberal Senator Paul Wellstone

(D-MN) introduced legislation that proposed expanding the coverage of the OSH Act to federal employees;[27] significantly increased the penalties for willful violations causing death to an employee by providing prison terms up to 10 years for initial violation and 20 years for subsequent violations, while at the same time removing the limits on the amount of financial penalty imposed;[28] and strengthened the whistleblower protections.[29]

He may have been sincere, but these proposals were treated mostly as just political blather. When Senator Ted Kennedy (D-MA), an original sponsor of Wellstone's bill, subsequently became chairman of the powerful Senate Labor Committee, this and similar legislation still went nowhere.

These proposals were not all mono-partisan. A Wyoming Republican, Senator Mike Enzi, prepared several OSHA reform bills, one of which would have increased the penalties for a fatal, willful violation of OSHA standards to 18 months in prison. However, this provision was missing from the legislation he subsequently introduced.[32]

As Senator Kennedy pointed out in a 2008 congressional hearing, it makes little sense that a violation of the South Pacific Tuna Act can be fined $325,000, while a worker fatality is capped under OSHA at only $7,000.[30] Yet, when the thousandth legislative iteration of the criminal penalties issue arose, in the 2013 draft Protecting America's Workers Act, it shared the terminal fate of all its predecessors.[31] And one can fearlessly predict that future congressional bills will share the same fate. That will probably change, if it ever does, only after some huge and highly publicized accident.

Notes

1. Mr. Miller is a partner with the Washington, D.C., law firm of Baise & Miller, P.C., available at www.miller@ baisemiller.com. A former firm associate, Michael Formica, was helpful in the preparation of this chapter.

2. 29 U.S.C. § 666(f).

3. 29 U.S.C. § 666(g).

4. 29 U.S.C. § 666(e).

5. All of these, of course, are situations that many victims might consider worse than death.

6. 505 US 88 (1992).

7. About half of the states currently have OSHA-approved state plans.

8. As discussed below, penalties are commonly increased by tying OSH Act prosecutions to additional prosecutions under existing environmental laws or raising more common criminal violations, such as conspiracy, at the same time.

9. *US v. Ladish Malting*, 135 F.3d 484 (7th Cir. 1998).

10. *Id.*

11. "Construction Worker's Death Results in Jail Time for Indiana Employer," OSHA National News Release, USDL 98-421, October 15, 1998.

12. *Ensign-Bickford Corporation v. OSHRC*, 717 F.2d 1419 (D.C. Cir. 1983).

13. For an interesting account of a case where a safety-careless company had a fatality and the Secretary of Labor himself became involved in the highly publicized enforcement—but one in which criminal charges were never ever considered—see Robert B. Reich, *Locked in the Cabinet* (N.Y.: Alfred A. Knopf, 1997), pp. 158-164, 228-231.

14. OSHA Interpretation Memo, Leo Carey to Regional Administrators, "Procedures For Tracking Criminal Referrals," May 31, 1990.

15. Fred Hosier, ed., quoting Jordan Schwartz and Eric Conn of Epstein Becker Green, "OSHA criminal charges on the rise", *Safety and OSHA News Alert*, 8 January 2013.

16. The number for 2010 was 14, 2011 was 10, 2012 was 13, and for 2013 only 3. OSHA, Occupational Safety and Health Administration Enforcement, Statistics. The elevated number in 2010 and perhaps 2011 was due to action against the refinery industry, particularly British Petroleum.

17. OSHA Regional News Release, 13-1717-DAK, 5 September 2013.

18. "Ca-OSHA [sic] Fines Against BART Bosses But Criminal Prosecution for Criminal Negligence and Deaths," transportworkers.org, 17 April 2014.

19. *IPB, Inc. v. Herman*, 144 F.3d 861 (D.C. Cir. 1998).

20. *People v. Chicago Magnet Wire Corp.*, 534 NE.2d 962 (Ill. 1989).

21. *People v. Pymm Thermometer Company*, 561 NYS.2d 687 (NY 1991).

22. *Gade* at 98.

23. *Id.*

24. *Id.* at 107.

25. *Id.*

26. *U.S. v. Union Foundry Co.* (N.D. Ala. No. 2.05-cr-00299, plea entered 6 September 2005), fine of $3.5 million and $0.75 million community service for RCRA violations, also OSHA willful violation; *U.S. v. McWane, Inc.* (N.D. Ala. No. CR-04-PR-199-S, 5 December 2005), jury found the company and three employees guilty of violating the Clean Water Act by illegally discharging industrial waste water through storm drains, making false statements, and so on; *U.S. v. Tyler Pipe* (E.D. Tex. No. 6:05-cr-00029, 22 March 2005), guilty pleas on felony violations of Clean Air Act, with fines of $4.5 million and mandatory $12 milion in emission control equipment; *U.S. v. McWane, Inc.* (D.Utah, No. 05-00811), indictment filed November 2005 for falsifying data and other violations of Clean Air Act.

27. Federal Employee Safety Act S. 650 106th Congress-1st Session.

28. Wrongful Death Accountability Act S. 651 106th Congress-1st Session.

29. Safety and Health Whistleblower Protection Act S. 652 106th Congress-1st Session.

30. Prepared statement, Hearing of the Committee on Health, Education, Labor, and Pension, U.S. Senate, p. 4.

31. H.R. 1648, introduced 18 April 2013, by Rep. George Miller (D-Cal.)

Chapter 13

Judicial Review of Enforcement Actions

Donelle R. Buratto
Ogletree, Deakins, Nash, Smoak & Stewart, P.C.
Detroit, MI

1.0 Overview

Judicial review plays an important, but purposefully limited, role in the adjudication of matters under the Occupational Safety and Health Act ("OSH Act"). The federal court system provides support to the main adjudicative body[1] of the Occupational Safety and Health Review Commission ("OSHRC" or "the Commission") and its Adminstrative Law Judges ("ALJs") primarily in the following circumstances:

- Appeals of final OSHRC orders[2] in enforcement matters;

- Pre-enforcement judicial review of health and safety standards promulgated under the OSH Act within the initial 60 days of enactment;[3]

- Actions by the Secretary of Labor to enforce administrative subpoenas, to enforce the antidiscrimination provisions of the OSH Act, or to restrain imminent dangers.[4]

OSHRC is an independent federal agency, separate and distinct from the Department of Labor's Occupational Safety and Health Administration ("OSHA"). OSHRC serves as an administrative court that holds hearings and provides appellate review of contested OSHA citations and penalties issued to an employer following an OSHA inspection. Once certain procedural steps are satisfied by a party seeking review of a final OSHRC order, the party may bring an action in federal court. Generally speaking, the federal court's review is restricted to issues preserved for appeal and decided on the written record from the below proceedings, briefing by the parties, and if needed, oral argument.

Federal courts are typically quite deferential to the decisions of OSHRC, as is common with the review of any administrative agency findings. However, final OSHRC orders will be overturned if the factual findings are not supported by substantial evidence in the record or if the OSHRC's legal conclusions are found to be arbitrary, capricious, an abuse of discretion, or otherwise not in accordance with the law. The federal court decision is final with two exceptions: first, if the court remands the case back to the OSHRC for further proceedings; and second, if the Supreme Court for the United States agrees to hear an appeal of the case.[5]

2.0 Jurisdiction

2.1 Who Can Bring An Appeal?

It is clear that employers and the Secretary of Labor may file a direct appeal of a final order of OSHRC. Specifically, the OSH Act provides that "any person adversely affected or aggrieved by an Order of the Commission" may obtain such review and that "the Secretary may also obtain review or enforcement of any final order of the Commission."[6] Any employer who has been found in violation of an OSHA standard or regulation; assessed a civil penalty; or ordered to take abatement actions would satisfy this requirement.

The most common other party who seeks to appeal a final order of OSHRC is an employee or his/her representative (typically, a union representative). This intervention is permissible, if the employee or representative has properly filed a Notice of Contest ("NOC"). The employee/representative's party status in these proceedings is limited to contesting whether the period of time fixed in the citation for the abatement of the violation is unreasonable.[7]

The OSH Act does not make clear whether OSHRC has standing to participate as a party in a matter reviewing one of its decisions, and the federal courts are divided on this issue. Three circuits—the Third, Sixth, and Ninth—have held that OSHRC is not a proper party, with the Second and Tenth circuits supporting that position in dictum.[8] The Fourth and Fifth circuits though have taken the opposite position, stating that OSHRC is similar to other administrative agencies and that it is appropriate for OSHRC to defend its policies that Congress empowered it to adopt.[9]

Manufacturers of products have also been found by courts to lack standing to challenge OSHA enforcement actions brought against their customers and to challenge safety regulations for the workplace.[10]

2.2 Which Courts Can Hear An Appeal?

Once a party has established that it has standing to appeal a final order issued by OSHRC, it will need to determine the appropriate court in which to file it. The OSH Act specifically provides for the appeal of final orders of OSHRC to the United States courts of appeals. The federal court system is divided into eleven circuits by region, plus the District of Columbia circuit. The proper federal court for an appeal is either the circuit in which the violation is alleged to have occurred or where the employer has its principal office, or in the court of Appeals for the District of Columbia Circuit.[11] Notably, the Secretary of Labor has similar jurisdictional options and may file an appeal of a final order in the circuit in which the violation is alleged to have occurred or where the employer has its principal office.[12] The proceedings will follow the same process whether the appeal is filed by an aggrieved party or the Secretary of Labor, and the judgment of the court will be final unless it remands the case back to OSHRC for further proceedings or review is granted by the Supreme Court.[13]

3.0 Timing

3.1 Requirements Before Filing An Appeal

An employer must exhaust certain administrative remedies before it can seek judicial review. First, and as discussed further in Chapter 11, the employer must contest the original citation, penalty, and/or abatement provisions by filing an NOC with the Secretary of Labor within 15 working days after receipt of the citation. The NOC must specifically indicate that the employer is contesting all of the citation, penalty, and abatement elements or else specify which specific part is being contested. Any uncontested portions of the citation will become binding upon the employer at the expiration of the 15-working day contest period and is not subject to judicial review.[14]

If an employer timely contests a citation, the challenged portions will be subject to review in a full evidentiary hearing before an ALJ, where the parties may present evidence through witness testimony and written or demonstrative exhibits. After the hearing, the ALJ will issue an opinion upholding or dismissing the contested portions of the citation. If an employer or the Secretary of Labor is dissatisfied with this opinion, they may file a Petition for Discretionary Review (PDR). A party does not have a right to review by OSHRC.[15] A Commissioner may exercise discretion and direct review on his or her own motion or based on the petition of a party.[16] The deadlines for filing a PDR are 10 days if it is filed with the ALJ or 20 days if the PDR is filed directly with the Executive Secretary.[17] However, federal regulations provide that "[t]he earlier a petition is filed, the more consideration it can be given."[18] If the Commission denies the PDR or if the employer or the Secretary of Labor remains unhappy with OSHRC's final order, the parties may then seek judicial review in the appropriate United States court of appeal.[19] The omission of any of these steps will prevent a party from access to judicial review.

3.2 When Do I Bring An Appeal?

The OSHA enforcement context has many strictly enforced time limits, and the area of judicial enforcement is no exception. Any party with standing (typically either the employer or the Secretary of Labor) may appeal a final order of OSHRC by filing a written petition seeking to have the order modified or set aside within sixty (60) days after the issuance of an order.[20] If a petition for review is not filed within the sixty day time period, OSHRC's findings of fact and order must be considered conclusive when the court considers any petition for enforcement. In these circumstances, as well as in the case of a uncontested citation which has become a final order of the Commission, the clerk of the court must enter a decree enforcing the order and transmit a copy of the decree to the Secretary of Labor and the employer named in the petition unless otherwise ordered by the Court.[21]

3.3 Filing for a Stay of Final Order

Simply seeking judicial review of an order of OSHRC does not stay the order.[22] Put another way, the Secretary of Labor may pursue payment of penalties or abatement compliance while an employer's appeal is pending and cite the employer for failure to satisfy these requirements if a stay is not in place. However, any party aggrieved by a final order of OSHRC may seek a stay from OSHRC while the matter is in its jurisdiction or from the court once the petition is filed.[23]

3.4 Pre-Enforcement Judicial Review

In addition to judicial review of enforcement actions, the OSH Act also allows any person who may be adversely affected by an issued safety standard to file a petition challenging the validity of the standard and for a judicial review of such standard.[24] The petition must be filed within 60 days after the standard is put into effect in the United States court of appeals for the circuit where the person resides or has his principal place of business.[25] Unless the court orders otherwise, the filing of this type of position will not operate as a stay of the standard. An employer challenging a standard in this regard faces an uphill battle as the determinations of the Secretary of Labor in promulgating the standard must be considered conclusive if they are supported by substantial evidence in the record considered as a whole.[26]

4.0 Scope of Judicial Review

4.1 Rules and Practice Guidelines

The procedural rules governing a review will vary depending on the stage of the challenge to the citation. Hearings before ALJs are governed by the Federal Rules of Civil Procedure ("FRCP") unless there is a specific provision to the contrary.[27] The FRCP are detailed rules that govern practice in the federal district courts and cover topics including the form, timing, and substance of actions taken with the courts. At the actual hearing before the ALJ, the ALJ will follow the Federal Rules of Evidence, which similarly set out specific rules regarding what evidence is and is not admissible in federal actions.[28] Chapter 11 provides a comprehensive discussion of the rules of procedure and evidence used in proceedings before OSHRC.

When an employer appeals a Commission decision to a federal court, yet another set of rules apply—the Federal Rules of Appellate Procedure ("FRAP"). Specifically, Rules 15 through 20 govern the review or enforcement of an order of an administrative agency, board, commission, or officer as follows: Rule 15—Review or Enforcement of an Agency Order—How Obtained; Intervention; Rule 16—The Record on Review or Enforcement; Rule 17—Filing the Record; Rule 18—Stay Pending Review; Rule 19—Settlement of a Judgment Enforcing an Agency Order in Part; and Rule 20—Applicability of Rules to the Review or Enforcement of an Agency Order.[29] Various other portions of the FRAP set standards as to when and how appeals must be brought, including detailed requirements on the fees, timing, and format of the filings. For example, Rule 27(d)(1) provides that a motion or a response must not exceed 20 pages, that any cover of the document used must be white, and that the document "must be bound in any manner that is secure, does not obscure the text, and permits the document to lie reasonably flat when open." In addition to the FRAP that apply to all federal circuit courts, each specific circuit court has its own rules that appealing parties must also follow. Failure to comply with either the FRAP or the relevant federal court's local rules may result in dismissal of the appeal.

Unlike the proceedings at the ALJ level, the judicial review at the federal court level is based on a review of the written record and limited to the issues raised on appeal before OSHRC. It typically will include oral argument, where the attorneys for the parties verbally present their positions to the judges. This oral argument is typically limited to 15 minutes per side and is focused on questions asked by the three-judge panel. It is not a full evidentiary hearing. If a party seeks to present any additional evidence to the panel, it must apply for leave and prove to the satisfaction of the court not only that the additional evidence is material, but also that there were reasonable grounds for failing to present the evidence in the hearing before OSHRC.[30] The court may order that the additional evidence be taken before OSHRC and be made a part of the written record.[31]

4.2 Standard of Review: Conclusions of Fact

Pursuant to the OSH Act and further endorsed by the courts, OSHRC's findings of fact are considered conclusive if they are supported by substantial evidence on the record considered as a whole.[32] The "substantial evidence" test applies whether OSHRC hears witness testimony itself or adopts the ALJ's findings of fact.[33] The circuits generally define "substantial evidence" as "such relevant evidence as a reasonable mind might accept as adequate to support a conclusion...."[34] This is seemingly a high standard to meet, with one court further describing the standard as "enough to warrant denial of a motion for a directed verdict in a civil case tried to a jury."[35] In reality though, courts presume OSHRC to have technical expertise and experience in the field of job safety, and the administrative adjudications by OSHRC are given great weight and deference.[36]

4.3 Standard of Review: Conclusion of Law

Unlike conclusions of fact, the OSH Act does not specify standards of review for conclusions of law. Therefore, the general rules set forth in the Administrative Procedures Act ("APA") govern the federal courts' review of

OSHRC decisions. Under section 10(e) of the APA, the reviewing court must decide all relevant questions of law, interpret statutory provisions, and determine the meaning or applicability of the terms of an agency decision.[37] The court may review either the whole record or only those parts of it cited by a party.[38] The court will find an agency decision, findings, and conclusions to be unlawful and set them aside only if they are found to be:

- Arbitrary, capricious, an abuse of discretion, or otherwise not in accordance with law;

- Contrary to constitutional right, power, privilege, or immunity;

- In excess of statutory jurisdiction, authority, or limitations, or short of statutory right;

- Without observance of procedure required by law;

- Unsupported by substantial evidence; or

- Unwarranted by the facts to the extent that the facts are subject to trial de novo by the reviewing court.[39]

Again as to conclusions of law, an employer challenging a decision at appellate court level faces an uphill battle. The above criteria set forth difficult standards to meet, particularly in light of the deference that is typically afforded to decisions of the OSHRC, OSHA, and other governmental agencies. For example, the "arbitrary and capricious" standard is a narrow one, requiring a reviewing court to evaluate whether the decision was based on a consideration of the relevant factors and whether there has been a clear error of judgment.[40] In doing so, the court is not allowed to substitute its judgment for the agency or supplement the reasoning provided by the agency. However, the agency must only prove a rational connection between the facts found and the choice made, not an ideal reasoning process.[41]

The highest court in the country, the United States Supreme Court, has also repeatedly stated that it is "well-established that an agency's construction of its own regulations is entitled to substantial deference," including in the OSHA context.[42] The OSH Act is somewhat unique from other regulatory schemes in that Congress divided the enforcement and rulemaking powers from the adjudicative power—assigning the former to the Secretary of Labor and the latter to OSHRC. By doing so, there are times when these entities may disagree on interpretations of the OSH Act regulations. The United States Supreme Court has held that in such circumstances, courts must favor the agency's interpretation as it has unique expertise and policymaking prerogatives related to the complex and changing circumstances; is better able to reconstruct the purpose of the regulations in question; and is versed in a much greater number of regulatory problems than OSHRC.[43]

4.4 Precedential Effect of Judicial Decisions

Unlike traditional civil or criminal litigation where the decisions of the appellate courts would be binding on the lower courts, OSHRC is generally not bound by rulings of the federal appellate courts and establishes its own precedent unless reversed by the United States Supreme Court. The reasoning behind this system is that Congress created the Commission to achieve uniformity in adjudications on national occupational safety and health policy.[44] However, in reality and as recognized by some of the federal courts, the Commission and its ALJs follow the law of the circuit to which the case would most likely be appealed.[45] OSHRC has also limited the precedential effect of unchallenged ALJ decisions, stating that these decisions are only binding on the parties to the case and have no broader precedential effect.[46]

5.0 Conclusion

An employer who is dissatisfied with decisions at the ALJ and OSHRC levels or with a newly promulgated standard may have an avenue to continue its challenge through judicial review. In doing so, the employer must be mindful of requirements such as exhaustion of administrative remedies and the procedural rules of the federal appellate court system. Any party pursuing judicial review must also be cognizant of the limited scope of review, the standards that must be met for the court to overturn a decision, and the deference typically afforded to OSHA in such proceedings. Even with these potential hurdles that parties must satisfy when seeking relief, the federal appellate courts remain a critical part of this country's occupational health and safety system and support the purposes underlying the OSH Act.

Notes

1.   29 U.S.C. §660.

2. For the purposes of this chapter, a "final order" of the Commission includes both decisions of the OSHRC and decisions of ALJs who conducted full evidentiary hearings in place of the Commission and which were not appealed. A decision of the ALJ will become a final order of the Commission if neither the Secretary of Labor nor the employer seeks review or if the OSHRC denies a request for discretionary review.

3. 29 U.S.C. §655(f).

4. *See* 29 U.S.C. §657(b) (actions by the Secretary to enforce administrative subpoenas); 29 U.S.C. §660(c)(2) (actions by Secretary to enforce the antidiscrimination provisions of the OSH Act); 29 U.S.C. §662(a), (d) (actions on behalf of Secretary to restrain imminent dangers). The OSH Act also provides a right to bring an action in federal court to any employee who was injured by reason of the failure of the Secretary of Labor to seek injunctive relief to restrain an imminent danger and actions on behalf of the United States to recover civil penalties. 29 U.S.C. §662(d); 29 U.S.C. §666(l).

5. 29 U.S.C. §660(a); 28 U.S.C. §1254.

6. 29 U.S.C. §660(a) and (b).

7. 29 U.S.C. §659(c); 29 U.S.C. §660(a).

8. *Marshall v. Occupational Safety & Health Review Comm'n and OCAW*, 635 F.2d 544 (6th Cir. 1980); Marshall v. Sun Petroleum, 622 F.2d 1176 (3rd Cir. 1980); *Dale Madden Construction, Inc. v. Hodgson*, 502 F.2d 278 (9th Cir. 1974).

9. *Brennan v. Gilles & Cotting, Inc.*, 504 F.2d 1255, 1267 (4th Cir. 1974); *Diamond Roofing Co. v. OSHRC*, 528 F.2d 645, 648 n.8 (5th Cir. 1976).

10. *R.T. Vanderbilt Co. v. Occupational Safety & Health Review Comm'n*, 728 F.2d 815, 818 (6th Cir. 1984); *Fire Equipment Mfrs. Ass'n v. Marshall*, 679 F.2d 679, 681 (7th Cir.1982), cert. denied, 459 U.S. 1105 (1983).

11. 29 U.S.C. §660(a).

12. 29 U.S.C. §660(b).

13. 29 U.S.C. §660(a).

14. 29 U.S.C. §659(a).

15. 29 C.F.R. §2200.91(a).

16. 29 C.F.R. §2200.91(a).

17. 29 C.F.R. §2200.91(b).

18. 29 C.F.R. §2200.91(b)

19. 29 U.S.C. §660(a).

20. 29 U.S.C. §660(a).

21. 29 U.S.C. §660(b).

22. 29 U.S.C. §660(a).

23. 29 U.S.C. §660(a); 29 C.F.R. §2200.94.

24. 29 U.S.C. §655(f).

25. 29 U.S.C. §655(f).

26. 29 U.S.C. §655(f).

27. 29 C.F.R. §2200.2.

28. 29 C.F.R. §2200.71.

29. Fed. R. App. P. 15-20.

30. 29 U.S.C. §660(a).

31. 29 U.S.C. §660(a).

32. 29 U.S.C. §660(a); *Modern Cont'l/ Obayashi v. Occupational Safety & Health Review Comm'n*, 196 F.3d 274 (1st Cir. 1999).

33. *Modern Cont'l/ Obayashi v. Occupational Safety & Health Review Comm'n*, 196 F.3d 274 (1st Cir. 1999).

34. *Northwood Stone & Asphalt, Inc. v. Occupational Safety & Health Review Comm'n*, 82 F.3d 418 (6th Cir. 1996); Compass Environmental, Inc. v. Occupational Safety & Health Review Comm'n, 663 F.3d 1164, 1167 (10th Cir. 2011).

35. *Martin Painting & Coating Co. v. Marshall*, 629 F.2d 437, 438 (6th Cir. 1980).

36. *Marshall v. Cities Service Oil Co.*, 577 F.2d 126, 131 (10th Cir. 1978).

37. 5 U.S.C. §706.

38. 5 U.S.C. §706.

39. 5 U.S.C. §706(2).

40. *Bowman Transportation, Inc. v. Arkansas-Best Freight System, Inc. et al.*, 419 U.S. 281, 285-286 (1974).

41. *Burlington Truck Lines v. United States*, 371 U.S. 156, 168 (1962); *Bowman Transportation, Inc. v. Arkansas-Best Freight System, Inc. et al.*, 419 U.S. 281, 285-286 (1974).

42. *Lyng v. Payne*, 476 U.S. 926, 939 (1986); *Martin v. Occupational Safety & Health Review Comm'n*, 499 U.S. 144, 149 (1991).

43. *Martin v. Occupational Safety & Health Review Comm'n*, 499 U.S. 144, 151-52 (1991).

44. 29 U.S.C. § 651(b)(3); S & H Rigers & Erectors, Inc., 1979 CCH OSHD 23,480; rev'd 1981 CCH OSHD 1 25,733; 659 F.2d 1273 (CA-5, 1981).

45. *ComTran Grp., Inc. v. U.S. Dep't of Labor*, 722 F.3d 1304, 1307 (11th Cir. 2013); *See also, e.g., Secretary of Labor v. Interstate Brands Corp.*, 20 O.S.H. Cas. (BNA) 1102, at *2 n. 7 (2003).

46. *See Leone Constr. Co.*, 3 O.S.H. Cas. (BNA) ¶ 1979 (O.S.H.R.C. Feb. 10, 1976).

Chapter 14

Imminent Danger Inspections

Frank D. Davis, Esq.
Ogletree, Deakins, Nash, Smoak & Stewart, P.C.
Dallas, Texas

1.0 Overview

• Worker's arm crushed in unguarded conveyor belt, causing amputation of her right arm below the elbow.

• Worker's leg crushed by falling steel beams forcing amputation when overloaded two-ton-rated crane broke and dropped almost three tons of steel less than three feet onto worker's leg.

• Worker loses hand to paper slicer because employer had disabled safety switch to increase production.

• Father and son carpenter team killed when they fell from mobile scaffolding because employer provided neither safety harnesses nor guardrails.

• All skin and muscle of worker torn from bone when worker tried to clean the rollers of a coating machine while they still rotated, which was consistent with plant procedures, instead of shutting down and de-energizing the machine first.

Each of these accounts represents a hazardous condition in the workplace that led to serious bodily injury or death. If the hazardous condition had been identified in advance, could these tragic consequences been avoided? The U.S. Congress thought so and passed legislation designed to identify and correct such hazards, or "imminent dangers," expeditiously and before tragic workplace injuries could result.

2.0 Imminent Danger Defined

Section 13(a) of the Occupational Safety and Health Act (OSH Act) defines imminent danger as "any conditions or practices in any place of employment which are such that a danger exists which could reasonably be expected to cause death or serious physical harm immediately or before the imminence of such danger can be eliminated through the enforcement procedures otherwise provided by this Act."[1] It is the duty of the Occupational Safety and Health Administration (OSHA or the Administration) to administer the OSH Act and enforce provisions such as the imminent danger legislation.

OSHA's front-line officer responsible for ensuring employer compliance with the OSH Act is the compliance safety and health officer (CSHO). The primary responsibility of the CSHO is to carry out the Administration's general mandate "to assure so far as possible every working man and woman in the nation safe and healthful working conditions . . . ."[2] The CSHO's primary means of identifying workplace hazards is through on-site inspections of the employer's business. During these inspections, CSHOs evaluate an employer's compliance with the OSH Act's standards by interviewing employees, walking through the employer's facility, and reviewing employer safety records, training, and safety policies.[3]

3.0 Nuts and Bolts of an Inspection

OSHA utilizes various inspections to enforce the OSH Act. Imminent danger inspections are unprogrammed inspections that the Administration schedules in response to alleged hazardous working conditions identified at a specific worksite.[4] OSHA usually learns of conditions leading to imminent danger inspections through employee reports, but sometimes a CSHO may notice and address an imminent danger situation during a programmed inspection.[5]

Because employee complaints of imminent dangers in the workplace receive top priority, OSHA immediately evaluates whether there is a reasonable basis to support the allegation.[9] To constitute an imminent danger that requires immediate inspection, the alleged hazard must satisfy two criteria:

1. It must be reasonably likely that a serious accident will occur immediately or if not immediately, before abatement would otherwise be required; and

2. The threatened harm is likely to cause death or serious physical injury. For a health hazard, exposure to the toxic substance or other health hazard must be so severe as to shorten life or cause substantial reduction in physical or mental efficiency, even if the resulting harm may not manifest itself immediately.[10]

OSHA also may evaluate the reliability of the employee complaint before initiating an inspection. For instance, while OSHA will conduct an imminent danger inspection during a strike or other labor dispute,[11] the area director for the local OSHA office that receives the complaint will evaluate the legitimacy of any complaint prior to scheduling such inspections during a strike or labor dispute "to insure as far as possible that the complaint reflects a good faith belief that a true hazard exists."[12]

Unprogrammed inspections, like imminent danger inspections, can be comprehensive or partial. A comprehensive inspection entails a complete inspection of the high-hazard areas of an employer's establishment. A partial inspection is one in which the focus is limited to certain, identified hazardous areas, operations, conditions, or practices. Usually, an imminent danger inspection will be limited to the specific hazard that originated the complaint, but a CSHO may expand a partial inspection based on information gathered during the course of the inspection, including violations of the OSH Act he witnesses during the inspection.[13]

4.0 The On-Site Visit

Upon receipt of an imminent danger complaint, OSHA will assign a CSHO and attempt to schedule an inspection of the facility within 24 hours of the complaint. While a copy of the complaint must be given to the employer no later than the time the CSHO begins his or her inspection of the employer's premises, the names of the employees referenced in the complaint and the name of the complainant him or herself may not be disclosed, directly or indirectly, to the employer without permission of the complainant.[14]

Generally, CSHOs conduct investigations without advance notice. In fact, alerting an employer without proper authorization in advance of an OSHA inspection can bring a fine of up to $1,000 and/or a six-month jail term. This general expectation and consequence applies to OSHA's CSHOs as well as compliance officers operating under individual state plans, which are discussed in Chapter 15.[16] Typically, OSHA does not give employer notice because it does not want employers to correct or change practices before a CSHO inspects the operation. In the case of an imminent danger inspection, however, OSHA may provide an employer advance notice up to twenty-four hours before the beginning of the inspection. OSHA does this hoping the employer will correct/address the hazardous condition immediately so as to abate employee exposure to such hazards.[17]

When the CSHO arrives to conduct the inspection, the CSHO must present official credentials.[18] Prior to allowing the CSHO to conduct an inspection, an employer should always ask to see the compliance officer's credentials.[19] If the employer questions the credentials, the employer can verify the credentials of federal and state compliance officers by calling the local federal or state office. As a point of practice, compliance officers can never collect a penalty at the time of inspection or promote the sale of a product or service.[20]

A CSHO is authorized by the OSH Act "to enter without delay and at reasonable times any factory, plant, establishment, construction site, or other area, workplace or environment where work is performed by an employee of an employer and to inspect and investigate during regular working hours and at other reasonable times, and within reasonable limits and in a reasonable manner, any such place of employment and all pertinent conditions, structures, machines, apparatus, devices, equipment, and materials therein, and to question privately any such employer, owner, operator, agent, or employee."[21] Even though this language seems to give OSHA very broad inspection and entrance authority, a CSHO cannot enter an employer's premises without a search warrant unless the employer allows the CSHO access. Specifically, in 1978, the U.S. Supreme Court ruled that the OSH Act violated the constitutional prohibition against unreasonable searches insofar as it purported to authorize warrantless, nonconsensual searches of commercial workplaces.[22] As a result, OSHA inspections generally must be conducted pursuant to a warrant if the employer does not agree to the inspection.

In most cases, an employer allows a CSHO access to inspect the employer's facility. Unless an employer is working on a military base or some other government facility, an employer may choose to deny the CSHO access to the plant because of trade secrets, labor relation problems, or any other reason.[23] In such instances, OSHA is authorized to seek a search warrant from a United States District Court overseeing the area in which the employer operates its business.[24]

An employer may not precondition a CSHO's entry upon signing any form or release or any type of waiver.[25] This standard applies to any employer forms controlling the release of trade secret information equally. If the employer insists that the CSHO sign a waiver before allowing entrance, OSHA will treat that situation as an employer's refusal to allow the CSHO entrance and will seek a search warrant. Notwithstanding the CSHO's right to enter a workplace without signing a waiver, a CSHO still may sign a visitor register or other registration book indicating his or her presence in the employer's facility.[26]

- If necessary, OSHA may obtain a warrant from by showing:

- The proposed scope of the inspection;

- A description of the alleged imminent danger situation;

- The date received and the source of the information;

- The original formal complaint;

- An explanation for the reasonable expectation of death or serious physical harm coupled with the immediacy of danger; and

- OSHA followed all current imminent danger processes and procedures.[27]

The warrant usually is narrowly construed, seeking to inspect only the condition at the employer's facility reported as an imminent danger.[28] This is especially true when the search of the workplace is triggered by an employee complaint only. Specifically, federal courts have held that search warrants covering an employer's entire facility are unreasonable and invalid when the breadth of the warrant exceeded the scope of the employee complaint that originated the inspection.[29]

Once the Administration secures a warrant, the inspection generally begins within twenty-four hours.[30]

In addition to a search warrant, the Administration may issue administrative subpoenas when a reasonable need for records, documents, testimony, or other supporting evidence exists. OSHA may issue administrative subpoenas as a matter of course, but usually refrains from doing so unless the employer is unwilling—or expected to be unwilling—to voluntarily disclose the information.[31]

Unlike search warrants, administrative subpoenas of employer records and policies may be broader than merely the area containing the reported hazard. To be enforceable, an OSHA subpoena for documents must be "reasonably relevant to the authorized inquiry."[32] Despite the plain language of the OSH Act, courts usually give OSHA significant deference, allowing broad discovery of employer records. In one such case, an employer operated a three-location facility. At one location, an employee submitted a complaint to OSHA regarding specific conditions in his workplace, which was limited to one of the employer's three locations. OSHA inspected the employee's workplace pursuant to a warrant narrowly tailored to suit the complaint and conducted an inspection of the area of concern. Pursuant to its authority to ascertain whether other locations experienced significant employee injuries and required inspection, OSHA subpoenaed the employer's safety and health records for all three locations. The employer objected, arguing that the subpoena exceeded the scope of the employee complaint and refused to produce the subpoenaed records. OSHA sought enforcement of its subpoena before a United States federal court, which concluded OSHA had authority to subpoena records from all three locations because the records were reasonably relevant to OSHA's authority to ascertain whether a broader inspection of the employer's operation was necessary.[33]

5.0 Employee Representatives

An employee representative is entitled to participate in the inspection of the employer's premises.[34] An employee representative "refers to (1) a representative of the certified or recognized bargaining agent, or, if none exists, (2) an employee member of a safety and health committee who has been chosen by the employees (employee committee members or employees at large) as their OSHA representative, or (3) an individual employee who has been selected as a walkaround representative by the employees of the establishment."[35] If an employer refuses to let an employee representative participate in the inspection, the CSHO will treat the situation as an employer's refusal to allow the inspection, and the Administration likely will seek a warrant, if it already had not done so, to pursue the imminent danger inspection.[36]

6.0 Opening Conference

Prior to walking the premises, the CSHO will convene an opening conference to explain how the employer was selected and to describe the expected scope of the inspection.[37] During the conference, the CSHO also will explain the specific purpose of the visit, the scope of the inspection, and the standards that apply.[38] In conducting this opening conference, OSHA "encourages employers and employees to meet together in the spirit of open communication."[39] If either the employer or the employee objects to engaging in a joint opening conference, the CSHO will conduct separate opening conferences with each party.[40] Following the opening conference, the CSHO will ask the employer to select an employer representative to accompany the CSHO during the walk around portion of the inspection.[41]

The OSH Act also provides that an authorized employee representative will accompany the CSHO during the inspection. If an employee representative is unavailable to attend the opening conference and to accompany the CSHO on the inspection, the CSHO will interview employees during the course of his inspection of facility.[42] CSHO interviews with employees are private; however, an employee may request that an employee representative participate in the interview as well.[43]

At establishments where more than one employer is present or in situations where groups of employees have different representatives, the CSHO may allow a different employer or employee representative for different phases of the inspection. More than one employer and/or employee representative may accompany the CSHO throughout or during any phase of an inspection if the CSHO determines such additional representatives will aid, and not interfere with, the inspection.[44]

7.0 The Walk Around

With employer and employee representatives selected, the CSHO begins the inspection of the relevant work areas for safety and health hazards. The CSHO will determine the course and scope of the walkaround inspection, but an imminent danger inspection should be limited to the specific hazard identified in the complaint. During the course of the walkaround inspection, the CSHO is authorized to observe safety and health conditions and practices in the facility; consult with employees privately; take photographs, videos, and instrument readings; examine records; collect air samples; measure noise levels; survey existing engineering controls; and monitor employee exposure to toxic fumes, gases, and dusts.[45]

During the course of the inspection, the CSHO may ask the employer to abate any hazard identified as an imminent danger and to remove employees from the area of a hazard identified as an imminent danger. Moreover, as soon as reasonably practicable after the conclusion of the walk around, OSHA will also contact the employer and ask it to immediately abate any hazardous situation identified as an imminent danger and/or remove employees from the area identified as an imminent danger until the identified hazard is abated.[46] The Administration encourages the employer to do whatever is possible to eliminate the identified hazard promptly and on a voluntary basis. When the employer voluntarily and permanently eliminates the imminent danger as soon as the Administration or CSHO identifies it to the employer, an appropriate citation and notification of penalty shall be issued but no further abatement will be necessary. If an employer fails or refuses to correct an imminent danger, OSHA will issue a citation and initiate imminent danger proceedings, including the posting of a Notice of Alleged Imminent Danger at the workplace.[47]

8.0 Notices of Imminent Danger & Temporary Restraining Orders

In a case where the employer cannot or does not eliminate the hazard or remove employees from exposure to an imminent danger, the Administration takes swift and definite steps. Specifically, the CSHO will post a Notice of Imminent Danger and call the area director, who will decide whether to obtain a temporary restraining order (TRO) from a federal court. The purpose of the TRO is to force an employer to immediately abate the imminent danger and/or to remove employees from the hazardous area. The CSHO, however, has no authority to order the closing of the operation or to direct employees to leave the area of the imminent danger or the workplace.[48] Rather, the CSHO shall notify employees and employee representatives of the posting of the Notice of Imminent Danger and shall advise them of their rights to report any other perceived violation without fear of discrimination or retaliation from the employer.[49] If an employee agrees that the workplace poses an imminent danger, he has a legal right to refuse to enter the area or perform the task that is the subject of the notice until the hazard is abated.[50]

In many cases, the CSHO may conclude no imminent danger exists at the time of the inspection. Further evaluation of the file, employer documents, and additional evidence may be required before OSHA can determine whether an imminent danger exists. In appropriate cases, a Notice of Imminent Danger may be posted later, after OSHA officials have an opportunity to further evaluate the evidence gathered by the CSHO during the inspection.[51]

9.0 Closing Conference

At the conclusion of an inspection, the CSHO conducts a closing conference with the employer and employee representatives, jointly or separately, as circumstances dictate. The CSHO may conduct the closing conference either on-site or by telephone.[52] In the event the employer declines to have a joint closing conference with the employee representative, the CSHO normally holds the conference with the employee representative first, unless the employee representative requests otherwise, in order to ensure that worker input, if any, is received and that any needed changes are made before the employer is informed of violations and proposed citations.[53]

During the closing conference, the CSHO will describe to the employer and employee representative all unsafe or unhealthful conditions observed during the inspection. At the same time, the CSHO will describe all apparent violations for which a citation and proposed penalty may be issued. Because the CSHO may not have all pertinent information at the time of the closing conference, a second closing conference may be held by telephone or in person to inform the employer and employee representative whether any additional concerns exist or whether additional citations may be issued or recommended.[54]

10.0 Citations and Penalties

Even if the CSHO posts a notice and/or OSHA seeks a TRO, the Administration still will issue citations to the employer if it finds violations of the OSH Act. The employer must post a copy of each citation at or near the place where the violation occurred for three days or until the violation is abated, whichever is longer.[55]

The ranges of penalties are extreme. For other-than-serious violations, the penalty can be as little at $0. In the case of a criminal willful violation, however, the penalty can be as much as $500,000. Accordingly, the classes of penalties attached to each alleged violation of OSHA standards is a significant aspect of the citation.

Other-than-serious violations are those that have a direct relationship to job safety and health, but which are unlikely to cause death or serious physical harm. Penalties for other-than-serious violations range from $0 to $7,000 for each violation assessed and may be reduced, depending on the employer's good faith, such as the employer's demonstrated efforts to comply with the OSH Act, history of previous violations, and business size.[56]

A serious violation exists when a there is a substantial probability that death or serious physical harm would result from the employer's violation of OSHA standards. OSHA may assess a penalty up to $7,000 per violation, depending on the gravity of the violation. Like an other-than-serious violation, the penalty for a serious violation may be adjusted based on the employer's good faith, history of previous violations, and size of business.[57]

An employer may be cited for a repeat violation if that employer has been cited previously for a substantially similar condition within the preceding five years, and the citation has become a final order of the Occupational Safety and Health Review Commission. Repeat violations can carry penalties up to $70,000 per instance.

If OSHA determines that the employer intentionally and knowingly violated an OSHA standard, it will issue a willful violation. In this case, OSHA must have concluded that (1) a hazardous condition exists; (2) the employer knew the condition violated a standard or other obligation of the OSH Act; and (3) the employer made no reasonable effort to eliminate it. Penalties for willful violations range from $5,000 to $70,000.

If an employer commits a willful violation that results in the death of an employee, the stakes are significantly raised. "An employer who is convicted in a criminal proceeding of a willful violation of a standard that has resulted in the death of an employee may be fined up to $250,000 (or $500,000 if the employer is a corporation)" and/or imprisoned up to six months.[58] A second conviction doubles the possible term of imprisonment.[59]

11.0 Abatement

Notwithstanding issuance of a TRO or a citation, OSHA also will direct the employer to abate any perceived violations of safety and health standards.[60] Abatement periods to correct deviations from standard requirements generally are short, less than thirty days. When the employer abates an identified hazard during the inspection, however, the abatement period listed on the citation shall note, "Corrected During Inspection."[61]

OSHA instructs its field offices that the abatement period for safety violations should be no more than thirty days. At the same time, the

Administration recognizes that certain situations may arise, such as for health violations, where extensive structural changes or new parts and equipment cannot be effected or obtained within thirty calendar days. When an identified imminent danger exists in the workplace, OSHA expects immediate abatement or that the employer will remove workers from the hazardous area.

In a situation where an employer contests either the period set for abatement or the citation itself, the abatement period usually will not begin until the employer and the Administration reach a settlement agreement or the matter is litigated and a decision is issued by the Occupational Safety and Health Review Commission, which is responsible for evaluating the legitimacy of OSHA citations appealed by an employer through a notice of contest after issuance. If the employer and the Administration reach a settlement agreement, the employer must abate any hazards within the period of that agreement. Alternatively, under the OSH Act, the abatement period begins when the Review Commission issues a final order regarding any challenge by the employer as to the abatement period or the citation, and the abatement period established by the Review Commission is not tolled while either the employer or the Administration appeals the decision, unless the employer was granted a stay. In situations where there is an employee contest of the abatement date, the abatement requirements of the citation remain unchanged. Also, where an employer contests only the proposed penalty, the abatement period established in the citation is unaffected by the employer's contest.[62]

The foregoing is the process for addressing abatement of all types of citations and also may be the process with regard to hazards and violations identified as part of an imminent danger in the workplace. If, however, the Administration established a short abatement period in an effort to immediately correct an imminent danger, the Administration may initiate appropriate imminent danger proceedings immediately and without regard for the employer's notice of contest.[63]

12.0 MSHA Imminent Danger Inspections

Few federal agencies are given such extensive power over an employer's operation as that bestowed upon the Mine Safety and Health Administration (MSHA). Under Section 107(a) of the Mine Safety and Health Act (the Mine Act), a mine inspector can shut down an entire mining operation if he identifies a hazardous condition that constitutes an imminent danger in a mine. Called a Section 107(a) Order or an Imminent

Danger Order, a mine inspector can force an employer to cease all mining operations and withdraw all employees from the affected area or mine. The order may be given orally or in writing. When spoken, the mine inspector will memorialize the oral imminent danger order, noting to whom he issued the order as well as the time, date, location, and reason for the order.

Like OSHA imminent danger inspections, MSHA learns of hazardous conditions that may trigger an imminent danger inspection in a variety ways. To facilitate employee complaints, for example, MSHA maintains a hotline and a Web site wherein employees may call in or file written complaints of hazardous conditions in the workplace.[64] As in the case of a report of an imminent danger to OSHA, MSHA will attempt to inspect the reported mine within 24 hours. Unlike OSHA, however, MSHA need not seek a search warrant to enter an employer's facility because the Mine Act is drafted narrowly so as to ensure entry without invading an employer's constitutional right against unlawful searches.[65] MSHA delays its inspection only if the complaint on its face, assuming all allegations were true, would not support the finding of an imminent danger.

MSHA also may discover hazardous conditions that constitute an imminent danger during one of multiple inspections conducted of mines during each calendar year. Specifically, the Mine Act requires MSHA to inspect underground mines four times per year. These inspections are not summary or short inspections. Rather, even a nominally sized mine may take several days to inspect. As such, mine inspectors may spend weeks and months inspecting one mine. If during that time the inspector finds any condition that warrants an imminent danger withdrawal order, the inspector has authority to shut down machinery, close mines, and order the withdrawal of personnel.[66]

MSHA can cite and fine mine operators in much the same way OSHA cites and fines employers. Of significance, however, is that the Mine Act gives MSHA authority to levy citations and penalties against individuals, not just employers. That is, under sections 110(c) and (d) of the Mine Act, MSHA is authorized to propose civil penalties against directors, officers, or agents of a corporation if the individual knowingly ordered, authorized, or carried out a violation of a mandatory safety or health standard. Likewise, MSHA may bring criminal charges against an operator or a corporate director, officer, or agent who willfully violates a provision of the Mine Act. Unlike the OSH Act, such penalties and criminal proceedings do not originate only from employee deaths caused by an employer's willful violation of a safety or health standard. Rather, MSHA may consider bringing such actions anytime it issues a section 107(a) imminent danger order of withdrawal. MSHA usually will recommend such proceedings within 240 days from the issuance of a citation or withdrawal order; however, the administration may take as long as 365 days to make such a determination.[67]

13.0 Individual Employee Rights and Labor Unions

Mindful of the dire consequences associated with hazardous job assignments, the U.S. Supreme Court in *Whirlpool Corp. v. Marshall* concluded that employees could refuse to perform some work if the employee reasonably believed the work would result in death or serious injury or illness. In that decision, a unanimous Court reversed decisions in two lower courts and found an employer could not discharge or otherwise take disciplinary action against any employee who refused to perform a job that was inherently unsafe—that is, constituted an imminent danger.[68]

The *Whirlpool* case involved two workers whom the employer directed to venture out onto a screen that was suspended 20 feet above a concrete floor to clean debris. The identical assignment had recently resulted in the death of another employee who had fallen through the screen to the floor below and died only weeks earlier. While the employer had reinforced certain portions of the screen and represented that they were safe, it could not represent that the entire screen was safe. When the employees refused to clean the screen out of fear for their own safety, the employer sent the workers home and withheld a day's pay. Upon review, the Court ruled the employees had the legal right to refuse the assignment based on their "reasonable concerns" for their own safety, and it warned that an employer may not discriminate against an employee for exercising this right. At the same time, the Court declined to conclude that an employer had to pay any employee for any period in which the employee refused to work, even in the face of a perceived imminent danger.[69]

While the Supreme Court elected to remain mute as to wages of employees involved in imminent danger scenarios, the Mine Act does not. Specifically, the Mine Act provides pay for employees when a mine is forced to close under a section 107(a) imminent danger withdrawal order. Section 111 of the Mine Act provides a statutory remedy for miners when their mine is idled or closed due to an MSHA order. In order to stake a claim to relief under this section, miners and/or their representatives must file a claim with the Federal Mine Safety and Health Review Commission.

Likewise, labor unions may seek through collective bargaining agreements to ensure their members receive pay for a scheduled shift even if the employee refuses to perform dangerous work. Regardless of their interest in remuneration, labor unions frequently inform their members of their rights under the OSH Act and *Whirlpool* through various publications. Such publications frequently outline the facts and findings of *Whirlpool* and provide recommended responses to hazardous work conditions. Some even advise as to how and when to report perceived imminent dangers and any discriminatory conduct by an employer to OSHA, the National Labor Relations Board, and the member's union.[70]

The OSH Act also provides relief for employees if the government does not act in the face of an alleged imminent danger. As discussed above, OSHA may petition a federal district court for a TRO to restrain an employer from engaging in operations that pose an imminent danger. The OSH Act does not give independent employees or their representatives this authority. Accordingly, in the event OSHA does not act, labor unions and employees have the right to seek a court order, called a writ of mandamus, that directs OSHA to seek a TRO and to stop an employer to correct a hazard in the workplace.[71]

14.0 Summary

In sum, Congress expanded the OSH Act and the Mine Act to include imminent danger inspections in an effort to avert workplace injuries that could be easily avoided by early identification and correction of workplace hazards. In accordance with this goal, Congress also afforded OSHA and MSHA broad authority to enter and inspect an employer's operation. In addition to OSHA and MSHA's authority to issue citations when imminent dangers are identified, both administrative agencies also have the authority to close an employer's operation until the hazardous condition is corrected.

An employer may avoid these actions, however, by accompanying the compliance officer on his or her inspection of the employer's facility and correcting any hazardous condition identified by the compliance officer during inspection. If an imminent danger cannot be corrected immediately, an employer may avoid a TRO or other consequences by voluntarily removing affected workers from the hazardous environment until the danger is corrected. In working with OSHA and MSHA cooperatively, an employer can ensure a safe work environment for its employees while also avoiding expensive and damaging conflicts with the administrative agencies.

Notes

1. 29 U.S.C. § 663(a).

2. OSHA Field Inspection Reference Manual (Ref. Man.) § 2.103, Ch. I(A)(4).

3. *Id.*

4. *Id.* at Ch. I(B).

5. 29 U.S.C. §657(f)(1); "Any employees or representatives of employees who believe that a violation of a safety or health hazard exists that threatens physical harm, or that an imminent danger exists, may request an inspection by giving notice . . . of such violation or danger." OSHA Inspections, U.S. Department of Labor, OSHA 2098 (2000), p. 3.

6. *Id.*

7. OSHA Inspections, U.S. Department of Labor, OSHA 2098 (2000), p. 3.

8. *Id.*

9. *Ref. Man.* at 2.103, Ch. I(C)(5).

10. *Id.*

11. *Ref. Man.* at Ch. II(A)(2).

12. *Id.*

13. *Ref. Man.* at 2.103, Ch. II(A).

14. *Id.*

15. *Id.*

16. OSHA 2098 at p. 1.

17. *Id.* at pp. 1–2; *Ref. Man.* at 2.103, Ch. I(E) ("Section 17(F) of the Act and 29 C.F.R. § 1903.6 contain a general prohibition against the giving of advance notice of inspections, except as authorized by the Secre- tary or the Secretary's designee").

18. 29 U.S.C. § 657(a).

19. OSHA 2098 at p. 5.

20. *Id.*

21. 29 U.S.C. § 657(a).

22. Marshall v. Barlow's, Inc., 436 U.S. 307 (1978).

23. *Id.*

24. *Ref. Man.* at 2.103 (A)(2)(c).

25. *Ref. Man.* at 2.103 (A)(2)(e).

26. *Id.*

27. *Id.* at Ch. II(A)(2)(c)(iv)(c).

28. See, for example, Donovan v. Sarasota Concrete Co., 693 F.2d 1061 (11th Cir. 1982); Marshall v. North American Car Co., 626 F.2d 320 (3rd Cir. 1980); Marshall v. Central Mine Equipment Co., 608 F.2d 719 (8th Cir. 1979).

29. *Id.*

30. *Id.* at Ch. II(A)(2)(c)(vi).

31. *Id.* at Ch. II(A)(2)(c)(vi).

32. See U.S. v. Morton Salt Co., 338 U.S. 632 (1950); U.S. v. Westinghouse Electric Co., 638 F.2d 570 (3rd Cir. 1980).

33. *Dole v. Trinity Indus. Inc.*, 904 F.2d 867 (1990) (citing *Endicott Johnson Corp. v. Perkins*, 317 U.S. 501 [1943] [where an OSHA subpoena "was not plainly incompetent or irrelevant to *any* lawful purpose of [OSHA] . . . , it was the duty of the district court to order its production."]).

34. 29 C.F.R. § 1903.8.

35. *Id.* at Ch. II(A)(2)(h).

36. *Id.*

37. *Id.* at Ch. II(A)(3); the Act 2098 at p. 5.

38. *Id.*

39. *Ref. Man.* at Ch. II(A)(3)(a). *Id.*

40. *Id.*

41. OSHA 2098 at p. 6.

42. *Id.*

43. *Id.*

44. 29 C.F.R. § 1903.8(A).

45. OSHA 2098 at p. 7.

46. *Id.*

47. *Id.* at p.8.

48. This authority is significantly different than that of a mines inspector, which is discussed below.

49. *Id.* at 2.103 Ch. II(B)(3)(c)(ii); 29 U.S.C. § 662.

50. See, for example, Whirlpool.

51. *Id.* at 2.103 Ch. II(B)(4).

52. *Ref. Man.* Ch. II(A)(5); Act 2098 at p. 8.

53. *Ref. Man.* at 2.103, Ch. II(A)(5).

54. *Id.*; Act 2098 at pp. 8–9.

55. OSHA 2098 at p. 10.

56. *Id.*

57. *Id.*

58. *Id.* at p. 11.

59. *Id.*

60. *Ref. Man.* at 2.102, Ch. IV(A).

61. *Id.*

62. *Id.*

63. *Id.*

64. http://www.msha.gov/codeaphone/codeaphonenew.htm.

65. 30 U.S.C. §813.

66. *Id.*

67. Interpretation and Guidelines on Enforcement or the 1977 Act, Dept. of Labor, Mine Safety and Health Administration, Vol. I (1996).

68. Whirlpool Corp. v. Marshall, 445 U.S. 1 (1980).

69. *Id.*

70. *See, for example, Right to Refuse Unsafe Work,* Communications Workers of America, Local 1101 (2000); *Un- derstanding the Facts About Imminent Danger,* International Association of Machinists and Aerospace Work- ers (2005).

71. 29 USC 662(d).

Appendix: The Occupational Safety and Health Act

OCCUPATIONAL SAFETY AND HEALTH
[OCCUPATIONAL SAFETY AND HEALTH ACT]

as amended[1]
29 U.S.C. § 651 et seq.

CONGRESSIONAL STATEMENT OF FINDINGS AND
DECLARATION OF PURPOSE AND POLICY

29 USC 651

(a) The Congress finds that personal injuries and illnesses arising out of work situations impose a substantial burden upon, and are a hindrance to, interstate commerce in terms of lost production, wage loss, medical expenses, and disability compensation payments.

(b) The Congress declares it to be its purpose and policy, through the exercise of its powers to regulate commerce among the several States and with foreign nations and to provide for the general welfare, to assure so far as possible every working man and woman in the Nation safe and healthful working conditions and to preserve our human resources–

   (1) by encouraging employers and employees in their efforts to reduce the number of occupational safety and health hazards at their places of employment, and to stimulate employers and employees to institute new and to perfect existing programs for providing safe and healthful working conditions;

   (2) by providing that employers and employees have separate but dependent responsibilities and rights with respect to achieving safe and healthful working conditions;

   (3) by authorizing the Secretary of Labor to set mandatory occupational safety and health standards applicable to businesses affecting interstate commerce, and by creating an Occupational Safety and Health Review Commission for carrying out adjudicatory functions under this chapter;

   (4) by building upon advances already made through employer and employee initiative for providing safe and healthful working conditions;

   (5) by providing for research in the field of occupational safety and health, including the psychological factors involved, and by developing innovative methods, techniques, and approaches for dealing with occupational safety and health problems;

   (6) by exploring ways to discover latent diseases, establishing causal connections between diseases and work in environmental conditions, and conducting other research relating to health problems, in recognition of the fact that occupational health standards present problems often different from those involved in occupational safety;

   (7) by providing medical criteria which will assure insofar as practicable that no employee will suffer diminished health, functional capacity, or life expectancy as a result of his work experience;

   (8) by providing for training programs to increase the number and competence of personnel engaged in the field of occupational safety and health;

   (9) by providing for the development and promulgation of occupational safety and health standards;

   (10) by providing an effective enforcement program which shall include a prohibition against giving advance notice of any inspection and sanctions for any individual violating this prohibition;

   (11) by encouraging the States to assume the fullest responsibility for the administration and enforcement of their occupational safety and health laws by providing grants to the States to assist in identifying their needs and responsibilities in the area of occupational safety and health, to develop plans in accordance with the provisions of this chapter, to improve the administration and enforcement of State occupational safety and health laws, and to conduct experimental and demonstration projects in connection therewith;

---

[1]Editor's note: Text from Title 29, Chapter 15 of U.S. Code, amended by Pub. L. 108-199, January 23, 2004; Pub. L. 108-447, December 8, 2004; Pub. L. 107-116, January 10, 2002; Pub. L. 106-554, December 21, 2000; Pub. L. 109-149, December 30, 2005; and Pub. L. 109-236, June 15, 2006..

(12) by providing for appropriate reporting procedures with respect to occupational safety and health which procedures will help achieve the objectives of this chapter and accurately describe the nature of the occupational safety and health problem;

(13) by encouraging joint labor-management efforts to reduce injuries and disease arising out of employment.

(Pub. L. 91-596, Sec. 2, Dec. 29, 1970, 84 Stat. 1590.)

References in Text

This chapter, referred to in subsec. (b)(3), (11), and (12), was in the original "this Act", meaning Pub. L. 91-596, Dec. 29, 1970, 84 Stat. 1590, as amended. For complete classification of this Act to the Code, see Short Title note set out under this section and Tables.

Effective Date

Section 34 of Pub. L. 91-596 provided that: "This Act [enacting this chapter and section 3142-1 of Title 42, The Public Health and Welfare, amending section 553 of this title, sections 5108, 5314, 5315, and 7902 of Title 5, Government Organization and Employees, sections 633 and 636 of Title 15, Commerce and Trade, section 1114 of Title 18, Crimes and Criminal Procedure, and section 1421 of former Title 49, Transportation, and enacting provisions set out as notes under this section and section 1114 of Title 18] shall take effect one hundred and twenty days after the date of its enactment [Dec. 29, 1970]."

Short Title of 1998 Amendment

Pub. L. 105-197, Sec. 1, July 16, 1998, 112 Stat. 638, provided that: "This Act [amending section 670 of this title] may be cited as the 'Occupational Safety and Health Administration Compliance Assistance Authorization Act of 1998'."

Short Title

Section 1 of Pub. L. 91-596 provided: "That this Act [enacting this chapter and section 3142-1 of Title 42, The Public Health and Welfare, amending section 553 of this title, sections 5108, 5314, 5315, and 7902 of Title 5, Government Organization and Employees, sections 633 and 636 of Title 15, Commerce and Trade, section 1114 of Title 18, Crimes and Criminal Procedure, and section 1421 of former Title 49, Transportation, and enacting provisions set out as notes under this section and section 1114 of Title 18] may be cited as the 'Occupational Safety and Health Act of 1970'."

Section Referred to in Other Sections

This section is referred to in section 671 of this title.

DEFINITIONS

29 USC 652

For the purposes of this chapter—

(1) The term "Secretary" means the Secretary of Labor.

(2) The term "Commission" means the Occupational Safety and Health Review Commission established under this chapter.

(3) The term "commerce" means trade, traffic, commerce, transportation, or communication among the several States, or between a State and any place outside thereof, or within the District of Columbia, or a possession of the United States (other than the Trust Territory of the Pacific Islands), or between points in the same State but through a point outside thereof.

(4) The term "person" means one or more individuals, partnerships, associations, corporations, business trusts, legal representatives, or any organized group of persons.

(5) The term "employer" means a person engaged in a business affecting commerce who has employees, but does not include the United States (not including the United States Postal Service) or any State or political subdivision of a State.

(6) The term "employee" means an employee of an employer who is employed in a business of his employer which affects commerce.

(7) The term "State" includes a State of the United States, the District of Columbia, Puerto Rico, the Virgin Islands, American Samoa, Guam, and the Trust Territory of the Pacific Islands.

(8) The term "occupational safety and health standard" means a standard which requires conditions, or the adoption or use of one or more practices, means, methods, operations, or processes, reasonably necessary or appropriate to provide safe or healthful employment and places of employment.

(9) The term "national consensus standard" means any occupational safety and health standard or modification thereof which (1), has been adopted and promulgated by a nationally recognized standards-producing organization under procedures whereby it can be determined by the Secretary that persons interested and affected by the scope or provisions of the standard have reached substantial agreement on its adoption, (2) was formulated in a manner which afforded an opportunity for diverse views to be considered and (3) has been designated as such a standard by the Secretary, after consultation with other appropriate Federal agencies.

(10) The term "established Federal standard" means any operative occupational safety and health standard established by any agency of the United States and presently in effect, or contained in any Act of Congress in force on December 29, 1970.

(11) The term "Committee" means the National Advisory Committee on Occupational Safety and Health established under this chapter.

(12) The term "Director" means the Director of the National Institute for Occupational Safety and Health.

(13) The term "Institute" means the National Institute for Occupational Safety and Health established under this chapter.

(14) The term "Workmen's Compensation Commission" means the National Commission on State Workmen's Compensation Laws established under this chapter.

(Pub. L. 91-596, Sec. 3, Dec. 29, 1970, 84 Stat. 1591; Pub. L. 105-241, Sec. 2(a), Sept. 28, 1998, 112 Stat. 1572.)

Amendments

1998–Par. (5). Pub. L. 105-241 inserted "(not including the United States Postal Service)" after "the United States".

Termination of Trust Territory of the Pacific Islands

For termination of Trust Territory of the Pacific Islands, see note set out preceding section 1681 of Title 48, Territories and Insular Possessions.

Termination of Advisory Committees

Advisory committees in existence on January 5, 1973, to terminate not later than the expiration of the 2-year period following January 5, 1973, unless, in the case of a committee established by the President or an officer of the Federal Government, such committee is renewed by appropriate action prior to the expiration of such 2-year period, or in the case of a committee established by the Congress, its duration is otherwise provided by law. See section 14 of Pub. L. 92-463, Oct. 6, 1972, 86 Stat. 776, set out in the Appendix to Title 5, Government Organization and Employees.

GEOGRAPHIC APPLICABILITY; JUDICIAL ENFORCEMENT; APPLICABILITY TO EXISTING STANDARDS; REPORT TO CONGRESS ON DUPLICATION AND COORDINATION OF FEDERAL LAWS; WORKMEN'S COMPENSATION LAW OR COMMON LAW OR STATUTORY RIGHTS, DUTIES, OR LIABILITIES OF EMPLOYERS AND EMPLOYEES UNAFFECTED

29 USC 653

(a) This chapter shall apply with respect to employment performed in a workplace in a State, the District of Columbia, the Commonwealth of Puerto Rico, the Virgin Islands, American Samoa, Guam, the Trust Territory of the Pacific Islands, Lake Island, Outer Continental Shelf lands defined in the Outer Continental Shelf Lands Act [43 U.S.C. 1331 et seq.], Johnston Island, and the Canal Zone. The Secretary of the Interior shall, by regulation, provide for judicial enforcement of this chapter by the courts established for areas in which there are no United States district courts having jurisdiction.

(b) (1) Nothing in this chapter shall apply to working conditions of employees with respect to which other Federal agencies, and State agencies acting under section 2021 of title 42, exercise statutory authority to prescribe or enforce standards or regulations affecting occupational safety or health.

(2) The safety and health standards promulgated under the Act of June 30, 1936, commonly known as the Walsh-Healey Act [41 U.S.C. 35 et seq.], the Service Contract Act of 1965 [41 U.S.C. 351 et seq.], Public Law 91-54, Act of August 9, 1969, Public Law 85-742, Act of August 23, 1958, and the National Foundation on Arts and Humanities Act [20 U.S.C. 951 et seq.] are superseded on the effective date of corresponding standards, promulgated under this chapter, which are determined by the Secretary to be more effective. Standards issued under the laws listed in this paragraph and in effect on or after the effective date of this chapter shall be deemed to be occupational safety and health standards issued under this chapter, as well as under such other Acts.

(3) The Secretary shall, within three years after the effective date of this chapter, report to the Congress his recommendations for legislation to avoid unnecessary duplication and to achieve coordination between this chapter and other Federal laws.

(4) Nothing in this chapter shall be construed to supersede or in any manner affect any workmen's compensation law or to enlarge or diminish or affect in any other manner the common law or statutory rights, duties, or liabilities of employers and employees under any law with respect to injuries, diseases, or death of employees arising out of, or in the course of, employment.

(Pub. L. 91-596, Sec. 4, Dec. 29, 1970, 84 Stat. 1592.)

References in Text

The Outer Continental Shelf Lands Act, referred to in subsec. (a), is act Aug. 7, 1953, ch. 345, 67 Stat. 462, as amended, which is classified generally to subchapter III (Sec. 1331 et seq.) of chapter 29 of Title 43, Public Lands. For complete classification of this Act to the Code, see Short Title note set out under section 1331 of Title 43 and Tables.

For definition of Canal Zone, referred to in subsec. (a), see section 3602(b) of Title 22, Foreign Relations and Intercourse.

Act of June 30, 1936, commonly known as the Walsh-Healey Act, referred to in subsec. (b)(2), is act June 30, 1936, ch. 881, 49 Stat. 2036, as amended, which is classified generally to section 35 et seq. of Title 41, Public Contracts. For complete classification of this Act to the Code, see Short Title note set out under section 35 of Title 41 and Tables. See section 262 of this title.

The Service Contract Act of 1965, referred to in subsec. (b)(2), is Pub. L. 89-286, Oct. 22, 1965, 79 Stat. 1034, as amended, which is classified generally to chapter 6 (Sec. 351 et seq.) of Title 41. For complete classification of this Act to the Code, see Short Title note set out under section 351 of Title 41 and Tables.

Public Law 91-54, Act of August 9, 1969, referred to in subsec. (b)(2), is Pub. L. 91-54, Aug. 9, 1969, 83 Stat. 96, which enacted section 333 of Title 40, Public Buildings, Property, and Works, and amended section 2 of Pub. L. 87-581, Aug. 13, 1962, 76 Stat. 357, set out as a note under section 327 of Title 40. For complete classification of this Act to the Code, see Tables.

Public Law 85-742, Act of August 23, 1958, referred to in subsec. (b)(2), is Pub. L. 85-742, Aug. 23, 1958, 72 Stat. 835, which amended section 941 of Title 33, Navigation and Navigable Waters, and enacted provisions set out as a note under section 941 of Title 33. For complete classification of this Act to the Code, see Tables.

The National Foundation on the Arts and the Humanities Act, referred to in subsec. (b)(2), is Pub. L. 89-209, Sept. 29, 1965, 79 Stat. 845, as amended, known as the National Foundation on the Arts and the Humanities Act of 1965, which is classified principally to subchapter I (Sec. 951 et seq.) of chapter 26 of Title 20, Education. For complete classification of this Act to the Code, see Short Title note set out under section 951 of Title 20 and Tables.

The effective date of this chapter, referred to in subsec. (b)(2), (3), is the effective date of Pub. L. 91-596, which is 120 days after Dec. 29, 1970, see section 34 of Pub. L. 91-596, set out as an Effective Date note under section 651 of this title.

Termination of Trust Territory of the Pacific Islands

For termination of Trust Territory of the Pacific Islands, see note set out preceding section 1681 of Title 48, Territories and Insular Possessions.

EPA Administrator Not Exercising "Statutory Authority" Under This Section in Exercising Any Authority Under Toxic Substances Control Act

In exercising any authority under the Toxic Substances Control Act (15 U.S.C. 2601 et seq.) in connection with amendment made by section 15(a) of Pub. L. 101-637, the Administrator of the Environmental Protection Agency not, for purposes of subsection (b)(1) of this section, to be considered to be exercising statutory authority to prescribe or enforce standards or regulations affecting occupational safety and health, see section 15(b) of Pub. L. 101-637, set out as a note under section 2646 of Title 15, Commerce and Trade.

Section Referred to in Other Sections

This section is referred to in section 673 of this title; title 15 section 2608; title 42 section 7412; title 49 section 5107.

DUTIES OF EMPLOYERS AND EMPLOYEES

29 USC 654

(a) Each employer–
   (1) shall furnish to each of his employees employment and a place of employment which are free from recognized hazards that are causing or are likely to cause death or serious physical harm to his employees;
   (2) shall comply with occupational safety and health standards promulgated under this chapter.
(b) Each employee shall comply with occupational safety and health standards and all rules, regulations, and orders issued pursuant to this chapter which are applicable to his own actions and conduct.

(Pub. L. 91-596, Sec. 5, Dec. 29, 1970, 84 Stat. 1593.)

Section Referred to in Other Sections

This section is referred to in sections 658, 666 of this title; title 2 section 1341; title 3 section 425; title 42 section 7412.

STANDARDS

29 USC 655

(a) Promulgation by Secretary of national consensus standards and established Federal standards; time for promulgation; conflicting standards
   Without regard to chapter 5 of title 5 or to the other subsections of this section, the Secretary shall, as soon as practicable during the period beginning with the effective date of this chapter and ending two years after such date, by rule promulgate as an occupational safety or health standard any national consensus standard, and any established Federal standard, unless he determines that the promulgation of such a standard would not result in improved safety or health for specifically designated employees. In the event of conflict among any such standards, the Secretary shall promulgate the standard which assures the greatest protection of the safety or health of the affected employees.
(b) Procedure for promulgation, modification, or revocation of standards
   The Secretary may by rule promulgate, modify, or revoke any occupational safety or health standard in the following manner:
   (1) Whenever the Secretary, upon the basis of information submitted to him in writing by an interested person, a representative of any organization of employers or employees, a nationally recognized standards-producing organization, the Secretary of Health and Human Services, the National Institute for Occupational Safety and Health, or a State or political subdivision, or on the basis of information developed by the Secretary or

otherwise available to him, determines that a rule should be promulgated in order to serve the objectives of this chapter, the Secretary may request the recommendations of an advisory committee appointed under section 656 of this title. The Secretary shall provide such an advisory committee with any proposals of his own or of the Secretary of Health and Human Services, together with all pertinent factual information developed by the Secretary or the Secretary of Health and Human Services, or otherwise available, including the results of research, demonstrations, and experiments. An advisory committee shall submit to the Secretary its recommendations regarding the rule to be promulgated within ninety days from the date of its appointment or within such longer or shorter period as may be prescribed by the Secretary, but in no event for a period which is longer than two hundred and seventy days.

(2) The Secretary shall publish a proposed rule promulgating, modifying, or revoking an occupational safety or health standard in the Federal Register and shall afford interested persons a period of thirty days after publication to submit written data or comments. Where an advisory committee is appointed and the Secretary determines that a rule should be issued, he shall publish the proposed rule within sixty days after the submission of the advisory committee's recommendations or the expiration of the period prescribed by the Secretary for such submission.

(3) On on before the last day of the period provided for the submission of written data or comments under paragraph (2), any interested person may file with the Secretary written objections to the proposed rule, stating the grounds therefor and requesting a public hearing on such objections. Within thirty days after the last day for filing such objections, the Secretary shall publish in the Federal Register a notice specifying the occupational safety or health standard to which objections have been filed and a hearing requested, and specifying a time and place for such hearing.

(4) Within sixty days after the expiration of the period provided for the submission of written data or comments under paragraph (2), or within sixty days after the completion of any hearing held under paragraph (3), the Secretary shall issue a rule promulgating, modifying, or revoking an occupational safety or health standard or make a determination that a rule should not be issued. Such a rule may contain a provision delaying its effective date for such period (not in excess of ninety days) as the Secretary determines may be necessary to insure that affected employers and employees will be informed of the existence of the standard and of its terms and that employers affected are given an opportunity to familiarize themselves and their employees with the existence of the requirements of the standard.

(5) The Secretary, in promulgating standards dealing with toxic materials or harmful physical agents under this subsection, shall set the standard which most adequately assures, to the extent feasible, on the basis of the best available evidence, that no employee will suffer material impairment of health or functional capacity even if such employee has regular exposure to the hazard dealt with by such standard for the period of his working life. Development of standards under this subsection shall be based upon research, demonstrations, experiments, and such other information as may be appropriate. In addition to the attainment of the highest degree of health and safety protection for the employee, other considerations shall be the latest available scientific data in the field, the feasibility of the standards, and experience gained under this and other health and safety laws. Whenever practicable, the standard promulgated shall be expressed in terms of objective criteria and of the performance desired.

(6) (A) Any employer may apply to the Secretary for a temporary order granting a variance from a standard or any provision thereof promulgated under this section. Such temporary order shall be granted only if the employer files an application which meets the requirements of clause (B) and establishes that (i) he is unable to comply with a standard by its effective date because of unavailability of professional or technical personnel or of materials and equipment needed to come into compliance with the standard or because necessary construction or alteration of facilities cannot be completed by the effective date, (ii) he is taking all available steps to safeguard his employees against the hazards covered by the standard, and (iii) he has an effective program for coming into compliance with the standard as quickly as practicable. Any temporary order issued under this paragraph shall prescribe the practices, means,

methods, operations, and processes which the employer must adopt and use while the order is in effect and state in detail his program for coming into compliance with the standard. Such a temporary order may be granted only after notice to employees and an opportunity for a hearing: Provided, That the Secretary may issue one interim order to be effective until a decision is made on the basis of the hearing. No temporary order may be in effect for longer than the period needed by the employer to achieve compliance with the standard or one year, whichever is shorter, except that such an order may be renewed not more than twice (I) so long as the requirements of this paragraph are met and (II) if an application for renewal is filed at least 90 days prior to the expiration date of the order. No interim renewal of an order may remain in effect for longer than 180 days.

(B) An application for a temporary order under this paragraph (6) shall contain:
    (i) a specification of the standard or portion thereof from which the employer seeks a variance,
    (ii) a representation by the employer, supported by representations from qualified persons having firsthand knowledge of the facts represented, that he is unable to comply with the standard or portion thereof and a detailed statement of the reasons therefor,
    (iii) a statement of the steps he has taken and will take (with specific dates) to protect employees against the hazard covered by the standard,
    (iv) a statement of when he expects to be able to comply with the standard and what steps he has taken and what steps he will take (with dates specified) to come into compliance with the standard, and
    (v) a certification that he has informed his employees of the application by giving a copy thereof to their authorized representative, posting a statement giving a summary of the application and specifying where a copy may be examined at the place or places where notices to employees are normally posted, and by other appropriate means.

A description of how employees have been informed shall be contained in the certification. The information to employees shall also inform them of their right to petition the Secretary for a hearing.

(C) The Secretary is authorized to grant a variance from any standard or portion thereof whenever he determines, or the Secretary of Health and Human Services certifies, that such variance is necessary to permit an employer to participate in an experiment approved by him or the Secretary of Health and Human Services designed to demonstrate or validate new and improved techniques to safeguard the health or safety of workers.

(7) Any standard promulgated under this subsection shall prescribe the use of labels or other appropriate forms of warning as are necessary to insure that employees are apprised of all hazards to which they are exposed, relevant symptoms and appropriate emergency treatment, and proper conditions and precautions of safe use or exposure. Where appropriate, such standard shall also prescribe suitable protective equipment and control or technological procedures to be used in connection with such hazards and shall provide for monitoring or measuring employee exposure at such locations and intervals, and in such manner as may be necessary for the protection of employees. In addition, where appropriate, any such standard shall prescribe the type and frequency of medical examinations or other tests which shall be made available, by the employer or at his cost, to employees exposed to such hazards in order to most effectively determine whether the health of such employees is adversely affected by such exposure. In the event such medical examinations are in the nature of research, as determined by the Secretary of Health and Human Services, such examinations may be furnished at the expense of the Secretary of Health and Human Services. The results of such examinations or tests shall be furnished only to the Secretary or the Secretary of Health and Human Services, and, at the request of the employee, to his physician. The Secretary, in consultation with the Secretary of Health and Human Services, may by rule promulgated pursuant to section 553 of title 5, make appropriate modifications in the foregoing requirements relating to the use of labels or other forms of warning, monitoring or measuring, and medical examinations, as may be warranted by experience, information, or medical or technological developments acquired subsequent to the promulgation of the relevant standard.

(8) Whenever a rule promulgated by the Secretary differs substantially from an existing national consensus standard, the Secretary shall, at the same time, publish in the Federal Register a statement of the reasons why the rule as adopted will better effectuate the purposes of this chapter than the national consensus standard.

(c) Emergency temporary standards

(1) The Secretary shall provide, without regard to the requirements of chapter 5 of title 5, for an emergency temporary standard to take immediate effect upon publication in the Federal Register if he determines (A) that employees are exposed to grave danger from exposure to substances or agents determined to be toxic or physically harmful or from new hazards, and (B) that such emergency standard is necessary to protect employees from such danger.

(2) Such standard shall be effective until superseded by a standard promulgated in accordance with the procedures prescribed in paragraph (3) of this subsection.

(3) Upon publication of such standard in the Federal Register the Secretary shall commence a proceeding in accordance with subsection (b) of this section, and the standard as published shall also serve as a proposed rule for the proceeding. The Secretary shall promulgate a standard under this paragraph no later than six months after publication of the emergency standard as provided in paragraph (2) of this subsection.

(d) Variances from standards; procedure

Any affected employer may apply to the Secretary for a rule or order for a variance from a standard promulgated under this section. Affected employees shall be given notice of each such application and an opportunity to participate in a hearing. The Secretary shall issue such rule or order if he determines on the record, after opportunity for an inspection where appropriate and a hearing, that the proponent of the variance has demonstrated by a preponderance of the evidence that the conditions, practices, means, methods, operations, or processes used or proposed to be used by an employer will provide employment and places of employment to his employees which are as safe and healthful as those which would prevail if he complied with the standard. The rule or order so issued shall prescribe the conditions the employer must maintain, and the practices, means, methods, operations, and processes which he must adopt and utilize to the extent they differ from the standard in question. Such a rule or order may be modified or revoked upon application by an employer, employees, or by the Secretary on his own motion, in the manner prescribed for its issuance under this subsection at any time after six months from its issuance.

(e) Statement of reasons for Secretary's determinations; publication in Federal Register

Whenever the Secretary promulgates any standard, makes any rule, order, or decision, grants any exemption or extension of time, or compromises, mitigates, or settles any penalty assessed under this chapter, he shall include a statement of the reasons for such action, which shall be published in the Federal Register.

(f) Judicial review

Any person who may be adversely affected by a standard issued under this section may at any time prior to the sixtieth day after such standard is promulgated file a petition challenging the validity of such standard with the United States court of appeals for the circuit wherein such person resides or has his principal place of business, for a judicial review of such standard. A copy of the petition shall be forthwith transmitted by the clerk of the court to the Secretary. The filing of such petition shall not, unless otherwise ordered by the court, operate as a stay of the standard. The determinations of the Secretary shall be conclusive if supported by substantial evidence in the record considered as a whole.

(g) Priority for establishment of standards

In determining the priority for establishing standards under this section, the Secretary shall give due regard to the urgency of the need for mandatory safety and health standards for particular industries, trades, crafts, occupations, businesses, workplaces or work environments. The Secretary shall also give due regard to the recommendations of the Secretary of Health and Human Services regarding the need for mandatory standards in determining the priority for establishing such standards.

(Pub. L. 91-596, Sec. 6, Dec. 29, 1970, 84 Stat. 1593; Pub. L. 96-88, title V, Sec. 509(b), Oct. 17, 1979, 93 Stat. 695.)

References in Text

The effective date of this chapter, referred to in subsec. (a), is the effective date of Pub. L. 91-596, Dec. 29, 1970, 84 Stat. 1590, which is 120 days after Dec. 29, 1970, see section 34 of Pub. L. 91-596, set out as an Effective Date note under section 651 of this title.

Change of Name

"Secretary of Health and Human Services" substituted for "Secretary of Health, Education, and Welfare" in subsecs. (b)(1), (6)(C), (7), and (g) pursuant to section 509(b) of Pub. L. 96-88 which is classified to section 3508(b) of Title 20, Education.

Termination of Advisory Committees

Advisory committees in existence on January 5, 1973, to terminate not later than the expiration of the 2-year period following January 5, 1973, unless, in the case of a committee established by the President or an officer of the Federal Government, such committee is renewed by appropriate action prior to the expiration of such 2-year period, or in the case of a committee established by the Congress, its duration is otherwise provided by law. See section 14 of Pub. L. 92-463, Oct. 6, 1972, 86 Stat. 776, set out in the Appendix to Title 5, Government Organization and Employees.

Prohibition on Exposure of Workers to Chemical or Other Hazards for Purpose of Conducting Experiments

Pub. L. 102-394, title I, Sec. 102, Oct. 6, 1992, 106 Stat. 1799, provided that: "None of the funds appropriated under this Act or subsequent Departments of Labor, Health and Human Services, and Education, and Related Agencies Appropriations Acts shall be used to grant variances, interim orders or letters of clarification to employers which will allow exposure of workers to chemicals or other workplace hazards in excess of existing Occupational Safety and Health Administration standards for the purpose of conducting experiments on workers' health or safety."

Similar provisions were contained in the following prior appropriation acts:

Pub. L. 102-170, title I, Sec. 102, Nov. 26, 1991, 105 Stat. 1114.
Pub. L. 101-517, title I, Sec. 102, Nov. 5, 1990, 104 Stat. 2196.
Pub. L. 101-166, title I, Sec. 102, Nov. 21, 1989, 103 Stat. 1165.
Pub. L. 100-202, Sec. 101(h) [title I, Sec. 102], Dec. 22, 1987, 101 Stat. 1329-256, 1329-263.
Pub. L. 99-500, Sec. 101(i) [H.R. 5233, title I, Sec. 102], Oct. 18, 1986, 100 Stat. 1783-287, and
Pub. L. 99-591, Sec. 101(i) [H.R. 5233, title I, Sec. 102], Oct. 30, 1986, 100 Stat. 3341-287.
Pub. L. 99-178, title I, Sec. 102, Dec. 12, 1985, 99 Stat. 1109.
Pub. L. 98-619, title I, Sec. 102, Nov. 8, 1984, 98 Stat. 3311.

Occupational Health Standard Concerning Exposure to Bloodborne Pathogens

Pub. L. 102-170, title I, Sec. 100, Nov. 26, 1991, 105 Stat. 1113, provided that:

"(a) Notwithstanding any other provision of law, on or before December 1, 1991, the Secretary of Labor, acting under the Occupational Safety and Health Act of 1970 [29 U.S.C. 651 et seq.], shall promulgate a final occupational health standard concerning occupational exposure to bloodborne pathogens. The final standard shall be based on the proposed standard as published in the Federal Register on May 30, 1989 (54 FR 23042), concerning occupational exposures to the hepatitis B virus, the human immunodeficiency virus and other bloodborne pathogens.

"(b) In the event that the final standard referred to in subsection (a) is not promulgated by the date required under such subsection, the proposed standard on occupational exposure to bloodborne pathogens as published in the Federal Register on May 30, 1989 (54 FR 23042) shall become effective as if such proposed standard had been promulgated as a final standard by the Secretary of Labor, and remain in effect until the date on which such Secretary promulgates the final standard referred to in subsection (a).

"(c) Nothing in this Act [enacting section 962 of Title 30, Mineral Lands and Mining, amending section 290b of Title 42, The Public Health and Welfare, enacting provisions set out as notes under section 1070a of Title 20, Education and section 1383 of Title 42, and amending provisions set out as notes under section 1255a of Title 8, Aliens and Nationality, and section 1221-1 of Title 20] shall be construed to require the Secretary of Labor (acting through the Occupational Safety and Health Administration) to revise the employment accident reporting regulations published at 29 C.F.R. 1904.8."

Retention of Markings and Placards

Pub. L. 101-615, Sec. 29, Nov. 16, 1990, 104 Stat. 3277, provided that: "Not later than 18 months after the date of enactment of this Act [Nov. 16, 1990], the Secretary of Labor, in consultation with the Secretary of Transportation and the Secretary of the Treasury, shall issue under section 6(b) of the Occupational Safety and Health Act of 1970 (29 U.S.C. 655(b)) standards requiring any employer who receives a package, container, motor vehicle, rail freight car, aircraft, or vessel which contains a hazardous material and which is required to be marked, placarded, or labeled in accordance with regulations issued under the Hazardous Materials Transportation Act [former 49 U.S.C. 1801 et seq.] to retain the markings, placards, and labels, and any other information as may be required by such regulations on the package, container, motor vehicle, rail freight car, aircraft, or vessel, until the hazardous materials have been removed therefrom."

Chemical Process Safety Management

Pub. L. 101-549, title III, Sec. 304, Nov. 15, 1990, 104 Stat. 2576, provided that:

"(a) Chemical Process Safety Standard.–The Secretary of Labor shall act under the Occupational Safety and Health Act of 1970 (29 U.S.C. 653) [29 U.S.C. 651 et seq.] to prevent accidental releases of chemicals which could pose a threat to employees. Not later than 12 months after the date of enactment of the Clean Air Act Amendments of 1990 [Nov. 15, 1990], the Secretary of Labor, in coordination with the Administrator of the Environmental Protection Agency, shall promulgate, pursuant to the Occupational Safety and Health Act, a chemical process safety standard designed to protect employees from hazards associated with accidental releases of highly hazardous chemicals in the workplace.

(b) List of Highly Hazardous Chemicals.–The Secretary shall include as part of such standard a list of highly hazardous chemicals, which include toxic, flammable, highly reactive and explosive substances. The list of such chemicals may include those chemicals listed by the Administrator under section 302 of the Emergency Planning and Community Right to Know Act of 1986 [42 U.S.C. 11002]. The Secretary may make additions to such list when a substance is found to pose a threat of serious injury or fatality in the event of an accidental release in the workplace.

(c) Elements of Safety Standard.–Such standard shall, at minimum, require employers to–

  (1) develop and maintain written safety information identifying workplace chemical and process hazards, equipment used in the processes, and technology used in the processes;

  (2) perform a workplace hazard assessment, including, as appropriate, identification of potential sources of accidental releases, an identification of any previous release within the facility which had a likely potential for catastrophic consequences in the workplace, estimation of workplace effects of a range of releases, estimation of the health and safety effects of such range on employees;

  (3) consult with employees and their representatives on the development and conduct of hazard assessments and the development of chemical accident prevention plans and provide access to these and other records required under the standard;

  (4) establish a system to respond to the workplace hazard assessment findings, which shall address prevention, mitigation, and emergency responses;

  (5) periodically review the workplace hazard assessment and response system;

  (6) develop and implement written operating procedures for the chemical process including procedures for each operating phase, operating limitations, and safety and health considerations;

  (7) provide written safety and operating information to employees and train employees in operating procedures, emphasizing hazards and safe practices;

  (8) ensure contractors and contract employees are provided appropriate information and training;

  (9) train and educate employees and contractors in emergency response in a manner as comprehensive and effective as that required by the regulation promulgated pursuant to section 126(d) of the Superfund Amendments and Reauthorization Act [of 1986] [Pub. L. 99-499, set out in a note below];

  (10) establish a quality assurance program to ensure that initial process related equipment, maintenance materials, and spare parts are fabricated and installed consistent with design specifications;

(11) establish maintenance systems for critical process related equipment including written procedures, employee training, appropriate inspections, and testing of such equipment to ensure ongoing mechanical integrity;

(12) conduct pre-start-up safety reviews of all newly installed or modified equipment;

(13) establish and implement written procedures to manage change to process chemicals, technology, equipment and facilities; and

(14) investigate every incident which results in or could have resulted in a major accident in the workplace, with any findings to be reviewed by operating personnel and modifications made if appropriate.

(d) State Authority.–Nothing in this section may be construed to diminish the authority of the States and political subdivisions thereof as described in section 112(r)(11) of the Clean Air Act [42 U.S.C. 7412(r)(11)]."

Worker Protection Standards

Pub. L. 99-499, title I, Sec. 126(a)-(f), Oct. 17, 1986, 100 Stat. 1690-1692, as amended by Pub. L. 100-202, Sec. 101(f) [title II, Sec. 201], Dec. 22, 1987, 101 Stat. 1329-187, 1329-198, provided:

"(a) Promulgation.–Within one year after the date of the enactment of this section [Oct. 17, 1986], the Secretary of Labor shall, pursuant to section 6 of the Occupational Safety and Health Act of 1970 [29 U.S.C. 655], promulgate standards for the health and safety protection of employees engaged in hazardous waste operations.

(b) Proposed Standards.–The Secretary of Labor shall issue proposed regulations on such standards which shall include, but need not be limited to, the following worker protection provisions:

(1) Site analysis.–Requirements for a formal hazard analysis of the site and development of a site specific plan for worker protection.

(2) Training.–Requirements for contractors to provide initial and routine training of workers before such workers are permitted to engage in hazardous waste operations which would expose them to toxic substances.

(3) Medical surveillance.–A program of regular medical examination, monitoring, and surveillance of workers engaged in hazardous waste operations which would expose them to toxic substances.

(4) Protective equipment.–Requirements for appropriate personal protective equipment, clothing, and respirators for work in hazardous waste operations.

(5) Engineering controls.–Requirements for engineering controls concerning the use of equipment and exposure of workers engaged in hazardous waste operations.

(6) Maximum exposure limits.–Requirements for maximum exposure limitations for workers engaged in hazardous waste operations, including necessary monitoring and assessment procedures.

(7) Informational program.–A program to inform workers engaged in hazardous waste operations of the nature and degree of toxic exposure likely as a result of such hazardous waste operations.

(8) Handling.–Requirements for the handling, transporting, labeling, and disposing of hazardous wastes.

(9) New technology program.–A program for the introduction of new equipment or technologies that will maintain worker protections.

(10) Decontamination procedures.–Procedures for decontamination.

(11) Emergency response.–Requirements for emergency response and protection of workers engaged in hazardous waste operations.

(c) Final Regulations.–Final regulations under subsection (a) shall take effect one year after the date they are promulgated. In promulgating final regulations on standards under subsection (a), the Secretary of Labor shall include each of the provisions listed in paragraphs (1) through (11) of subsection (b) unless the Secretary determines that the evidence in the public record considered as a whole does not support inclusion of any such provision.

(d) Specific Training Standards.–

(1) Offsite instruction; field experience.–Standards promulgated under subsection (a) shall include training standards requiring that general site workers (such as equipment operators, general laborers, and other supervised personnel) engaged in hazardous substance

removal or other activities which expose or potentially expose such workers to hazardous substances receive a minimum of 40 hours of initial instruction off the site, and a minimum of three days of actual field experience under the direct supervision of a trained, experienced supervisor, at the time of assignment. The requirements of the preceding sentence shall not apply to any general site worker who has received the equivalent of such training. Workers who may be exposed to unique or special hazards shall be provided additional training.

(2) Training of supervisors.–Standards promulgated under subsection (a) shall include training standards requiring that onsite managers and supervisors directly responsible for the hazardous waste operations (such as foremen) receive the same training as general site workers set forth in paragraph (1) of this subsection and at least eight additional hours of specialized training on managing hazardous waste operations. The requirements of the preceding sentence shall not apply to any person who has received the equivalent of such training.

(3) Certification; enforcement.–Such training standards shall contain provisions for certifying that general site workers, onsite managers, and supervisors have received the specified training and shall prohibit any individual who has not received the specified training from engaging in hazardous waste operations covered by the standard. The certification procedures shall be no less comprehensive than those adopted by the Environmental Protection Agency in its Model Accreditation Plan for Asbestos Abatement Training as required under the Asbestos Hazard Emergency Response Act of 1986 [Pub. L. 99-519, see Short Title of 1986 Amendment note, set out under section 2601 of Title 15, Commerce and Trade].

(4) Training of emergency response personnel.–Such training standards shall set forth requirements for the training of workers who are responsible for responding to hazardous emergency situations who may be exposed to toxic substances in carrying out their responsibilities.

(e) Interim Regulations.–The Secretary of Labor shall issue interim final regulations under this section within 60 days after the enactment of this section [Oct. 17, 1986] which shall provide no less protection under this section for workers employed by contractors and emergency response workers than the protections contained in the Environmental Protection Agency Manual (1981) 'Health and Safety Requirements for Employees Engaged in Field Activities' and existing standards under the Occupational Safety and Health Act of 1970 [29 U.S.C. 651 et seq.] found in subpart C of part 1926 of title 29 of the Code of Federal Regulations. Such interim final regulations shall take effect upon issuance and shall apply until final regulations become effective under subsection (c).

(f) Coverage of Certain State and Local Employees.–Not later than 90 days after the promulgation of final regulations under subsection (a), the Administrator shall promulgate standards identical to those promulgated by the Secretary of Labor under subsection (a). Standards promulgated under this subsection shall apply to employees of State and local governments in each State which does not have in effect an approved State plan under section 18 of the Occupational Safety and Health Act of 1970 [29 U.S.C. 667] providing for standards for the health and safety protection of employees engaged in hazardous waste operations."

Section Referred to in Other Sections

This section is referred to in sections 656, 657, 658, 666, 667, 668, 669 of this title; title 2 section 1341; title 3 section 425; title 7 section 1942; title 42 section 4853.

ADMINISTRATION

29 USC 656

(a) National Advisory Committee on Occupational Safety and Health; establishment; membership; appointment; Chairman; functions; meetings; compensation; secretarial and clerical personnel

(1) There is hereby established a National Advisory Committee on Occupational Safety and Health consisting of twelve members appointed by the Secretary, four of whom are to be designated by the Secretary of Health and Human Services, without regard to the

provisions of title 5 governing appointments in the competitive service, and composed of representatives of management, labor, occupational safety and occupational health professions, and of the public. The Secretary shall designate one of the public members as Chairman. The members shall be selected upon the basis of their experience and competence in the field of occupational safety and health.

(2) The Committee shall advise, consult with, and make recommendations to the Secretary and the Secretary of Health and Human Services on matters relating to the administration of this chapter. The Committee shall hold no fewer than two meetings during each calendar year. All meetings of the Committee shall be open to the public and a transcript shall be kept and made available for public inspection.

(3) The members of the Committee shall be compensated in accordance with the provisions of section 3109 of title 5.

(4) The Secretary shall furnish to the Committee an executive secretary and such secretarial, clerical, and other services as are deemed necessary to the conduct of its business.

(b) Advisory committees; appointment; duties; membership; compensation; reimbursement to member's employer; meetings; availability of records; conflict of interest

An advisory committee may be appointed by the Secretary to assist him in his standard-setting functions under section 655 of this title. Each such committee shall consist of not more than fifteen members and shall include as a member one or more designees of the Secretary of Health and Human Services, and shall include among its members an equal number of persons qualified by experience and affiliation to present the viewpoint of the employers involved, and of persons similarly qualified to present the viewpoint of the workers involved, as well as one or more representatives of health and safety agencies of the States. An advisory committee may also include such other persons as the Secretary may appoint who are qualified by knowledge and experience to make a useful contribution to the work of such committee, including one or more representatives of professional organizations of technicians or professionals specializing in occupational safety or health, and one or more representatives of nationally recognized standards-producing organizations, but the number of persons so appointed to any such advisory committee shall not exceed the number appointed to such committee as representatives of Federal and State agencies. Persons appointed to advisory committees from private life shall be compensated in the same manner as consultants or experts under section 3109 of title 5. The Secretary shall pay to any State which is the employer of a member of such a committee who is a representative of the health or safety agency of that State, reimbursement sufficient to cover the actual cost to the State resulting from such representative's membership on such committee. Any meeting of such committee shall be open to the public and an accurate record shall be kept and made available to the public. No member of such committee (other than representatives of employers and employees) shall have an economic interest in any proposed rule.

(c) Use of services, facilities, and personnel of Federal, State, and local agencies; reimbursement; employment of experts and consultants or organizations; renewal of contracts; compensation; travel expenses

In carrying out his responsibilities under this chapter, the Secretary is authorized to—

(1) use, with the consent of any Federal agency, the services, facilities, and personnel of such agency, with or without reimbursement, and with the consent of any State or political subdivision thereof, accept and use the services, facilities, and personnel of any agency of such State or subdivision with reimbursement; and

(2) employ experts and consultants or organizations thereof as authorized by section 3109 of title 5, except that contracts for such employment may be renewed annually; compensate individuals so employed at rates not in excess of the rate specified at the time of service for grade GS-18 under section 5332 of title 5, including traveltime, and allow them while away from their homes or regular places of business, travel expenses (including per diem in lieu of subsistence) as authorized by section 5703 of title 5 for persons in the Government service employed intermittently, while so employed.

(Pub. L. 91-596, Sec. 7, Dec. 29, 1970, 84 Stat. 1597; Pub. L. 96-88, title V, Sec. 509(b), Oct. 17, 1979, 93 Stat. 695.)

Change of Name

"Secretary of Health and Human Services" substituted for "Secretary of Health, Education, and Welfare" in subsecs. (a)(1), (2) and (b) pursuant to section 509(b) of Pub. L. 96-88 which is classified to section 3508(b) of Title 20, Education.

Termination of Advisory Committees

Advisory committees in existence on January 5, 1973, to terminate not later than the expiration of the 2-year period following January 5, 1973, unless, in the case of a committee established by the President or an officer of the Federal Government, such committee is renewed by appropriate action prior to the expiration of such 2-year period, or in the case of a committee established by the Congress, its duration is otherwise provided by law. See section 14 of Pub. L. 92-463, Oct. 6, 1972, 86 Stat. 776, set out in the Appendix to Title 5, Government Organization and Employees.

References in Other Laws to GS-16, 17, or 18 Pay Rates

References in laws to the rates of pay for GS-16, 17, or 18, or to maximum rates of pay under the General Schedule, to be considered references to rates payable under specified sections of Title 5, Government Organization and Employees, see section 529 [title I, Sec. 101(c)(1)] of Pub. L. 101-509, set out in a note under section 5376 of Title 5.

Section Referred to in Other Sections

This section is referred to in section 655 of this title.

INSPECTIONS, INVESTIGATIONS, AND RECORDKEEPING

29 USC 657

(a) Authority of Secretary to enter, inspect, and investigate places of employment; time and manner

In order to carry out the purposes of this chapter, the Secretary, upon presenting appropriate credentials to the owner, operator, or agent in charge, is authorized—

  (1) to enter without delay and at reasonable times any factory, plant, establishment, construction site, or other area, workplace or environment where work is performed by an employee of an employer; and

  (2) to inspect and investigate during regular working hours and at other reasonable times, and within reasonable limits and in a reasonable manner, any such place of employment and all pertinent conditions, structures, machines, apparatus, devices, equipment, and materials therein, and to question privately any such employer, owner, operator, agent, or employee.

(b) Attendance and testimony of witnesses and production of evidence; enforcement of subpoena

In making his inspections and investigations under this chapter the Secretary may require the attendance and testimony of witnesses and the production of evidence under oath. Witnesses shall be paid the same fees and mileage that are paid witnesses in the courts of the United States. In case of a contumacy, failure, or refusal of any person to obey such an order, any district court of the United States or the United States courts of any territory or possession, within the jurisdiction of which such person is found, or resides or transacts business, upon the application by the Secretary, shall have jurisdiction to issue to such person an order requiring such person to appear to produce evidence if, as, and when so ordered, and to give testimony relating to the matter under investigation or in question, and any failure to obey such order of the court may be punished by said court as a contempt thereof.

(c) Maintenance, preservation, and availability of records; issuance of regulations; scope of records; periodic inspections by employer; posting of notices by employer; notification of employee of corrective action

  (1) Each employer shall make, keep and preserve, and make available to the Secretary or the Secretary of Health and Human Services, such records regarding his activities relating to this chapter as the Secretary, in cooperation with the Secretary of Health and Human Services, may prescribe by regulation as necessary or appropriate for the enforcement of this chapter or for developing information regarding the causes and prevention of occupa-

tional accidents and illnesses. In order to carry out the provisions of this paragraph such regulations may include provisions requiring employers to conduct periodic inspections. The Secretary shall also issue regulations requiring that employers, through posting of notices or other appropriate means, keep their employees informed of their protections and obligations under this chapter, including the provisions of applicable standards.

(2) The Secretary, in cooperation with the Secretary of Health and Human Services, shall prescribe regulations requiring employers to maintain accurate records of, and to make periodic reports on, work-related deaths, injuries and illnesses other than minor injuries requiring only first aid treatment and which do not involve medical treatment, loss of consciousness, restriction of work or motion, or transfer to another job.

(3) The Secretary, in cooperation with the Secretary of Health and Human Services, shall issue regulations requiring employers to maintain accurate records of employee exposures to potentially toxic materials or harmful physical agents which are required to be monitored or measured under section 655 of this title. Such regulations shall provide employees or their representatives with an opportunity to observe such monitoring or measuring, and to have access to the records thereof. Such regulations shall also make appropriate provision for each employee or former employee to have access to such records as will indicate his own exposure to toxic materials or harmful physical agents. Each employer shall promptly notify any employee who has been or is being exposed to toxic materials or harmful physical agents in concentrations or at levels which exceed those prescribed by an applicable occupational safety and health standard promulgated under section 655 of this title, and shall inform any employee who is being thus exposed of the corrective action being taken.

(d) Obtaining of information
Any information obtained by the Secretary, the Secretary of Health and Human Services, or a State agency under this chapter shall be obtained with a minimum burden upon employers, especially those operating small businesses. Unnecessary duplication of efforts in obtaining information shall be reduced to the maximum extent feasible.

(e) Employer and authorized employee representatives to accompany Secretary or his authorized representative on inspection of workplace; consultation with employees where no authorized employee representative is present
Subject to regulations issued by the Secretary, a representative of the employer and a representative authorized by his employees shall be given an opportunity to accompany the Secretary or his authorized representative during the physical inspection of any workplace under subsection (a) of this section for the purpose of aiding such inspection. Where there is no authorized employee representative, the Secretary or his authorized representative shall consult with a reasonable number of employees concerning matters of health and safety in the workplace.

(f) Request for inspection by employees or representative of employees; grounds; procedure; determination of request; notification of Secretary or representative prior to or during any inspection of violations; procedure for review of refusal by representative of Secretary to issue citation for alleged violations
(1) Any employees or representative of employees who believe that a violation of a safety or health standard exists that threatens physical harm, or that an imminent danger exists, may request an inspection by giving notice to the Secretary or his authorized representative of such violation or danger. Any such notice shall be reduced to writing, shall set forth with reasonable particularity the grounds for the notice, and shall be signed by the employees or representative of employees, and a copy shall be provided the employer or his agent no later than at the time of inspection, except that, upon the request of the person giving such notice, his name and the names of individual employees referred to therein shall not appear in such copy or on any record published, released, or made available pursuant to subsection (g) of this section. If upon receipt of such notification the Secretary determines there are reasonable grounds to believe that such violation or danger exists, he shall make a special inspection in accordance with the provisions of this section as soon as practicable, to determine if such violation or danger exists. If the Secretary determines there are no reasonable grounds to believe that a violation or danger exists he shall notify the employees or representative of the employees in writing of such determination.

(2) Prior to or during any inspection of a workplace, any employees or representative of employees employed in such workplace may notify the Secretary or any representative of the Secretary responsible for conducting the inspection, in writing, of any violation of this chapter which they have reason to believe exists in such workplace. The Secretary shall, by regulation, establish procedures for informal review of any refusal by a representative of the Secretary to issue a citation with respect to any such alleged violation and shall furnish the employees or representative of employees requesting such review a written statement of the reasons for the Secretary's final disposition of the case.

(g) Compilation, analysis, and publication of reports and information; rules and regulations

    (1) The Secretary and Secretary of Health and Human Services are authorized to compile, analyze, and publish, either in summary or detailed form, all reports or information obtained under this section.

    (2) The Secretary and the Secretary of Health and Human Services shall each prescribe such rules and regulations as he may deem necessary to carry out their responsibilities under this chapter, including rules and regulations dealing with the inspection of an employer's establishment.

(h) Use of results of enforcement activities

The Secretary shall not use the results of enforcement activities, such as the number of citations issued or penalties assessed, to evaluate employees directly involved in enforcement activities under this chapter or to impose quotas or goals with regard to the results of such activities.

(Pub. L. 91-596, Sec. 8, Dec. 29, 1970, 84 Stat. 1598; Pub. L. 96-88, title V, Sec. 509(b), Oct. 17, 1979, 93 Stat. 695; Pub. L. 105-198, Sec. 1, July 16, 1998, 112 Stat. 640.)

Amendments

1998–Subsec. (h). Pub. L. 105-198 added subsec. (h).

Change of Name

"Secretary of Health and Human Services" substituted for "Secretary of Health, Education, and Welfare" in subsecs. (c), (d), and (g) pursuant to section 509(b) of Pub. L. 96-88 which is classified to section 3508(b) of Title 20, Education.

Section Referred to in Other Sections

This section is referred to in sections 667, 669, 670, 673 of this title; title 2 section 1341; title 3 section 425.

CITATIONS

29 USC 658

(a) Authority to issue; grounds; contents; notice in lieu of citation for de minimis violations

If, upon inspection or investigation, the Secretary or his authorized representative believes that an employer has violated a requirement of section 654 of this title, of any standard, rule or order promulgated pursuant to section 655 of this title, or of any regulations prescribed pursuant to this chapter, he shall with reasonable promptness issue a citation to the employer. Each citation shall be in writing and shall describe with particularity the nature of the violation, including a reference to the provision of the chapter, standard, rule, regulation, or order alleged to have been violated. In addition, the citation shall fix a reasonable time for the abatement of the violation. The Secretary may prescribe procedures for the issuance of a notice in lieu of a citation with respect to de minimis violations which have no direct or immediate relationship to safety or health.

(b) Posting

Each citation issued under this section, or a copy or copies thereof, shall be prominently posted, as prescribed in regulations issued by the Secretary, at or near each place a violation referred to in the citation occurred.

(c) Time for issuance

No citation may be issued under this section after the expiration of six months following the occurrence of any violation.

(Pub. L. 91-596, Sec. 9, Dec. 29, 1970, 84 Stat. 1601.)

Section Referred to in Other Sections
This section is referred to in sections 659, 666, 667 of this title; title 2 section 1341; title 3 section 425.

ENFORCEMENT PROCEDURES

29 USC 659

(a) Notification of employer of proposed assessment of penalty subsequent to issuance of citation; time for notification of Secretary by employer of contest by employer of citation or proposed assessment; citation and proposed assessment as final order upon failure of employer to notify of contest and failure of employees to file notice

If, after an inspection or investigation, the Secretary issues a citation under section 658(a) of this title, he shall, within a reasonable time after the termination of such inspection or investigation, notify the employer by certified mail of the penalty, if any, proposed to be assessed under section 666 of this title and that the employer has fifteen working days within which to notify the Secretary that he wishes to contest the citation or proposed assessment of penalty. If, within fifteen working days from the receipt of the notice issued by the Secretary the employer fails to notify the Secretary that he intends to contest the citation or proposed assessment of penalty, and no notice is filed by any employee or representative of employees under subsection (c) of this section within such time, the citation and the assessment, as proposed, shall be deemed a final order of the Commission and not subject to review by any court or agency.

(b) Notification of employer of failure to correct in allotted time period violation for which citation was issued and proposed assessment of penalty for failure to correct; time for notification of Secretary by employer of contest by employer of notification of failure to correct or proposed assessment; notification or proposed assessment as final order upon failure of employer to notify of contest

If the Secretary has reason to believe that an employer has failed to correct a violation for which a citation has been issued within the period permitted for its correction (which period shall not begin to run until the entry of a final order by the Commission in the case of any review proceedings under this section initiated by the employer in good faith and not solely for delay or avoidance of penalties), the Secretary shall notify the employer by certified mail of such failure and of the penalty proposed to be assessed under section 666 of this title by reason of such failure, and that the employer has fifteen working days within which to notify the Secretary that he wishes to contest the Secretary's notification or the proposed assessment of penalty. If, within fifteen working days from the receipt of notification issued by the Secretary, the employer fails to notify the Secretary that he intends to contest the notification or proposed assessment of penalty, the notification or proposed assessment of penalty, the notification and assessment, as proposed, shall be deemed a final order of the Commission and not subject to review by any court or agency.

(c) Advisement of Commission by Secretary of notification of contest by employer of citation or notification or of filing of notice by any employee or representative of employees; hearing by Commission; orders of Commission and Secretary; rules of procedure

If an employer notifies the Secretary that he intends to contest a citation issued under section 658(a) of this title or notification issued under subsection (a) or (b) of this section, or if, within fifteen working days of the issuance of a citation under section 658(a) of this title, any employee or representative of employees files a notice with the Secretary alleging that the period of time fixed in the citation for the abatement of the violation is unreasonable, the Secretary shall immediately advise the Commission of such notification, and the Commission shall afford an opportunity for a hearing (in accordance with section 554 of title 5 but without regard to subsection (a)(3) of such section). The Commission shall thereafter issue an order, based on findings of fact, affirming, modifying, or vacating the Secretary's citation or proposed penalty, or directing other appropriate relief, and such order shall become final thirty days after its issuance. Upon a showing by an employer of a good faith effort to comply with the abatement requirements of a citation, and that abatement has not been completed because of factors beyond his reasonable control, the Secretary, after an opportunity for a hearing as provided in this subsection, shall issue an order affirming or modifying the abatement requirements in such citation. The rules of procedure prescribed by the Commission shall provide affected employees or representatives of affected employees an opportunity to participate as parties to hearings under this subsection.

(Pub. L. 91-596, Sec. 10, Dec. 29, 1970, 84 Stat. 1601.)

Section Referred to in Other Sections

This section is referred to in sections 660, 666, 667 of this title; title 2 section 1341; title 3 section 425.

JUDICIAL REVIEW

29 USC 660

(a) Filing of petition by persons adversely affected or aggrieved; orders subject to review; jurisdiction; venue; procedure; conclusiveness of record and findings of Commission; appropriate relief; finality of judgment

Any person adversely affected or aggrieved by an order of the Commission issued under subsection (c) of section 659 of this title may obtain a review of such order in any United States court of appeals for the circuit in which the violation is alleged to have occurred or where the employer has its principal office, or in the Court of Appeals for the District of Columbia Circuit, by filing in such court within sixty days following the issuance of such order a written petition praying that the order be modified or set aside. A copy of such petition shall be forthwith transmitted by the clerk of the court to the Commission and to the other parties, and thereupon the Commission shall file in the court the record in the proceeding as provided in section 2112 of title 28. Upon such filing, the court shall have jurisdiction of the proceeding and of the question determined therein, and shall have power to grant such temporary relief or restraining order as it deems just and proper, and to make and enter upon the pleadings, testimony, and proceedings set forth in such record a decree affirming, modifying, or setting aside in whole or in part, the order of the Commission and enforcing the same to the extent that such order is affirmed or modified. The commencement of proceedings under this subsection shall not, unless ordered by the court, operate as a stay of the order of the Commission. No objection that has not been urged before the Commission shall be considered by the court, unless the failure or neglect to urge such objection shall be excused because of extraordinary circumstances. The findings of the Commission with respect to questions of fact, if supported by substantial evidence on the record considered as a whole, shall be conclusive. If any party shall apply to the court for leave to adduce additional evidence and shall show to the satisfaction of the court that such additional evidence is material and that there were reasonable grounds for the failure to adduce such evidence in the hearing before the Commission, the court may order such additional evidence to be taken before the Commission and to be made a part of the record. The Commission may modify its findings as to the facts, or make new findings, by reason of additional evidence so taken and filed, and it shall file such modified or new findings, which findings with respect to questions of fact, if supported by substantial evidence on the record considered as a whole, shall be conclusive, and its recommendations, if any, for the modification or setting aside of its original order. Upon the filing of the record with it, the jurisdiction of the court shall be exclusive and its judgment and decree shall be final, except that the same shall be subject to review by the Supreme Court of the United States, as provided in section 1254 of title 28.

(b) Filing of petition by Secretary; orders subject to review; jurisdiction; venue; procedure; conclusiveness of record and findings of Commission; enforcement of orders; contempt proceedings

The Secretary may also obtain review or enforcement of any final order of the Commission by filing a petition for such relief in the United States court of appeals for the circuit in which the alleged violation occurred or in which the employer has its principal office, and the provisions of subsection (a) of this section shall govern such proceedings to the extent applicable. If no petition for review, as provided in subsection (a) of this section, is filed within sixty days after service of the Commission's order, the Commission's findings of fact and order shall be conclusive in connection with any petition for enforcement which is filed by the Secretary after the expiration of such sixty-day period. In any such case, as well as in the case of a noncontested citation or notification by the Secretary which has become a final order of the Commission under subsection (a) or (b) of section 659 of this title, the clerk of the court, unless otherwise ordered by the court, shall forthwith enter a decree enforcing the order and shall transmit a copy of such decree to the Secretary and the employer named in the petition. In any contempt proceeding brought to enforce a decree of a court of appeals entered pursuant to this subsection or subsection (a) of this section, the court of appeals may assess the penalties provided in section 666 of this title, in addition to invoking any other available remedies.

(c) Discharge or discrimination against employee for exercise of rights under this chapter; prohibition; procedure for relief

(1) No person shall discharge or in any manner discriminate against any employee because such employee has filed any complaint or instituted or caused to be instituted any proceeding under or related to this chapter or has testified or is about to testify in any such proceeding or because of the exercise by such employee on behalf of himself or others of any right afforded by this chapter.

(2) Any employee who believes that he has been discharged or otherwise discriminated against by any person in violation of this subsection may, within thirty days after such violation occurs, file a complaint with the Secretary alleging such discrimination. Upon receipt of such complaint, the Secretary shall cause such investigation to be made as he deems appropriate. If upon such investigation, the Secretary determines that the provisions of this subsection have been violated, he shall bring an action in any appropriate United States district court against such person. In any such action the United States district courts shall have jurisdiction, for cause shown to restrain violations of paragraph (1) of this subsection and order all appropriate relief including rehiring or reinstatement of the employee to his former position with back pay.

(3) Within 90 days of the receipt of a complaint filed under this subsection the Secretary shall notify the complainant of his determination under paragraph (2) of this subsection.

(Pub. L. 91-596, Sec. 11, Dec. 29, 1970, 84 Stat. 1602; Pub. L. 98-620, title IV, Sec. 402(32), Nov. 8, 1984, 98 Stat. 3360.)

Amendments

1984—Subsec. (a). Pub. L. 98-620 struck out provision requiring expeditious hearing of petitions filed under this subsection.

Effective Date of 1984 Amendment

Amendment by Pub. L. 98-620 not applicable to cases pending on Nov. 8, 1984, see section 403 of Pub. L. 98-620, set out as a note under section 1657 of Title 28, Judiciary and Judicial Procedure.

Section Referred to in Other Sections

This section is referred to in title 15 section 2651.

OCCUPATIONAL SAFETY AND HEALTH REVIEW COMMISSION

29 USC 661

(a) Establishment; membership; appointment; Chairman
The Occupational Safety and Health Review Commission is hereby established. The Commission shall be composed of three members who shall be appointed by the President, by and with the advice and consent of the Senate, from among persons who by reason of training, education, or experience are qualified to carry out the functions of the Commission under this chapter. The President shall designate one of the members of the Commission to serve as Chairman.

(b) Terms of office; removal by President
The terms of members of the Commission shall be six years except that (1) the members of the Commission first taking office shall serve, as designated by the President at the time of appointment, one for a term of two years, one for a term of four years, and one for a term of six years, and (2) a vacancy caused by the death, resignation, or removal of a member prior to the expiration of the term for which he was appointed shall be filled only for the remainder of such unexpired term. A member of the Commission may be removed by the President for inefficiency, neglect of duty, or malfeasance in office.

(c) Omitted

(d) Principal office; hearings or other proceedings at other places
The principal office of the Commission shall be in the District of Columbia. Whenever the Commission deems that the convenience of the public or of the parties may be promoted, or delay or expense may be minimized, it may hold hearings or conduct other proceedings at any other place.

(e) Functions and duties of Chairman; appointment and compensation of administrative law judges and other employees
The Chairman shall be responsible on behalf of the Commission for the administrative operations of the Commission and shall appoint such administrative law judges and other employees as he deems necessary to assist in the performance of the Commission's functions and to fix their compensation in accordance with the provisions of chapter 51 and subchapter III of chapter 53 of title 5 relating to classification and General Schedule pay rates: Provided, That assignment, removal and compensation of administrative law judges shall be in accordance with sections 3105, 3344, 5372, and 7521 of title 5.

(f) Quorum; official action
For the purpose of carrying out its functions under this chapter, two members of the Commission shall constitute a quorum and official action can be taken only on the affirmative vote of at least two members.

(g) Hearings and records open to public; promulgation of rules; applicability of Federal Rules of Civil Procedure
Every official act of the Commission shall be entered of record, and its hearings and records shall be open to the public. The Commission is authorized to make such rules as are necessary for the orderly transaction of its proceedings. Unless the Commission has adopted a different rule, its proceedings shall be in accordance with the Federal Rules of Civil Procedure.

(h) Depositions and production of documentary evidence; fees
The Commission may order testimony to be taken by deposition in any proceeding pending before it at any state of such proceeding. Any person may be compelled to appear and depose, and to produce books, papers, or documents, in the same manner as witnesses may be compelled to appear and testify and produce like documentary evidence before the Commission. Witnesses whose depositions are taken under this subsection, and the persons taking such depositions, shall be entitled to the same fees as are paid for like services in the courts of the United States.

(i) Investigatory powers
For the purpose of any proceeding before the Commission, the provisions of section 161 of this title are hereby made applicable to the jurisdiction and powers of the Commission.

(j) Administrative law judges; determinations; report as final order of Commission
A[1] administrative law judge appointed by the Commission shall hear, and make a determination upon, any proceeding instituted before the Commission and any motion in connection therewith, assigned to such administrative law judge by the Chairman of the Commission, and shall make a report of any such determination which constitutes his final disposition of the proceedings. The report of the administrative law judge shall become the final order of the Commission within thirty days after such report by the administrative law judge, unless within such period any Commission member has directed that such report shall be reviewed by the Commission.

(k) Appointment and compensation of administrative law judges
Except as otherwise provided in this chapter, the administrative law judges shall be subject to the laws governing employees in the classified civil service, except that appointments shall be made without regard to section 5108 of title 5. Each administrative law judge shall receive compensation at a rate not less than that prescribed for GS-16 under section 5332 of title 5.

(Pub. L. 91-596, Sec. 12, Dec. 29, 1970, 84 Stat. 1603; Pub. L. 95-251, Sec. 2(a)(7), Mar. 27, 1978, 92 Stat. 183.)

---

[1]So in original. Probably should be "An".

References in Text

The General Schedule, referred to in subsec. (e), is set out under section 5332 of Title 5, Government Organization and Employees.

The Federal Rules of Civil Procedure, referred to in subsec. (g), are set out in the Appendix to Title 28, Judiciary and Judicial Procedure.

Codification

Subsec. (c) of this section amended sections 5314 and 5315 of Title 5, Government Organization and Employees.

In subsec. (e), reference to section 5372 of title 5 was substituted for section 5362 on authority of Pub. L. 95-454, Sec. 801(a)(3)(A)(ii), Oct. 13, 1978, 92 Stat. 1221, which redesignated sections 5361 through 5365 of title 5 as sections 5371 through 5375.

Amendments

1978–Subsecs. (e), (j), (k). Pub. L. 95-251 substituted "administrative law judge" and "administrative law judges" for "hearing examiner" and "hearing examiners", respectively, wherever appearing.

References in Other Laws to GS-16, 17, or 18 Pay Rates

References in laws to the rates of pay for GS-16, 17, or 18, or to maximum rates of pay under the General Schedule, to be considered references to rates payable under specified sections of Title 5, Government Organization and Employees, see section 529 [title I, Sec. 101(c)(1)] of Pub. L. 101-509, set out in a note under section 5376 of Title 5.

INJUNCTION PROCEEDINGS

29 USC 662

(a) Petition by Secretary to restrain imminent dangers; scope of order
The United States district courts shall have jurisdiction, upon petition of the Secretary, to restrain any conditions or practices in any place of employment which are such that a danger exists which could reasonably be expected to cause death or serious physical harm immediately or before the imminence of such danger can be eliminated through the enforcement procedures otherwise provided by this chapter. Any order issued under this section may require such steps to be taken as may be necessary to avoid, correct, or remove such imminent danger and prohibit the employment or presence of any individual in locations or under conditions where such imminent danger exists, except individuals whose presence is necessary to avoid, correct, or remove such imminent danger or to maintain the capacity of a continuous process operation to resume normal operations without a complete cessation of operations, or where a cessation of operations is necessary, to permit such to be accomplished in a safe and orderly manner.

(b) Appropriate injunctive relief or temporary restraining order pending outcome of enforcement proceeding; applicability of Rule 65 of Federal Rules of Civil Procedure
Upon the filing of any such petition the district court shall have jurisdiction to grant such injunctive relief or temporary restraining order pending the outcome of an enforcement proceeding pursuant to this chapter. The proceeding shall be as provided by Rule 65 of the Federal Rules, Civil Procedure, except that no temporary restraining order issued without notice shall be effective for a period longer than five days.

(c) Notification of affected employees and employers by inspector of danger and of recommendation to Secretary to seek relief
Whenever and as soon as an inspector concludes that conditions or practices described in subsection (a) of this section exist in any place of employment, he shall inform the affected employees and employers of the danger and that he is recommending to the Secretary that relief be sought.

(d) Failure of Secretary to seek relief; writ of mandamus
If the Secretary arbitrarily or capriciously fails to seek relief under this section, any employee who may be injured by reason of such failure, or the representative of such employees, might

bring an action against the Secretary in the United States district court for the district in which the imminent danger is alleged to exist or the employer has its principal office, or for the District of Columbia, for a writ of mandamus to compel the Secretary to seek such an order and for such further relief as may be appropriate.

(Pub. L. 91-596, Sec. 13, Dec. 29, 1970, 84 Stat. 1605.)

References in Text

Rule 65 of the Federal Rules of Civil Procedures, referred to in subsec. (b), is set out in the Appendix to Title 28, Judiciary and Judicial Procedure.

Federal Rules of Civil Procedure

Writ of mandamus abolished in United States district courts, but relief available by appropriate action or motion, see rule 81, Title 28, Appendix, Judiciary and Judicial Procedure.

Section Referred to in Other Sections

This section is referred to in title 2 section 1341; title 3 section 425.

REPRESENTATION IN CIVIL LITIGATION

29 USC 663

Except as provided in section 518(a) of title 28 relating to litigation before the Supreme Court, the Solicitor of Labor may appear for and represent the Secretary in any civil litigation brought under this chapter but all such litigations shall be subject to the direction and control of the Attorney General.

(Pub. L. 91-596, Sec. 14, Dec. 29, 1970, 84 Stat. 1606.)

DISCLOSURE OF TRADE SECRETS; PROTECTIVE ORDERS

29 USC 664

All information reported to or otherwise obtained by the Secretary or his representative in connection with any inspection or proceeding under this chapter which contains or which might reveal a trade secret referred to in section 1905 of title 18 shall be considered confidential for the purpose of that section, except that such information may be disclosed to other officers or employees concerned with carrying out this chapter or when relevant in any proceeding under this chapter. In any such proceeding the Secretary, the Commission, or the court shall issue such orders as may be appropriate to protect the confidentiality of trade secrets.

(Pub. L. 91-596, Sec. 15, Dec. 29, 1970, 84 Stat. 1606.)

VARIATIONS, TOLERANCES, AND EXEMPTIONS FROM REQUIRED PROVISIONS; PROCEDURE; DURATION

29 USC 665

The Secretary, on the record, after notice and opportunity for a hearing may provide such reasonable limitations and may make such rules and regulations allowing reasonable variations, tolerances, and exemptions to and from any or all provisions of this chapter as he may find necessary and proper to avoid serious impairment of the national defense. Such action shall not be in effect for more than six months without notification to affected employees and an opportunity being afforded for a hearing.

(Pub. L. 91-596, Sec. 16, Dec. 29, 1970, 84 Stat. 1606.)

CIVIL AND CRIMINAL PENALTIES

29 USC 666

(a) Willful or repeated violation

Any employer who willfully or repeatedly violates the requirements of section 654 of this title, any standard, rule, or order promulgated pursuant to section 655 of this title, or regulations prescribed pursuant to this chapter may be assessed a civil penalty of not more than $70,000 for each violation, but not less than $5,000 for each willful violation.

(b) Citation for serious violation

Any employer who has received a citation for a serious violation of the requirements of section 654 of this title, of any standard, rule, or order promulgated pursuant to section 655 of this title, or of any regulations prescribed pursuant to this chapter, shall be assessed a civil penalty of up to $7,000 for each such violation.

(c) Citation for violation determined not serious

Any employer who has received a citation for a violation of the requirements of section 654 of this title, of any standard, rule, or order promulgated pursuant to section 655 of this title, or of regulations prescribed pursuant to this chapter, and such violation is specifically determined not to be of a serious nature, may be assessed a civil penalty of up to $7,000 for each such violation.

(d) Failure to correct violation

Any employer who fails to correct a violation for which a citation has been issued under section 658(a) of this title within the period permitted for its correction (which period shall not begin to run until the date of the final order of the Commission in the case of any review proceeding under section 659 of this title initiated by the employer in good faith and not solely for delay or avoidance of penalties), may be assessed a civil penalty of not more than $7,000 for each day during which such failure or violation continues.

(e) Willful violation causing death to employee

Any employer who willfully violates any standard, rule, or order promulgated pursuant to section 655 of this title, or of any regulations prescribed pursuant to this chapter, and that violation caused death to any employee, shall, upon conviction, be punished by a fine of not more than $10,000 or by imprisonment for not more than six months, or by both; except that if the conviction is for a violation committed after a first conviction of such person, punishment shall be by a fine of not more than $20,000 or by imprisonment for not more than one year, or by both.

(f) Giving advance notice of inspection

Any person who gives advance notice of any inspection to be conducted under this chapter, without authority from the Secretary or his designees, shall, upon conviction, be punished by a fine of not more than $1,000 or by imprisonment for not more than six months, or by both.

(g) False statements, representations or certification

Whoever knowingly makes any false statement, representation, or certification in any application, record, report, plan, or other document filed or required to be maintained pursuant to this chapter shall, upon conviction, be punished by a fine of not more than $10,000, or by imprisonment for not more than six months, or by both.

(h) Omitted

(i) Violation of posting requirements

Any employer who violates any of the posting requirements, as prescribed under the provisions of this chapter, shall be assessed a civil penalty of up to $7,000 for each violation.

(j) Authority of Commission to assess civil penalties

The Commission shall have authority to assess all civil penalties provided in this section, giving due consideration to the appropriateness of the penalty with respect to the size of the business of the employer being charged, the gravity of the violation, the good faith of the employer, and the history of previous violations.

(k) Determination of serious violation

For purposes of this section, a serious violation shall be deemed to exist in a place of employment if there is a substantial probability that death or serious physical harm could result from a condition which exists, or from one or more practices, means, methods, operations, or processes which have been adopted or are in use, in such place of employment unless the employer did not, and could not with the exercise of reasonable diligence, know of the presence of the violation.

(l) Procedure for payment of civil penalties
Civil penalties owned under this chapter shall be paid to the Secretary for deposit into the Treasury of the United States and shall accrue to the United States and may be recovered in a civil action in the name of the United States brought in the United States district court for the district where the violation is alleged to have occurred or where the employer has its principal office.

(Pub. L. 91-596, Sec. 17, Dec. 29, 1970, 84 Stat. 1606, 1607; Pub. L. 101-508, title III, Sec. 3101, Nov. 5, 1990, 104 Stat. 1388-29.)

Codification

Subsec. (h) of this section amended section 1114 of Title 18, Crimes and Criminal Procedure, and enacted note set out thereunder.

Amendments

1990–Subsec. (a). Pub. L. 101-508, Sec. 3101(1), substituted "$70,000 for each violation, but not less than $5,000 for each willful violation" for "$10,000 for each violation".

Subsecs. (b) to (d), (i). Pub. L. 101-508, Sec. 3101(2), substituted "$7,000" for "$1,000".

Section Referred to in Other Sections

This section is referred to in sections 659, 660, 667 of this title.

STATE JURISDICTION AND PLANS

29 USC 667

(a) Assertion of State standards in absence of applicable Federal standards
Nothing in this chapter shall prevent any State agency or court from asserting jurisdiction under State law over any occupational safety or health issue with respect to which no standard is in effect under section 655 of this title.

(b) Submission of State plan for development and enforcement of State standards to preempt applicable Federal standards
Any State which, at any time, desires to assume responsibility for development and enforcement therein of occupational safety and health standards relating to any occupational safety or health issue with respect to which a Federal standard has been promulgated under section 655 of this title shall submit a State plan for the development of such standards and their enforcement.

(c) Conditions for approval of plan
The Secretary shall approve the plan submitted by a State under subsection (b) of this section, or any modification thereof, if such plan in his judgment—

(1) designates a State agency or agencies as the agency or agencies responsible for administering the plan throughout the State,

(2) provides for the development and enforcement of safety and health standards relating to one or more safety or health issues, which standards (and the enforcement of which standards) are or will be at least as effective in providing safe and healthful employment and places of employment as the standards promulgated under section 655 of this title which relate to the same issues, and which standards, when applicable to products which are distributed or used in interstate commerce, are required by compelling local conditions and do not unduly burden interstate commerce,

(3) provides for a right of entry and inspection of all workplaces subject to this chapter which is at least as effective as that provided in section 657 of this title, and includes a prohibition on advance notice of inspections,

(4) contains satisfactory assurances that such agency or agencies have or will have the legal authority and qualified personnel necessary for the enforcement of such standards,

(5) gives satisfactory assurances that such State will devote adequate funds to the administration and enforcement of such standards,

(6) contains satisfactory assurances that such State will, to the extent permitted by its law, establish and maintain an effective and comprehensive occupational safety and health program applicable to all employees of public agencies of the State and its political subdivisions, which program is as effective as the standards contained in an approved plan,

(7) requires employers in the State to make reports to the Secretary in the same manner and to the same extent as if the plan were not in effect, and

(8) provides that the State agency will make such reports to the Secretary in such form and containing such information, as the Secretary shall from time to time require.

(d) Rejection of plan; notice and opportunity for hearing
If the Secretary rejects a plan submitted under subsection (b) of this section, he shall afford the State submitting the plan due notice and opportunity for a hearing before so doing.

(e) Discretion of Secretary to exercise authority over comparable standards subsequent to approval of State plan; duration; retention of jurisdiction by Secretary upon determination of enforcement of plan by State
After the Secretary approves a State plan submitted under subsection (b) of this section, he may, but shall not be required to, exercise his authority under sections 657, 658, 659, 662, and 666 of this title with respect to comparable standards promulgated under section 655 of this title, for the period specified in the next sentence. The Secretary may exercise the authority referred to above until he determines, on the basis of actual operations under the State plan, that the criteria set forth in subsection (c) of this section are being applied, but he shall not make such determination for at least three years after the plan's approval under subsection (c) of this section. Upon making the determination referred to in the preceding sentence, the provisions of sections 654(a)(2), 657 (except for the purpose of carrying out subsection (f) of this section), 658, 659, 662, and 666 of this title, and standards promulgated under section 655 of this title, shall not apply with respect to any occupational safety or health issues covered under the plan, but the Secretary may retain jurisdiction under the above provisions in any proceeding commenced under section 658 or 659 of this title before the date of determination.

(f) Continuing evaluation by Secretary of State enforcement of approved plan; withdrawal of approval of plan by Secretary; grounds; procedure; conditions for retention of jurisdiction by State
The Secretary shall, on the basis of reports submitted by the State agency and his own inspections make a continuing evaluation of the manner in which each State having a plan approved under this section is carrying out such plan. Whenever the Secretary finds, after affording due notice and opportunity for a hearing, that in the administration of the State plan there is a failure to comply substantially with any provision of the State plan (or any assurance contained therein), he shall notify the State agency of his withdrawal of approval of such plan and upon receipt of such notice such plan shall cease to be in effect, but the State may retain jurisdiction in any case commenced before the withdrawal of the plan in order to enforce standards under the plan whenever the issues involved do not relate to the reasons for the withdrawal of the plan.

(g) Judicial review of Secretary's withdrawal of approval or rejection of plan; jurisdiction; venue; procedure; appropriate relief; finality of judgment
The State may obtain a review of a decision of the Secretary withdrawing approval of or rejecting its plan by the United States court of appeals for the circuit in which the State is located by filing in such court within thirty days following receipt of notice of such decision a petition to modify or set aside in whole or in part the action of the Secretary. A copy of such petition shall forthwith be served upon the Secretary, and thereupon the Secretary shall certify and file in the court the record upon which the decision complained of was issued as provided in section 2112 of title 28. Unless the court finds that the Secretary's decision in rejecting a proposed State plan or withdrawing his approval of such a plan is not supported by substantial evidence the court shall affirm the Secretary's decision. The judgment of the court shall be subject to review by the Supreme Court of the United States upon certiorari or certification as provided in section 1254 of title 28.

(h) Temporary enforcement of State standards

The Secretary may enter into an agreement with a State under which the State will be permitted to continue to enforce one or more occupational health and safety standards in effect in such State until final action is taken by the Secretary with respect to a plan submitted by a State under subsection (b) of this section, or two years from December 29, 1970, whichever is earlier.

(Pub. L. 91-596, Sec. 18, Dec. 29, 1970, 84 Stat. 1608.)

Section Referred to in Other Sections

This section is referred to in sections 670, 671a, 672 of this title; title 7 section 1942.

PROGRAMS OF FEDERAL AGENCIES

29 USC 668

(a) Establishment, development, and maintenance by head of each Federal agency

It shall be the responsibility of the head of each Federal agency (not including the United States Postal Service) to establish and maintain an effective and comprehensive occupational safety and health program which is consistent with the standards promulgated under section 655 of this title. The head of each agency shall (after consultation with representatives of the employees thereof)–

(1) provide safe and healthful places and conditions of employment, consistent with the standards set under section 655 of this title;

(2) acquire, maintain, and require the use of safety equipment, personal protective equipment, and devices reasonably necessary to protect employees;

(3) keep adequate records of all occupational accidents and illnesses for proper evaluation and necessary corrective action;

(4) consult with the Secretary with regard to the adequacy as to form and content of records kept pursuant to subsection (a)(3) of this section; and

(5) make an annual report to the Secretary with respect to occupational accidents and injuries and the agency's program under this section. Such report shall include any report submitted under section 7902(e)(2) of title 5.

(b) Report by Secretary to President

The Secretary shall report to the President a summary or digest of reports submitted to him under subsection (a)(5) of this section, together with his evaluations of and recommendations derived from such reports.

(c) Omitted

(d) Access by Secretary to records and reports required of agencies

The Secretary shall have access to records and reports kept and filed by Federal agencies pursuant to subsections (a)(3) and (5) of this section unless those records and reports are specifically required by Executive order to be kept secret in the interest of the national defense or foreign policy, in which case the Secretary shall have access to such information as will not jeopardize national defense or foreign policy.

(Pub. L. 91-596, Sec. 19, Dec. 29, 1970, 84 Stat. 1609; Pub. L. 97-375, title I, Sec. 110(c), Dec. 21, 1982, 96 Stat. 1821; Pub. L. 105-241, Sec. 2(b)(1), Sept. 28, 1998, 112 Stat. 1572.)

Codification

Subsec. (c) of this section amended section 7902 of Title 5, Government Organization and Employees.

Amendments

1998–Subsec. (a). Pub. L. 105-241 inserted "(not including the United States Postal Service)" after "each Federal agency".

1982–Subsec. (b). Pub. L. 97-375 struck out direction that the President transmit annually to the Senate and House a report of the activities of Federal agencies under this section.

Occupational Safety and Health Programs for Federal Employees

Occupational safety and health programs for Federal employees and continuation of Federal Advisory Council on Occupational Safety and Health, see Ex. Ord. No. 12196, Feb. 26, 1980, 45 F.R. 12769, set out as a note under section 7902 of Title 5, Government Organization and Employees.

Section Referred to in Other Sections

This section is referred to in title 3 section 425; title 39 section 410.

RESEARCH AND RELATED ACTIVITIES

29 USC 669

(a) Authority of Secretary of Health and Human Services to conduct research, experiments, and demonstrations, develop plans, establish criteria, promulgate regulations, authorize programs, and publish results and industrywide studies; consultations

(1) The Secretary of Health and Human Services, after consultation with the Secretary and with other appropriate Federal departments or agencies, shall conduct (directly or by grants or contracts) research, experiments, and demonstrations relating to occupational safety and health, including studies of psychological factors involved, and relating to innovative methods, techniques, and approaches for dealing with occupational safety and health problems.

(2) The Secretary of Health and Human Services shall from time to time consult with the Secretary in order to develop specific plans for such research, demonstrations, and experiments as are necessary to produce criteria, including criteria identifying toxic substances, enabling the Secretary to meet his responsibility for the formulation of safety and health standards under this chapter; and the Secretary of Health and Human Services, on the basis of such research, demonstrations, and experiments and any other information available to him, shall develop and publish at least annually such criteria as will effectuate the purposes of this chapter.

(3) The Secretary of Health and Human Services, on the basis of such research, demonstrations, and experiments, and any other information available to him, shall develop criteria dealing with toxic materials and harmful physical agents and substances which will describe exposure levels that are safe for various periods of employment, including but not limited to the exposure levels at which no employee will suffer impaired health or functional capacities or diminished life expectancy as a result of his work experience.

(4) The Secretary of Health and Human Services shall also conduct special research, experiments, and demonstrations relating to occupational safety and health as are necessary to explore new problems, including those created by new technology in occupational safety and health, which may require ameliorative action beyond that which is otherwise provided for in the operating provisions of this chapter. The Secretary of Health and Human Services shall also conduct research into the motivational and behavioral factors relating to the field of occupational safety and health.

(5) The Secretary of Health and Human Services, in order to comply with his responsibilities under paragraph (2), and in order to develop needed information regarding potentially toxic substances or harmful physical agents, may prescribe regulations requiring employers to measure, record, and make reports on the exposure of employees to substances or physical agents which the Secretary of Health and Human Services reasonably believes may endanger the health or safety of employees. The Secretary of Health and Human Services also is authorized to establish such programs of medical examinations and tests as may be necessary for determining the incidence of occupational illnesses and the susceptibility of employees to such illnesses. Nothing in this or any other provision of this chapter shall be deemed to authorize or require medical examination, immunization, or treatment for those who object thereto on religious grounds, except where such is necessary for the protection of the health or safety of others. Upon the request of any employer who is required to measure and record exposure of employees to substances or physical agents as provided under this subsection, the Secretary of Health and Human Services shall furnish full financial or other assistance to such employer for the purpose of defraying any additional expense incurred by him in carrying out the measuring and recording as provided in this subsection.

(6) The Secretary of Health and Human Services shall publish within six months of December 29, 1970, and thereafter as needed but at least annually a list of all known toxic substances by generic family or other useful grouping, and the concentrations at which such toxicity is known to occur. He shall determine following a written request by any employer or authorized representative of employees, specifying with reasonable particularity the grounds on which the request is made, whether any substance normally found in the place of employment has potentially toxic effects in such concentrations as used or found; and shall submit such determination both to employers and affected employees as soon as possible. If the Secretary of Health and Human Services determines that any substance is potentially toxic at the concentrations in which it is used or found in a place of employment, and such substance is not covered by an occupational safety or health standard promulgated under section 655 of this title, the Secretary of Health and Human Services shall immediately submit such determination to the Secretary, together with all pertinent criteria.

(7) Within two years of December 29, 1970, and annually thereafter the Secretary of Health and Human Services shall conduct and publish industrywide studies of the effect of chronic or low-level exposure to industrial materials, processes, and stresses on the potential for illness, disease, or loss of functional capacity in aging adults.

(b) Authority of Secretary of Health and Human Services to make inspections and question employers and employees
The Secretary of Health and Human Services is authorized to make inspections and question employers and employees as provided in section 657 of this title in order to carry out his functions and responsibilities under this section.

(c) Contracting authority of Secretary of Labor; cooperation between Secretary of Labor and Secretary of Health and Human Services
The Secretary is authorized to enter into contracts, agreements, or other arrangements with appropriate public agencies or private organizations for the purpose of conducting studies relating to his responsibilities under this chapter. In carrying out his responsibilities under this subsection, the Secretary shall cooperate with the Secretary of Health and Human Services in order to avoid any duplication of efforts under this section.

(d) Dissemination of information to interested parties
Information obtained by the Secretary and the Secretary of Health and Human Services under this section shall be disseminated by the Secretary to employers and employees and organizations thereof.

(e) Delegation of functions of Secretary of Health and Human Services to Director of the National Institute for Occupational Safety and Health
The functions of the Secretary of Health and Human Services under this chapter shall, to the extent feasible, be delegated to the Director of the National Institute for Occupational Safety and Health established by section 671 of this title.

(Pub. L. 91-596, Sec. 20, Dec. 29, 1970, 84 Stat. 1610; Pub. L. 96-88, title V, Sec. 509(b), Oct. 17, 1979, 93 Stat. 695.)

Change of Name

"Secretary of Health and Human Services" substituted in text for "Secretary of Health, Education, and Welfare" pursuant to section 509(b) of Pub. L. 96-88 which is classified to section 3508(b) of Title 20, Education.

Section Referred to in Other Sections

This section is referred to in section 671 of this title.

EXPANDED RESEARCH ON WORKER HEALTH AND SAFETY

29 USC 669a

The Secretary of Health and Human Services (referred to in this section as the "Secretary"), acting through the Director of the National Institute of Occupational Safety and Health, shall enhance and expand research as deemed appropriate on the health and safety of workers who are at risk for bioterrorist threats or attacks in the workplace, including research on the health effects of measures taken to treat or protect such workers for diseases or disorders resulting from a bioterrorist threat or attack. Nothing in this section may be construed as establishing new regulatory authority for the Secretary or the Director to issue or modify any occupational safety and health rule or regulation.

TRAINING AND EMPLOYEE EDUCATION

29 USC 670

(a) Authority of Secretary of Health and Human Services to conduct education and informational programs; consultations
The Secretary of Health and Human Services, after consultation with the Secretary and with other appropriate Federal departments and agencies, shall conduct, directly or by grants or contracts (1) education programs to provide an adequate supply of qualified personnel to carry out the purposes of this chapter, and (2) informational programs on the importance of and proper use of adequate safety and health equipment.

(b) Authority of Secretary of Labor to conduct short-term training of personnel
The Secretary is also authorized to conduct, directly or by grants or contracts, short-term training of personnel engaged in work related to his responsibilities under this chapter.

(c) Authority of Secretary of Labor to establish and supervise education and training programs and consult and advise interested parties
The Secretary, in consultation with the Secretary of Health and Human Services, shall (1) provide for the establishment and supervision of programs for the education and training of employers and employees in the recognition, avoidance, and prevention of unsafe or unhealthful working conditions in employments covered by this chapter, and (2) consult with and advise employers and employees, and organizations representing employers and employees as to effective means of preventing occupational injuries and illnesses.

(d) Compliance assistance program
(1) The Secretary shall establish and support cooperative agreements with the States under which employers subject to this chapter may consult with State personnel with respect to–
  (A) the application of occupational safety and health requirements under this chapter or under State plans approved under section 667 of this title; and
  (B) voluntary efforts that employers may undertake to establish and maintain safe and healthful employment and places of employment.
  Such agreements may provide, as a condition of receiving funds under such agreements, for contributions by States towards meeting the costs of such agreements.
(2) Pursuant to such agreements the State shall provide on-site consultation at the employer's worksite to employers who request such assistance. The State may also provide other education and training programs for employers and employees in the State. The State shall ensure that on-site consultations conducted pursuant to such agreements include provision for the participation by employees.
(3) Activities under this subsection shall be conducted independently of any enforcement activity. If an employer fails to take immediate action to eliminate employee exposure to an imminent danger identified in a consultation or fails to correct a serious hazard so identified within a reasonable time, a report shall be made to the appropriate enforcement authority for such action as is appropriate.
(4) The Secretary shall, by regulation after notice and opportunity for comment, establish rules under which an employer–
  (A) which requests and undergoes an on-site consultative visit provided under this subsection;
  (B) which corrects the hazards that have been identified during the visit within the time frames established by the State and agrees to request a subsequent consultative visit if major changes in working conditions or work processes occur which introduce new hazards in the workplace; and
  (C) which is implementing procedures for regularly identifying and preventing hazards regulated under this chapter and maintains appropriate involvement of, and training

for, management and non-management employees in achieving safe and healthful working conditions,

may be exempt from an inspection (except an inspection requested under section 657(f) of this title or an inspection to determine the cause of a workplace accident which resulted in the death of one or more employees or hospitalization for three or more employees) for a period of 1 year from the closing of the consultative visit.

(5) A State shall provide worksite consultations under paragraph (2) at the request of an employer. Priority in scheduling such consultations shall be assigned to requests from small businesses which are in higher hazard industries or have the most hazardous conditions at issue in the request.

(Pub. L. 91-596, Sec. 21, Dec. 29, 1970, 84 Stat. 1612; Pub. L. 96-88, title V, Sec. 509(b), Oct. 17, 1979, 93 Stat. 695; Pub. L. 105-197, Sec. 2, July 16, 1998, 112 Stat. 638.)

Amendments

1998–Subsec. (d). Pub. L. 105-197 added subsec. (d).

Change of Name

"Secretary of Health and Human Services" substituted for "Secretary of Health, Education, and Welfare" in subsecs. (a) and (c) pursuant to section 509(b) of Pub. L. 96-88 which is classified to section 3508(b) of Title 20, Education.

Retention of Training Institute Course Tuition Fees by OSHA

Pub. L. 107-116, title I, Jan 10, 2002, 115 Stat. 2182, provided in part that: "notwithstanding 31 U.S.C. 3302, the Occupational Safety and Health Administration may retain up to $750,000 per fiscal year of training institute course tuition fees, otherwise authorized by law to be collected, and may utilize such sums for occupational safety and health training and education grants".

Similar provisions were contained in the following prior appropriation acts:

Pub. L. 106-554, title I, Sec. 1(a)(1), Dec. 21, 2000, 114 Stat. 2763A-8.

Pub. L. 106-113, div. B, Sec. 1000(a)(4) [title I], Nov. 29, 1999, 113 Stat. 1535, 1501A-222.

Pub. L. 105-277, div. A, Sec. 101(f) [title I], Oct. 21, 1998, 112 Stat. 2681-337, 2681-343.

Pub. L. 105-78, title I, Nov. 13, 1997, 111 Stat. 1474.

Pub. L. 104-208, div. A, title I, Sec. 101(e) [title I], Sept. 30, 1996, 110 Stat. 3009-233, 3009-239.

Pub. L. 104-134, title I, Sec. 101(d) [title I], Apr. 26, 1996, 110 Stat. 1321-211, 1321-217; renumbered title I, Pub. L. 104-140, Sec. 1(a), May 2, 1996, 110 Stat. 1327.

Pub. L. 103-333, title I, Sept. 30, 1994, 108 Stat. 2544.

Section Referred to in Other Sections

This section is referred to in section 671 of this title.

---

29 USC 670 Note

(a) For necessary expenses for the Occupational Safety and Health Administration, $460,786,000, including not to exceed $92,505,000 which shall be the maximum amount available for grants to States under section 23(g) of the Occupational Safety and Health Act (the "Act"), which grants shall be no less than 50 percent of the costs of State occupational safety and health programs required to be incurred under plans approved by the Secretary under section 18 of the Act; and, in addition, notwithstanding 31 U.S.C. 3302, the Occupational Safety and Health Administration may retain up to $750,000 per fiscal year of training institute course tuition fees, otherwise authorized by law to be collected, and may utilize such sums for occupational safety and health training and education grants: Provided, That, notwithstanding 31 U.S.C. 3302, the Secretary of Labor is authorized, during the fiscal year ending September 30, 2004, to collect and retain fees for services provided to Nationally Recognized Testing Laboratories, and may utilize such sums, in accordance with the provisions of 29 U.S.C. 9a, to administer national and international laboratory recognition programs that ensure the safety of equipment and products used by workers in the workplace:

Provided further, that none of the funds appropriated under this paragraph shall be obligated or expended to prescribe, issue, administer, or enforce any standard, rule, regulation, or order under the Act which is applicable to any person who is engaged in a farming operation which does not maintain a temporary labor camp and employs 10 or fewer employees:

Provided further, that no funds appropriated under this paragraph shall be obligated or expended to administer or enforce any standard, rule, regulation, or order under the Act with respect to any employer of 10 or fewer employees who is included within a category having an occupational injury lost workday case rate, at the most precise Standard Industrial Classification Code for which such data are published, less than the national average rate as such rates are most recently published by the Secretary, acting through the Bureau of Labor Statistics, in accordance with section 24 of that Act (29 U.S.C. 673), except–

(1) to provide, as authorized by such Act, consultation, technical assistance, educational and training services, and to conduct surveys and studies;

(2) to conduct an inspection or investigation in response to an employee complaint, to issue a citation for violations found during such inspection, and to assess a penalty for violations which are not corrected within a reasonable abatement period and for any willful violations found;

(3) to take any action authorized by such Act with respect to imminent dangers;

(4) to take any action authorized by such Act with respect to health hazards;

(5) to take any action authorized by such Act with respect to a report of an employment accident which is fatal to one or more employees or which results in hospitalization of two or more employees, and to take any action pursuant to such investigation authorized by such Act; and

(6) to take any action authorized by such Act with respect to complaints of discrimination against employees for exercising rights under such Act:

Provided further, that the foregoing proviso shall not apply to any person who is engaged in a farming operation which does not maintain a temporary labor camp and employs 10 or fewer employees:

Provided further, that not less than $3,200,000 shall be used to extend funding for the Institutional Competency Building training grants which commenced in September 2000, for program activities for the period of September 30, 2003 to September 30, 2004, provided that a grantee has demonstrated satisfactory performance.

(b) For necessary expenses for the Occupational Safety and Health Administration, $468,109,000, including not to exceed $91,747,000 which shall be the maximum amount available for grants to States under section 23(g) of the Occupational Safety and Health Act (the "Act"), which grants shall be no less than 50 percent of the costs of State occupational safety and health programs required to be incurred under plans approved by the Secretary under section 18 of the Act; and, in addition, notwithstanding 31 U.S.C. 3302, the Occupational Safety and Health Administration may retain up to $750,000 per fiscal year of training institute course tuition fees, otherwise authorized by law to be collected, and may utilize such sums for occupational safety and health training and education grants:

Provided, that notwithstanding 31 U.S.C. 3302, the Secretary of Labor is authorized, during the fiscal year ending September 30, 2005, to collect and retain fees for services provided to Nationally Recognized Testing Laboratories, and may utilize such sums, in accordance with the provisions of 29 U.S.C. 9a, to administer national and international laboratory recognition programs that ensure the safety of equipment and products used by workers in the workplace:

Provided further, that none of the funds appropriated under this paragraph shall be obligated or expended to prescribe, issue, administer, or enforce any standard, rule, regulation, or order under the Act which is applicable to any person who is engaged in a farming operation which does not maintain a temporary labor camp and employs 10 or fewer employees:

Provided further, that no funds appropriated under this paragraph shall be obligated or expended to administer or enforce any standard, rule, regulation, or order under the Act with respect to any employer of 10 or fewer employees who is included within a category having a Days Away, Restricted, or Transferred (DART) occupational injury and illness rate, at the most precise industrial classification code for which such data are published, less than the national average rate as such rates are most recently published by the Secretary, acting through the Bureau of Labor Statistics, in accordance with section 24 of that Act (29 U.S.C. 673), except–

(1) to provide, as authorized by such Act, consultation, technical assistance, educational and training services, and to conduct surveys and studies;

(2) to conduct an inspection or investigation in response to an employee complaint, to issue a citation for violations found during such inspection, and to assess a penalty for violations which are not corrected within a reasonable abatement period and for any willful violations found;

(3) to take any action authorized by such Act with respect to imminent dangers;

(4) to take any action authorized by such Act with respect to health hazards;

(5) to take any action authorized by such Act with respect to a report of an employment accident which is fatal to one or more employees or which results in hospitalization of two or more employees, and to take any action pursuant to such investigation authorized by such Act; and

(6) to take any action authorized by such Act with respect to complaints of discrimination against employees for exercising rights under such Act:

Provided further, that the foregoing proviso shall not apply to any person who is engaged in a farming operation which does not maintain a temporary labor camp and employs 10 or fewer employees:

Provided further, that not less than $3,200,000 shall be used to extend funding for the Institutional Competency Building training grants which commenced in September 2000, for program activities for the period of September 30, 2005 to September 30, 2006, provided that a grantee has demonstrated satisfactory performance:

Provided further, that none of the funds appropriated under this paragraph shall be obligated or expended to administer or enforce the provisions of 29 CFR 1910.134(f)(2) (General Industry Respiratory Protection Standard) to the extent that such provisions require the annual fit testing (after the initial fit testing) of respirators for occupational exposure to tuberculosis.

Salaries and Expenses

For necessary expenses for the Occupational Safety and Health Administration, $477,199,000, including not to exceed $92,013,000 which shall be the maximum amount available for grants to States under section 23(g) of the Occupational Safety and Health Act (the "Act"), which grants shall be no less than 50 percent of the costs of State occupational safety and health programs required to be incurred under plans approved by the Secretary under section 18 of the Act; and, in addition, notwithstanding 31 U.S.C. 3302, the Occupational Safety and Health Administration may retain up to $750,000 per fiscal year of training institute course tuition fees, otherwise authorized by law to be collected, and may utilize such sums for occupational safety and health training and education grants: Provided, That, notwithstanding 31 U.S.C. 3302, the Secretary of Labor is authorized, during the fiscal year ending September 30, 2006, to collect and retain fees for services provided to Nationally Recognized Testing Laboratories, and may utilize such sums, in accordance with the provisions of 29 U.S.C. 9a, to administer national and international laboratory recognition programs that ensure the safety of equipment and products used by workers in the workplace:

Provided further, That none of the funds appropriated under this paragraph shall be obligated or expended to prescribe, issue, administer, or enforce any standard, rule, regulation, or order under the Act which is applicable to any person who is engaged in a farming operation which does not maintain a temporary labor camp and employs 10 or fewer employees:

Provided further, That no funds appropriated under this paragraph shall be obligated or expended to administer or enforce any standard, rule, regulation, or order under the Act with respect to any employer of 10 or fewer employees who is included within a category having a Days Away, Restricted, or Transferred (DART) occupational injury and illness rate, at the most precise industrial classification code for which such data are published, less than the national average rate as such rates are most recently published by the Secretary, acting through the Bureau of Labor Statistics, in accordance with section 24 of that Act (29 U.S.C. 673), except–

(1) to provide, as authorized by such Act, consultation, technical assistance, educational and training services, and to conduct surveys and studies;

(2) to conduct an inspection or investigation in response to an employee complaint, to issue a citation for violations found during such inspection, and to assess a penalty for violations which are not corrected within a reasonable abatement period and for any willful violations found;

(3) to take any action authorized by such Act with respect to imminent dangers;

322 ❖ Occupational Safety and Health Law Handbook

(4) to take any action authorized by such Act with respect to health hazards;
(5) to take any action authorized by such Act with respect to a report of an employment accident which is fatal to one or more employees or which results in hospitalization of two or more employees, and to take any action pursuant to such investigation authorized by such Act; and
(6) to take any action authorized by such Act with respect to complaints of discrimination against employees for exercising rights under such Act:

Provided further, That the foregoing proviso shall not apply to any person who is engaged in a farming operation which does not maintain a temporary labor camp and employs 10 or fewer employees:

Provided further, That not less than $3,200,000 shall be used to extend funding for the Institutional Competency Building training grants which commenced in September 2000, for program activities for the period of September 30, 2006, to September 30, 2007, provided that a grantee has demonstrated satisfactory performance:

Provided further, That none of the funds appropriated under this paragraph shall be obligated or expended to administer or enforce the provisions of 29 CFR 1910.134(f)(2) (General Industry Respiratory Protection Standard) to the extent that such provisions require the annual fit testing (after the initial fit testing) of respirators for occupational exposure to tuberculosis.

NATIONAL INSTITUTE FOR OCCUPATIONAL SAFETY AND HEALTH

29 USC 671

(a) Statement of purpose
It is the purpose of this section to establish a National Institute for Occupational Safety and Health in the Department of Health and Human Services in order to carry out the policy set forth in section 651 of this title and to perform the functions of the Secretary of Health and Human Services under sections 669 and 670 of this title.

(b) Establishment; Director; appointment; term
There is hereby established in the Department of Health and Human Services a National Institute for Occupational Safety and Health. The Institute shall be headed by a Director who shall be appointed by the Secretary of Health and Human Services, and who shall serve for a term of six years unless previously removed by the Secretary of Health and Human Services.

(c) Development and establishment of standards; performance of functions of Secretary of Health and Human Services
The Institute is authorized to–
(1) develop and establish recommended occupational safety and health standards; and
(2) perform all functions of the Secretary of Health and Human Services under sections 669 and 670 of this title.

(d) Authority of Director
Upon his own initiative, or upon the request of the Secretary or the Secretary of Health and Human Services, the Director is authorized (1) to conduct such research and experimental programs as he determines are necessary for the development of criteria for new and improved occupational safety and health standards, and (2) after consideration of the results of such research and experimental programs make recommendations concerning new or improved occupational safety and health standards. Any occupational safety and health standard recommended pursuant to this section shall immediately be forwarded to the Secretary of Labor, and to the Secretary of Health and Human Services.

(e) Additional authority of Director
In addition to any authority vested in the Institute by other provisions of this section, the Director, in carrying out the functions of the Institute, is authorized to–
(1) prescribe such regulations as he deems necessary governing the manner in which its functions shall be carried out;
(2) receive money and other property donated, bequeathed, or devised, without condition or restriction other than that it be used for the purposes of the Institute and to use, sell, or otherwise dispose of such property for the purpose of carrying out its functions;

(3) receive (and use, sell, or otherwise dispose of, in accordance with paragraph (2)), money and other property donated, bequeathed or devised to the Institute with a condition or restriction, including a condition that the Institute use other funds of the Institute for the purposes of the gift;

(4) in accordance with the civil service laws, appoint and fix the compensation of such personnel as may be necessary to carry out the provisions of this section;

(5) obtain the services of experts and consultants in accordance with the provisions of section 3109 of title 5;

(6) accept and utilize the services of voluntary and noncompensated personnel and reimburse them for travel expenses, including per diem, as authorized by section 5703 of title 5;

(7) enter into contracts, grants or other arrangements, or modifications thereof to carry out the provisions of this section, and such contracts or modifications thereof may be entered into without performance or other bonds, and without regard to section 5 of title 41, or any other provision of law relating to competitive bidding;

(8) make advance, progress, and other payments which the Director deems necessary under this title without regard to the provisions of section 3324(a) and (b) of title 31; and

(9) make other necessary expenditures.

(f) Annual reports
The Director shall submit to the Secretary of Health and Human Services, to the President, and to the Congress an annual report of the operations of the Institute under this chapter, which shall include a detailed statement of all private and public funds received and expended by it, and such recommendations as he deems appropriate.

(g) Lead-based paint activities

(1) Training grant program

(A) The Institute, in conjunction with the Administrator of the Environmental Protection Agency, may make grants for the training and education of workers and supervisors who are or may be directly engaged in lead-based paint activities.

(B) Grants referred to in subparagraph (A) shall be awarded to nonprofit organizations (including colleges and universities, joint labor-management trust funds, States, and nonprofit government employee organizations)–

(i) which are engaged in the training and education of workers and supervisors who are or who may be directly engaged in lead-based paint activities (as defined in title IV of the Toxic Substances Control Act [15 U.S.C. 2681 et seq.]),

(ii) which have demonstrated experience in implementing and operating health and safety training and education programs, and

(iii) with a demonstrated ability to reach, and involve in lead-based paint training programs, target populations of individuals who are or will be engaged in lead-based paint activities.

Grants under this subsection shall be awarded only to those organizations that fund at least 30 percent of their lead-based paint activities training programs from non-Federal sources, excluding in-kind contributions. Grants may also be made to local governments to carry out such training and education for their employees.

(C) There are authorized to be appropriated, at a minimum, $10,000,000 to the Institute for each of the fiscal years 1994 through 1997 to make grants under this paragraph.

(2) Evaluation of programs
The Institute shall conduct periodic and comprehensive assessments of the efficacy of the worker and supervisor training programs developed and offered by those receiving grants under this section. The Director shall prepare reports on the results of these assessments addressed to the Administrator of the Environmental Protection Agency to include recommendations as may be appropriate for the revision of these programs. The sum of $500,000 is authorized to be appropriated to the Institute for each of the fiscal years 1994 through 1997 to carry out this paragraph.

(h) Office of Mine Safety and Health.

(1) In general.
There shall be permanently established within the Institute an Office of Mine Safety and Health which shall be administered by an Associate Director to be appointed by the Director.

(2) Purpose.--The purpose of the Office is to enhance the development of new mine safety technology and technological applications and to expedite the commercial availability and implementation of such technology in mining environments.

(3) Functions.--In addition to all purposes and authorities provided for under this section, the Office of Mine Safety and Health shall be responsible for research, development, and testing of new technologies and equipment designed to enhance mine safety and health. To carry out such functions the Director of the Institute, acting through the Office, shall have the authority to--

(A) award competitive grants to institutions and private entities to encourage the development and manufacture of mine safety equipment;

(B) award contracts to educational institutions or private laboratories for the performance of product testing or related work with respect to new mine technology and equipment; and

(C) establish an interagency working group as provided for in paragraph (5).

(4) Grant authority.--To be eligible to receive a grant under the authority provided for under paragraph (3)(A), an entity or institution shall--

(A) submit to the Director of the Institute an application at such time, in such manner, and containing such information as the Director may require; and

(B) include in the application under subparagraph (A), a description of the mine safety equipment to be developed and manufactured under the grant and a description of the reasons that such equipment would otherwise not be developed or manufactured, including reasons relating to the limited potential commercial market for such equipment.

(5) Interagency working group.--

(A) Establishment.--The Director of the Institute, in carrying out paragraph (3)(D) shall establish an interagency working group to share technology and technological research and developments that could be utilized to enhance mine safety and accident response.

(B) Membership.--The working group under subparagraph (A) shall be chaired by the Associate Director of the Office who shall appoint the members of the working group, which may include representatives of other Federal agencies or departments as determined appropriate by the Associate Director.

(C) Duties.--The working group under subparagraph (A) shall conduct an evaluation of research conducted by, and the technological developments of, agencies and departments who are represented on the working group that may have applicability to mine safety and accident response and make recommendations to the Director for the further development and eventual implementation of such technology.

(6) Annual report.--Not later than 1 year after the establishment of the Office under this subsection, and annually thereafter, the Director of the Institute shall submit to the Committee on Health, Education, Labor, and Pensions of the Senate and the Committee on Education and the Workforce of the House of Representatives a report that, with respect to the year involved, describes the new mine safety technologies and equipment that have been studied, tested, and certified for use, and with respect to those instances of technologies and equipment that have been considered but not yet certified for use, the reasons therefore.

(7) Authorization of appropriations.--There is authorized to be appropriated, such sums as may be necessary to enable the Institute and the Office of Mine Safety and Health to carry out this subsection.

(Pub. L. 91-596, Sec. 22, Dec. 29, 1970, 84 Stat. 1612; Pub. L. 96-88, title V, Sec. 509(b), Oct. 17, 1979, 93 Stat. 695; Pub. L. 102-550, title X, Sec. 1033, Oct. 28, 1992, 106 Stat. 3924.)

References in Text

The civil service laws, referred to in subsec. (e)(4), are set forth in Title 5, Government Organization and Employees. See, particularly, section 3301 et seq. of Title 5.

The Toxic Substances Control Act, referred to in subsec. (g)(1)(B)(i), is Pub. L. 94-469, Oct. 11, 1976, 90 Stat. 2003, as amended. Title IV of the Act is classified generally to subchapter IV (Sec. 2681 et seq.) of chapter 53 of Title 15, Commerce and Trade. For complete classification of this Act to the Code, see Short Title note set out under section 2601 of Title 15 and Tables.

Codification

In subsec. (e)(8), "section 3324(a) and (b) of title 31" substituted for "section 3648 of the Revised Statutes, as amended (31 U.S.C. 529)" on authority of Pub. L. 97-258, Sec. 4(b), Sept. 13, 1982, 96 Stat. 1067, the first section of which enacted Title 31, Money and Finance.

Amendments

1992–Subsec. (g). Pub. L. 102-550 added subsec. (g).

Change of Name

"Secretary of Health and Human Services" substituted for "Secretary of Health, Education, and Welfare" in subsecs. (a) to (d) and (f) pursuant to section 509(b) of Pub. L. 96-88 which is classified to section 3508(b) of Title 20, Education.

Section Referred to in Other Sections

This section is referred to in section 669 of this title.

WORKERS' FAMILY PROTECTION

29 USC 671a

(a) Short title
   This section may be cited as the "Workers' Family Protection Act".
(b) Findings and purpose
   (1) Findings
      Congress finds that–
      (A) hazardous chemicals and substances that can threaten the health and safety of workers are being transported out of industries on workers' clothing and persons;
      (B) these chemicals and substances have the potential to pose an additional threat to the health and welfare of workers and their families;
      (C) additional information is needed concerning issues related to employee transported contaminant releases; and
      (D) additional regulations may be needed to prevent future releases of this type.
   (2) Purpose
      It is the purpose of this section to–
      (A) increase understanding and awareness concerning the extent and possible health impacts of the problems and incidents described in paragraph (1);
      (B) prevent or mitigate future incidents of home contamination that could adversely affect the health and safety of workers and their families;
      (C) clarify regulatory authority for preventing and responding to such incidents; and
      (D) assist workers in redressing and responding to such incidents when they occur.
(c) Evaluation of employee transported contaminant releases
   (1) Study
      (A) In general
         Not later than 18 months after October 26, 1992, the Director of the National Institute for Occupational Safety and Health (hereafter in this section referred to as the "Director"), in cooperation with the Secretary of Labor, the Administrator of the Environmental Protection Agency, the Administrator of the Agency for Toxic Substances and Disease Registry, and the heads of other Federal Government agencies as determined to be appropriate by the Director, shall conduct a study to evaluate the potential for, the prevalence of, and the issues related to the contamination of workers' homes with hazardous chemicals and substances, including infectious agents, transported from the workplaces of such workers.
      (B) Matters to be evaluated
         In conducting the study and evaluation under subparagraph (A), the Director shall–
         (i) conduct a review of past incidents of home contamination through the utilization of literature and of records concerning past investigations and enforcement actions undertaken by–
            (I) the National Institute for Occupational Safety and Health;

(II) the Secretary of Labor to enforce the Occupational Safety and Health Act of 1970 (29 U.S.C. 651 et seq.);

(III) States to enforce occupational safety and health standards in accordance with section 18 of such Act (29 U.S.C. 667); and

(IV) other government agencies (including the Department of Energy and the Environmental Protection Agency), as the Director may determine to be appropriate;

(ii) evaluate current statutory, regulatory, and voluntary industrial hygiene or other measures used by small, medium and large employers to prevent or remediate home contamination;

(iii) compile a summary of the existing research and case histories conducted on incidents of employee transported contaminant releases, including–

(I) the effectiveness of workplace housekeeping practices and personal protective equipment in preventing such incidents;

(II) the health effects, if any, of the resulting exposure on workers and their families;

(III) the effectiveness of normal house cleaning and laundry procedures for removing hazardous materials and agents from workers' homes and personal clothing;

(IV) indoor air quality, as the research concerning such pertains to the fate of chemicals transported from a workplace into the home environment; and

(V) methods for differentiating exposure health effects and relative risks associated with specific agents from other sources of exposure inside and outside the home;

(iv) identify the role of Federal and State agencies in responding to incidents of home contamination;

(v) prepare and submit to the Task Force established under paragraph (2) and to the appropriate committees of Congress, a report concerning the results of the matters studied or evaluated under clauses (i) through (iv); and

(vi) study home contamination incidents and issues and worker and family protection policies and practices related to the special circumstances of firefighters and prepare and submit to the appropriate committees of Congress a report concerning the findings with respect to such study.

(2) Development of investigative strategy

(A) Task Force

Not later than 12 months after October 26, 1992, the Director shall establish a working group, to be known as the "Workers' Family Protection Task Force". The Task Force shall–

(i) be composed of not more than 15 individuals to be appointed by the Director from among individuals who are representative of workers, industry, scientists, industrial hygienists, the National Research Council, and government agencies, except that not more than one such individual shall be from each appropriate government agency and the number of individuals appointed to represent industry and workers shall be equal in number;

(ii) review the report submitted under paragraph (1)(B)(v);

(iii) determine, with respect to such report, the additional data needs, if any, and the need for additional evaluation of the scientific issues related to and the feasibility of developing such additional data; and

(iv) if additional data are determined by the Task Force to be needed, develop a recommended investigative strategy for use in obtaining such information.

(B) Investigative strategy

(i) Content

The investigative strategy developed under subparagraph (A)(iv) shall identify data gaps that can and cannot be filled, assumptions and uncertainties associated with various components of such strategy, a timetable for the implementation of such strategy, and methodologies used to gather any required data.

(ii) Peer review

The Director shall publish the proposed investigative strategy under subparagraph (A)(iv) for public comment and utilize other methods, including technical conferences or seminars, for the purpose of obtaining comments concerning the proposed strategy.

(iii) Final strategy

After the peer review and public comment is conducted under clause (ii), the Director, in consultation with the heads of other government agencies, shall propose a final strategy for investigating issues related to home contamination that shall be implemented by the National Institute for Occupational Safety and Health and other Federal agencies for the period of time necessary to enable such agencies to obtain the information identified under subparagraph (A)(iii).

(C) Construction

Nothing in this section shall be construed as precluding any government agency from investigating issues related to home contamination using existing procedures until such time as a final strategy is developed or from taking actions in addition to those proposed in the strategy after its completion.

(3) Implementation of investigative strategy

Upon completion of the investigative strategy under subparagraph (B)(iii), each Federal agency or department shall fulfill the role assigned to it by the strategy.

(d) Regulations

(1) In general

Not later than 4 years after October 26, 1992, and periodically thereafter, the Secretary of Labor, based on the information developed under subsection (c) of this section and on other information available to the Secretary, shall—

(A) determine if additional education about, emphasis on, or enforcement of existing regulations or standards is needed and will be sufficient, or if additional regulations or standards are needed with regard to employee transported releases of hazardous materials; and

(B) prepare and submit to the appropriate committees of Congress a report concerning the result of such determination.

(2) Additional regulations or standards

If the Secretary of Labor determines that additional regulations or standards are needed under paragraph (1), the Secretary shall promulgate, pursuant to the Secretary's authority under the Occupational Safety and Health Act of 1970 (29 U.S.C. 651 et seq.), such regulations or standards as determined to be appropriate not later than 3 years after such determination.

(e) Authorization of appropriations

There are authorized to be appropriated from sums otherwise authorized to be appropriated, for each fiscal year such sums as may be necessary to carry out this section.

(Pub. L. 102-522, title II, Sec. 209, Oct. 26, 1992, 106 Stat. 3420.)

References in Text

The Occupational Safety and Health Act of 1970, referred to in subsecs. (c)(1)(B)(i)(II) and (d)(2), is Pub. L. 91-596, Dec. 29, 1970, 84 Stat. 1590, as amended, which is classified principally to this chapter. For complete classification of this Act to the Code, see Short Title note set out under section 651 of this title and Tables.

Codification

Section was enacted as part of the Fire Administration Authorization Act of 1992, and not as part of the Occupational Safety and Health Act of 1970 which comprises this chapter.

GRANTS TO STATES

29 USC 672

(a) Designation of State agency to assist State in identifying State needs and responsibilities and in developing State plans

The Secretary is authorized, during the fiscal year ending June 30, 1971, and the two succeeding fiscal years, to make grants to the States which have designated a State agency under section 667 of this title to assist them—

(1) in identifying their needs and responsibilities in the area of occupational safety and health,

(2) in developing State plans under section 667 of this title, or

(3) in developing plans for–

   (A) establishing systems for the collection of information concerning the nature and frequency of occupational injuries and diseases;

   (B) increasing the expertise and enforcement capabilities of their personnel engaged in occupational safety and health programs; or

   (C) otherwise improving the administration and enforcement of State occupational safety and health laws, including standards thereunder, consistent with the objectives of this chapter.

(b) Experimental and demonstration projects
The Secretary is authorized, during the fiscal year ending June 30, 1971, and the two succeeding fiscal years, to make grants to the States for experimental and demonstration projects consistent with the objectives set forth in subsection (a) of this section.

(c) Designation by Governor of appropriate State agency for receipt of grant
The Governor of the State shall designate the appropriate State agency for receipt of any grant made by the Secretary under this section.

(d) Submission of application
Any State agency designated by the Governor of the State desiring a grant under this section shall submit an application therefor to the Secretary.

(e) Approval or rejection of application
The Secretary shall review the application, and shall, after consultation with the Secretary of Health and Human Services, approve or reject such application.

(f) Federal share
The Federal share for each State grant under subsection (a) or (b) of this section may not exceed 90 per centum of the total cost of the application. In the event the Federal share for all States under either such subsection is not the same, the differences among the States shall be established on the basis of objective criteria.

(g) Administration and enforcement of programs contained in approved State plans; Federal share
The Secretary is authorized to make grants to the States to assist them in administering and enforcing programs for occupational safety and health contained in State plans approved by the Secretary pursuant to section 667 of this title. The Federal share for each State grant under this subsection may not exceed 50 per centum of the total cost to the State of such a program. The last sentence of subsection (f) of this section shall be applicable in determining the Federal share under this subsection.

(h) Report to President and Congress
Prior to June 30, 1973, the Secretary shall, after consultation with the Secretary of Health and Human Services, transmit a report to the President and to the Congress, describing the experience under the grant programs authorized by this section and making any recommendations he may deem appropriate.

(Pub. L. 91-596, Sec. 23, Dec. 29, 1970, 84 Stat. 1613; Pub. L. 96-88, title V, Sec. 509(b), Oct. 17, 1979, 93 Stat. 695.)

Change of Name

"Secretary of Health and Human Services" substituted for "Secretary of Health, Education, and Welfare" in subsec. (c), pursuant to section 509(b) of Pub. L. 96-88 which is classified to section 3508(b) of Title 20, Education.

STATISTICS

29 USC 673

(a) Development and maintenance of program of collection, compilation, and analysis; employments subject to coverage; scope

In order to further the purposes of this chapter, the Secretary, in consultation with the Secretary of Health and Human Services, shall develop and maintain an effective program of collection, compilation, and analysis of occupational safety and health statistics. Such program may cover all employments whether or not subject to any other provisions of this chapter but shall not cover employments excluded by section 653 of this title. The Secretary shall compile accurate statistics on work injuries and illnesses which shall include all disabling, serious, or significant injuries and illnesses, whether or not involving loss of time from work, other than minor injuries requiring only first aid treatment and which do not involve medical treatment, loss of consciousness, restriction of work or motion, or transfer to another job.

(b) Authority of Secretary to promote, encourage, or engage in programs, make grants, and grant or contract for research and investigations

To carry out his duties under subsection (a) of this section, the Secretary may—

(1) promote, encourage, or directly engage in programs of studies, information and communication concerning occupational safety and health statistics;

(2) make grants to States or political subdivisions thereof in order to assist them in developing and administering programs dealing with occupational safety and health statistics; and

(3) arrange, through grants or contracts, for the conduct of such research and investigations as give promise of furthering the objectives of this section.

(c) Federal share for grants

The Federal share for each grant under subsection (b) of this section may be up to 50 per centum of the State's total cost.

(d) Utilization by Secretary of State or local services, facilities, and employees; consent; reimbursement

The Secretary may, with the consent of any State or political subdivision thereof, accept and use the services, facilities, and employees of the agencies of such State or political subdivision, with or without reimbursement, in order to assist him in carrying out his functions under this section.

(e) Reports by employers

On the basis of the records made and kept pursuant to section 657(c) of this title, employers shall file such reports with the Secretary as he shall prescribe by regulation, as necessary to carry out his functions under this chapter.

(f) Supersedure of agreements between Department of Labor and States for collection of statistics

Agreements between the Department of Labor and States pertaining to the collection of occupational safety and health statistics already in effect on the effective date of this chapter shall remain in effect until superseded by grants or contracts made under this chapter.

(Pub. L. 91-596, Sec. 24, Dec. 29, 1970, 84 Stat. 1614; Pub. L. 96-88, title V, Sec. 509(b), Oct. 17, 1979, 93 Stat. 695.)

References in Text

The effective date of this chapter, referred to in subsec. (f), means the effective date of Pub. L. 91-596, Dec. 29, 1970, 84 Stat. 1590, which is 120 days after Dec. 29, 1970, see section 34 of Pub. L. 91-596, set out as an Effective Date note under section 651 of this title.

Change of Name

"Secretary of Health and Human Services" substituted for "Secretary of Health, Education, and Welfare" in subsec. (a) pursuant to section 509(b) of Pub. L. 96-88 which is classified to section 3508(b) of Title 20, Education.

AUDIT OF GRANT RECIPIENT; MAINTENANCE OF RECORDS; CONTENTS OF RECORDS; ACCESS TO BOOKS, ETC.

29 USC 674

(a) Each recipient of a grant under this chapter shall keep such records as the Secretary or the Secretary of Health and Human Services shall prescribe, including records which fully disclose the amount and disposition by such recipient of the proceeds of such grant, the total cost of the project or undertaking in connection with which such grant is made or used, and the amount of that portion of the cost of the project or undertaking supplied by other sources, and such other records as will facilitate an effective audit.

(b) The Secretary or the Secretary of Health and Human Services, and the Comptroller General of the United States, or any of their duly authorized representatives, shall have access for the purpose of audit and examination to any books, documents, papers, and records of the recipients of any grant under this chapter that are pertinent to any such grant.

(Pub. L. 91-596, Sec. 25, Dec. 29, 1970, 84 Stat. 1615; Pub. L. 96-88, title V, Sec. 509(b), Oct. 17, 1979, 93 Stat. 695.)

Change of Name

"Secretary of Health and Human Services" substituted in text for "Secretary of Health, Education, and Welfare" pursuant to section 509(b) of Pub. L. 96-88 which is classified to section 3508(b) of Title 20, Education.

ANNUAL REPORTS BY SECRETARY OF LABOR AND SECRETARY OF HEALTH AND HUMAN SERVICES; CONTENTS

29 USC 675

Within one hundred and twenty days following the convening of each regular session of each Congress, the Secretary and the Secretary of Health and Human Services shall each prepare and submit to the President for transmittal to the Congress a report upon the subject matter of this chapter, the progress toward achievement of the purpose of this chapter, the needs and requirements in the field of occupational safety and health, and any other relevant information. Such reports shall include information regarding occupational safety and health standards, and criteria for such standards, developed during the preceding year; evaluation of standards and criteria previously developed under this chapter, defining areas of emphasis for new criteria and standards; an evaluation of the degree of observance of applicable occupational safety and health standards, and a summary of inspection and enforcement activity undertaken; analysis and evaluation of research activities for which results have been obtained under governmental and nongovernmental sponsorship; an analysis of major occupational diseases; evaluation of available control and measurement technology for hazards for which standards or criteria have been developed during the preceding year; description of cooperative efforts undertaken between Government agencies and other interested parties in the implementation of this chapter during the preceding year; a progress report on the development of an adequate supply of trained manpower in the field of occupational safety and health, including estimates of future needs and the efforts being made by Government and others to meet those needs; listing of all toxic substances in industrial usage for which labeling requirements, criteria, or standards have not yet been established; and such recommendations for additional legislation as are deemed necessary to protect the safety and health of the worker and improve the administration of this chapter.

(Pub. L. 91-596, Sec. 26, Dec. 29, 1970, 84 Stat. 1615; Pub. L. 96-88, title V, Sec. 509(b), Oct. 17, 1979, 93 Stat. 695.)

Change of Name

"Secretary of Health and Human Services" substituted in text for "Secretary of Health, Education, and Welfare" in text pursuant to section 509(b) of Pub. L. 96-88 which is classified to section 3508(b) of Title 20, Education.

Study of Occupationally Related Pulmonary and Respiratory Diseases; Study To Be Completed and Report Submitted by September 1, 1979

Pub. L. 95-239, Sec. 17, Mar. 1, 1978, 92 Stat. 105, authorized Secretary of Labor, in cooperation with Director of National Institute for Occupational Safety and Health, to conduct a study of occupationally related pulmonary and respiratory diseases and to complete such study and report findings to President and Congress not later than 18 months after Mar. 1, 1978.

29 USC 676
_____

Omitted.

Codification

Section, Pub. L. 91-596, Sec. 27, Dec. 29, 1970, 84 Stat. 1616, provided for establishment of a National Commission on State Workmen's Compensation Laws to make an effective study and evaluation of State workmen's compensation laws to determine whether such laws provide an adequate, prompt, and equitable system of compensation for injury or death, with a final report to be transmitted to President and Congress not later than July 31, 1972, ninety days after which the Commission ceased to exist.

SEPARABILITY

29 USC 677
_____

If any provision of this chapter, or the application of such provision to any person or circumstance, shall be held invalid, the remainder of this chapter, or the application of such provision to persons or circumstances other than those as to which it is held invalid, shall not be affected thereby.

(Pub. L. 91-596, Sec. 32, Dec. 29, 1970, 84 Stat. 1619.)

AUTHORIZATION OF APPROPRIATIONS

29 USC 678
_____

There are authorized to be appropriated to carry out this chapter for each fiscal year such sums as the Congress shall deem necessary.

(Pub. L. 91-596, Sec. 33, Dec. 29, 1970, 84 Stat. 1620.)